Advances in Intelligent Systems and Computing

Volume 1353

The series "Advances in Intelligent Systems and Computing" contains publications on theory, applications, and design methods of Intelligent Systems and Intelligent Computing. Virtually all disciplines such as engineering, natural sciences, computer and information science, ICT, economics, business, e-commerce, environment, healthcare, life science are covered. The list of topics spans all the areas of modern intelligent systems and computing such as: computational intelligence, soft computing including neural networks, fuzzy systems, evolutionary computing and the fusion of these paradigms, social intelligence, ambient intelligence, computational neuroscience, artificial life, virtual worlds and society, cognitive science and systems, Perception and Vision, DNA and immune based systems, self-organizing and adaptive systems, e-Learning and teaching, human-centered and human-centric computing, recommender systems, intelligent control, robotics and mechatronics including human-machine teaming, knowledge-based paradigms, learning paradigms, machine ethics, intelligent data analysis, knowledge management, intelligent agents, intelligent decision making and support, intelligent network security, trust management, interactive entertainment, Web intelligence and multimedia.

The publications within "Advances in Intelligent Systems and Computing" are primarily proceedings of important conferences, symposia and congresses. They cover significant recent developments in the field, both of a foundational and applicable character. An important characteristic feature of the series is the short publication time and world-wide distribution. This permits a rapid and broad dissemination of research results.

Indexed by DBLP, EI Compendex, INSPEC, WTI Frankfurt eG, zbMATH, Japanese Science and Technology Agency (JST).
All books published in the series are submitted for consideration in Web of Science.

More information about this series at http://www.springer.com/series/11156

Marcin Paprzycki · Sabu M. Thampi ·
Sushmita Mitra · Ljiljana Trajkovic ·
El-Sayed M. El-Alfy
Editors

Intelligent Systems, Technologies and Applications

Proceedings of Sixth ISTA 2020, India

Springer

Editors
Marcin Paprzycki
Systems Research Institute
Polish Academy of Sciences
Warsaw, Poland

Sabu M. Thampi
Indian Institute of Information Technology
and Management Kerala
Trivandrum, Kerala, India

Sushmita Mitra
Machine Intelligence Unit
Indian Statistical Institute
Kolkata, West Bengal, India

Ljiljana Trajkovic
Simon Fraser University
Burnaby
BC, Canada

El-Sayed M. El-Alfy
College of Computer Sciences
and Engineering
King Fahd University of Petroleum
and Minerals
Dhahran, Saudi Arabia

ISSN 2194-5357 ISSN 2194-5365 (electronic)
Advances in Intelligent Systems and Computing
ISBN 978-981-16-0729-5 ISBN 978-981-16-0730-1 (eBook)
https://doi.org/10.1007/978-981-16-0730-1

This Springer imprint is published by the registered company Springer Nature Singapore Pte Ltd.
The registered company address is: 152 Beach Road, #21-01/04 Gateway East, Singapore 189721, Singapore

Conference Organization

Chief Patron

G. Viswanathan, Chancellor, Vellore Institute of Technology

Patrons

Sankar Viswanathan, Vice-President, Vellore Institute of Technology
Sekar Viswanathan, Vice-President, Vellore Institute of Technology
G. V. Selvam, Vice-President, Vellore Institute of Technology
Sandhya Pentareddy, Executive Director, Vellore Institute of Technology
Kadhambari S. Viswanathan, Assistant Vice-President, Vellore Institute of Technology
Rambabu Kodali, Vice Chancellor, Vellore Institute of Technology
S. Narayanan, Pro-Vice Chancellor, Vellore Institute of Technology, Vellore
V. S. Kanchana Bhaaskaran, Pro-Vice Chancellor, Vellore Institute of Technology
P. K. Manoharan, Additional Registrar, Vellore Institute of Technology

General Chairs

Janusz Kacprzyk, Systems Research Institute, Polish Academy of Sciences, Poland
Marcin Paprzycki, Systems Research Institute, Polish Academy of Sciences, Poland
Sushmita Mitra, Machine Intelligence Unit, Indian Statistical Institute, Kolkata, India

General Executive Chair

Sabu M. Thampi, Indian Institute of Information Technology and Management-Kerala (IIITM-K), Trivandrum

Program Chairs

Juan Manuel Corchado Rodriguez, University of Salamanca, Spain
El-Sayed M. El-Alfy, King Fahd University of Petroleum and Minerals, Saudi Arabia
Ljiljana Trajkovic, Simon Fraser University, Canada

Organizing Chairs

S. Geetha, VIT, Chennai
R. Jagadeesh Kannan, VIT, Chennai

Organizing Secretaries

C. Sweetlin Hemalatha, VIT, Chennai
G. Suganya, VIT, Chennai
R. Kumar, VIT, Chennai

Organizing Co-chairs

S. Asha, VIT, Chennai
R. Pattabiraman, VIT, Chennai
V. Viswanathan, VIT, Chennai

TPC Members

http://www.acn-conference.org/2020/ista2020/committee.html

Organized by

Vellore Institute of Technology (VIT), Chennai, India

VIT®

Vellore Institute of Technology

(Deemed to be University under section 3 of UGC Act, 1956)

Preface

This volume of proceedings offers to readers a selection of refereed papers that were presented at the Sixth International Symposium on Intelligent Systems Technologies and Applications (ISTA'20). ISTA'20 brought together researchers in related fields to explore and discuss various aspects of intelligent systems technologies and their applications. The symposium was organized by Vellore Institute of Technology (VIT), Chennai, India, during October 14–17, 2020. Due to the recent pandemic situation, it was conducted as a virtual event. ISTA'20 was co-located with the International Conference on Applied Soft Computing and Communication Networks (ACN'20).

All submissions were evaluated on the basis of their significance, novelty, and technical quality. A double-blind review process was conducted to ensure that the names and affiliations of authors were unknown to TPC. This volume consists of 28 papers (19 regular and 9 short papers) that were virtually presented at the symposium. The papers cover different areas such as big data analytics, security and privacy, Internet of Things, machine and deep learning, health informatics, visual computing, signal processing, and natural language processing.

We are very grateful to the many people who contributed to the organization of the symposium. Our heartfelt thanks go to all the authors for submitting their research results to the symposium and to the members of the program committee for their careful and insightful reviews. We are grateful to Vellore Institute of Technology (VIT), Chennai, for organizing the symposium. It could not have taken place without the commitment of the local organizing committee, which in many ways helped to effectively coordinate the event. We are grateful to General Chairs and Program Chairs for their great support. We express our most sincere thanks to all keynote speakers who have shared their expertise and knowledge with us. Finally, we would like to thank our publisher Springer, in particular Senior Editor Aninda Bose, for their collaboration.

Warsaw, Poland Marcin Paprzycki
Trivandrum, India Sabu M. Thampi
Kolkata, India Sushmita Mitra
Burnaby, Canada Ljiljana Trajkovic
Dhahran, Saudi Arabia El-Sayed M. El-Alfy
October 2020

Contents

About the Editors

Marcin Paprzycki is Associate Professor in the Systems Research Institute, Polish Academy of Sciences, Warsaw, Poland. He has an M.S. from Adam Mickiewicz University in Poznań, Poland, a Ph.D. from Southern Methodist University in Dallas, Texas, and a Doctor of Science from the Bulgarian Academy of Sciences. He is a senior member of IEEE, a senior member of ACM, Senior Fulbright Lecturer, and IEEE CS Distinguished Visitor. He has contributed to more than 500 publications and was invited to the program committees of over 800 international conferences. He is on the editorial boards of 12 journals and a book series.

Sabu M. Thampi is Professor at the Indian Institute of Information Technology and Management-Kerala (IIITM-K), Technopark Campus, Trivandrum, Kerala, India. His current research interests include cognitive computing, Internet of things (IoT), authorship analysis, trust management, biometrics, social networks, nature-inspired computing, and video surveillance. He has published papers in book chapters, journals, and conference proceedings. He has authored and edited a few books. Sabu has served as Guest Editor for special issues in few journals and a program committee member for many international conferences and workshops. He has co-chaired several international workshops and conferences. He has initiated and is also involved in the organization of several annual conferences/symposiums. Sabu is currently serving as Editor for *Journal of Network and Computer Applications* (JNCA), Elsevier; *Connection Science*, Taylor & Francis; Associate Editor for *IEEE Access* and *International Journal of Embedded Systems*, Inderscience, UK, and Reviewer for several reputed international journals. Sabu is a senior member of IEEE and ACM. He is Founding Chair of ACM Trivandrum Professional Chapter.

Sushmita Mitra is full Professor at the Machine Intelligence Unit (MIU), Indian Statistical Institute, Kolkata. From 1992 to 1994 she was in the RWTH, Aachen, Germany as a DAAD Fellow. She was a Visiting Professor in the Computer Science Departments of the University of Alberta, Edmonton, Canada in 2004, 2007; Meiji University, Japan in 1999, 2004, 2005, 2007; and Aalborg University Esbjerg, Denmark in 2002, 2003. Dr. Mitra received the National Talent Search Scholarship (1978–1983) from NCERT, India, the University Gold Medal in 1988, the IEEE

TNN Outstanding Paper Award in 1994 for her pioneering work in neuro-fuzzy computing, the CIMPA-INRIA-UNESCO Fellowship in 1996, and Fulbright-Nehru Senior Research Fellowship in 2018–2020. She was the INAE Chair Professor during 2018–2020. Dr. Mitra is the author of the books *Neuro-Fuzzy Pattern Recognition: Methods in Soft Computing* and *Data Mining: Multimedia, Soft Computing, and Bioinformatics* published by John Wiley, and *Introduction to Machine Learning and Bioinformatics*, Chapman & Hall/CRC Press, beside a host of other edited books. Dr. Mitra has guestedited special issues of several journals, is an Associate Editor of *IEEE/ACM Transactions on Computational Biology and Bioinformatics*, *Information Sciences*, *Fundamenta Informatica*, and is a Founding Associate Editor of *Wiley Interdisciplinary Reviews: Data Mining and Knowledge Discovery* (WIRE DMKD). She has more than 150 research publications in refereed international journals. Dr. Mitra is a Fellow of the IEEE, Indian National Science Academy (INSA), International Association for Pattern Recognition (IAPR), and Fellow of the Indian National Academy of Engineering (INAE) and The National Academy of Sciences, India (NASI). She is an IEEE CIS Distinguished Lecturer and the current Chair-Elect, IEEE Kolkata Section. She has visited more than 30 countries as a Plenary/Invited Speaker or an academic visitor, and served as General Chair and/or Program Chair of several international conferences. Her current research interests include data mining, pattern recognition, soft computing, medical image processing, and Bioinformatics.

Ljiljana Trajkovic received the Dipl. Ing. degree from University of Pristina, Yugoslavia, in 1974, the M.Sc. degrees in electrical engineering and computer engineering from Syracuse University, Syracuse, NY, in 1979 and 1981, respectively, and the Ph.D. degree in electrical engineering from University of California at Los Angeles, in 1986.

She is currently a Professor in the School of Engineering Science at Simon Fraser University, Burnaby, British Columbia, Canada. From 1995 to 1997, she was a National Science Foundation (NSF) Visiting Professor in the Electrical Engineering and Computer Sciences Department, University of California, Berkeley. She was a Research Scientist at Bell Communications Research, Morristown, NJ, from 1990 to 1997, and a Member of the Technical Staff at AT&T Bell Laboratories, Murray Hill, NJ, from 1988 to 1990. Her research interests include high-performance communication networks, control of communication systems, computer-aided circuit analysis and design, and theory of nonlinear circuits and dynamical systems.

Dr. Trajkovic serves as IEEE Division X Delegate/Director (2019–2020) and served as IEEE Division X Delegate-Elect/Director-Elect (2018). She served as Senior Past President (2018–2019), Junior Past President (2016–2017), President (2014–2015), President-Elect (2013), Vice President Publications (2012–2013, 2010–2011), Vice President Long-Range Planning and Finance (2008–2009), and a Member at Large of the Board of Governors (2004–2006) of the IEEE Systems, Man, and Cybernetics Society. She served as 2007 President of the IEEE Circuits and Systems Society and a member of its Board of Governors (2004–2005, 2001–2003). She is Chair of the IEEE Circuits and Systems Society joint Chapter of the Vancouver/Victoria Sections. She was Chair of the IEEE Technical Committee on

Nonlinear Circuits and Systems (1998). She is General Co-chair of SMC 2020 and SMC 2020 Workshop on BMI Systems and served as General Co-chair of SMC 2019 and SMC 2018 Workshops on BMI Systems, SMC 2016, and HPSR 2014, Special Sessions Co-chair of SMC 2017, Technical Program Chair of SMC 2017 and SMC 2016 Workshops on BMI Systems, Technical Program Co-chair of ISCAS 2005, and Technical Program Chair and Vice General Co-Chair of ISCAS 2004. She served as an Associate Editor of the *IEEE Transactions on Circuits and Systems (Part I)* (2004–2005, 1993–1995), the *IEEE Transactions on Circuits and Systems (Part II)* (2018, 2002–2003, 1999–2001), and the *IEEE Circuits and Systems Magazine* (2001–2003). She is a Distinguished Lecturer of the *IEEE Systems, Man, and Cybernetics Society* (2020–2021) and the *IEEE Circuits and Systems Society* (2010–2011, 2002–2003). She is a Professional Member of IEEE-HKN and a Life Fellow of the IEEE.

El-Sayed M. El-Alfy is a Professor in Information and Computer Science Department, King Fahd University of Petroleum and Minerals (KFUPM), Saudi Arabia. He has more than 25 years of experience in industry and academia involving research, teaching, supervision, curriculum design, program assessment and quality assurance in higher education. He is approved ABET/CSAB Program Evaluator (PEV), IEEE Senior Member, and reviewer and consultant for NCAAA and several universities and research agents in various countries. He is an active researcher with interests in fields related to machine learning and nature-inspired computing and applications to data science and cybersecurity analytics, pattern recognition, multimedia forensics and security systems. He has published numerously in peer-reviewed international journals and conferences, edited a number of books published by reputable international publishers, attended and contributed in the organization of many world-class international conferences, supervised master and Ph.D. students, served as a guest editor for a number of special issues in international journals, and been in the editorial board of a number of premium international journals including *IEEE/CAA Journal of Automatica Sinica*, *IEEE Transactions on Neural Networks and Learning Systems*, *International Journal on Trust Management in Computing and Communications*, and *Journal of Emerging Technologies in Web Intelligence* (JETWI). He is also a member of *IEEE Computational Intelligence Society*, and ex-member of *ACM and IEEE Computer Society*. His work has been internationally recognized and received a number of awards.

Probability of Loan Default—Applying Data Analytics to Financial Credit Risk Prediction

Aleksandra Łuczak, Maria Ganzha, and Marcin Paprzycki

Abstract In the banking industry, one of the important issues is how to establish credit worthiness of potential clients. With the possibility of collecting digital records of results of past credit applications (of all clients), it can be stipulated that machine learning techniques can be used in "credit decision support" systems. There exists a substantial body of literature devoted to this subject. Moreover, benchmark datasets have been proposed, to establish effectiveness of proposed credit risk assessment approaches. The aim of this work is to compare performance of *seven* different classifiers, applied to *two* different benchmark datasets. Moreover, capabilities of, recently introduced, methods for combining results from multiple classifiers, into a meta-classifier, will be evaluated.

Keywords Meta-classification · Neural network · Random forest · XGBoost · Adaboost · K-nearest neighbours · Support vector machine · Parameter optimization · Credit risk · Credit risk prediction · Loan default · Data analytics · Machine learning · Performance comparison

1 Introduction

Banking industry of today collects "all possible" data concerning all of its customers. Moreover, for all practical purposes, such data is never discarded (even if clients close their accounts [1]). One of the areas, where collected data is expected to be of great value, is to provide support for the so-called credit risk assessment. Specifically, it is assumed that application of data analytics methods, to the collected data, can help correctly assess probability of loan defaults. This belief can be observed, among others,

A. Łuczak · M. Ganzha
Warsaw University of Technology, Warsaw, Poland
e-mail: M.Ganzha@mini.pw.edu.pl

M. Paprzycki (✉)
Systems Research Institute Polish Academy of Sciences, Warsaw, Poland
e-mail: marcin.paprzycki@ibspan.waw.pl

M. Paprzycki et al. (eds.), *Intelligent Systems, Technologies and Applications*, Advances in Intelligent Systems and Computing 1353,
https://doi.org/10.1007/978-981-16-0730-1_1

in a large body of literature devoted to the subject (see, Sect. 3). However, existing work is, mostly, focused on specific classifiers, with the goal of improving their *individual* performance. The aim of our work is to *compare* performance of different credit risk assessment methods, when applied to two standard "benchmark datasets". Moreover, we will investigate usefulness of a, recently introduced, meta-classifiers, which "merge" predictions of individual classifiers to provide a "combined score".

To this effect, we proceed as follows. We start from a brief summary of necessary background knowledge related to the banking industry (Sect. 2). Next, in Sect. 3, we present an overview of related literature. We follow, in Sects. 4 and 5, with an overview of the two datasets used in our work and the experimental setup. This allows us, in Sect. 6, to summarize results of performed experiments. We conclude our contribution in Sect. 7.

2 How Banks Assess Customers—Short Introduction

In all banks, there exist complicated procedures, used to comprehensively evaluate credit worthiness of their (potential) customers. Moreover, each bank has it own method to do so, and such methods are a closely guarded intellectual property. While one of the authors worked in a credit department of a large bank, for obvious reasons, we will be able to share only "common open knowledge/information" about the process of assessment of probability of loan default. Keeping this in mind, let us start from basics of credit risk assessment.

2.1 Need of Credit Risk Assessment

One of the most sensitive stages, of a bank processing a credit application, is the decision-making process. Following the digital transformation, banks introduced online lending and borrowing mechanisms. Here, the key element is a semi-automatic credit decision process, which is based, among others, on the demographic and the financial data. Here, the *semi-automatic process* means, usually that automatic positive decisions are accepted, while possible "problems" are remanded to human employees for final decisions.

The possibility of applying for loans online is one of the reasons that banks are granting more and more credits. Here, note that, at the same time, the total number of physical bank branches is systematically decreasing (as predicted, for instance, in [2]). The fact that the increase in number of credits is a global phenomenon is confirmed, among others, by the Polish Financial Supervision Authority [3]. According to the available data, at the end of December 2019, household debt, to the banks, amounted to 671 billion PLN. In addition, as much as 15.44 million, or 40.6% of Poles were actively repaying a loan, or a credit.

With the increasing prevalence of loans, and the decreasing requirements of banks (which earn money on the credits), there is also a recognized risk of approving loans to persons who cannot cope with their repayment, hence the need to automatically (as the total number of bank employees is decreasing) evaluate credit risk of an increasing number of loan applicants.

In this context, the *Probability of Default* (PD) indicates the probability of a loss, associated with a given credit. It is, usually, considered within a certain time-horizon (usually one year). Simply said, high value of PD indicates that customer may stop paying back the loan (temporarily, or permanently). The probability of customer's load default is the primary parameter in calculating creditworthiness. This factor plays a major role in loss control and income maximization [4]. PD can be estimated with the use of a number of analytical tools, and majority of such tools are based on some form of, broadly understood, "data analytics". In practice, in majority of banks, linear discriminate analysis [5] or logistic regression [6] is applied, because these algorithms combine simplicity and efficiency and are easy to explain (high interpretability) [7]. The latter makes then also slightly more desirable than neural network-based approaches, which are not explainable and thus raise variety of ethical issues.

Separately, in scientific literature, a number of approaches have been studied. Moreover, to support this research, at least two open, tagged, datasets have been created, which can be used to compare performance of considered methods.

3 Related Work

This being the case, let us now briefly present state of the art of analytic credit risk assessment, as represented in scientific publications. Since, as mentioned above, each bank has its own credit risk assessment procedures, which are a closely guarded intellectual property, only published literature can be discussed. We split the material into two parts. First, we summarize the proposed approaches, dataset(s) that they were applied to, and the reported quality of the loan default prediction. The latter will be used to compare results of our experiments to. Second, for the standard methods of data analytics, described in the reviewed literature, we provide short descriptions and basic references.

3.1 Data Analytics for Credit Risk Analysis

In the credit risk assessment, the most common method used to predict the PD is a scoring card [8]. It consists of assessing (weighting) selected characteristics of the potential borrower. After assigning a score to each one of them, these ratings are combined (summed up). Here, if the total score is higher than the (financial institution dependent) cut-off point, client receives a credit [9]. This is a very simple method and is very easy to operationalize, but has many disadvantages. For example, it has a relatively weak predictive power and cannot handle large datasets and features with complicated relationships [10].

Nowadays, more focus is devoted to the predictive power of classification models, used to control credit risk, because even a small increase in performance results in a large increase in profits [11, 12]. Therefore, with the development of machine learning methods, the improvement of the predictive power of scoring models is within reach of financial institutions, and they are taking up this approach [13–15]. In most cases, algorithms from the supervised ML family are used in the credit risk assessment, to find a correlation between the clients attributes, and their expected loan repayment. Here, note that in many articles, proposed approaches are confirmed by high accuracy [16–18].

Recently, many researchers have focused their attention on ensemble methods, and use of different base classifiers, to make the accuracy of prediction "as high as possible". Such heterogeneous classifiers can, for instance, ensemble three state-of-the art classifiers: logistic regression (LR), artificial neural network (ANN), and support vector machines (SVM) [19]. Hybrid ensemble methods were also based on Random Forest, Support Vector Machines, Decision Trees, Artificial Neural Network, Multidimensional Neural Network [20]. The method based on a hybrid associative memory with translation was reported in [21]. There exists a study, from 2015, that compares 41 different classification algorithms [22] and a study that compares Bagging DT, Random Subspace DT, Random Forest and Rotation Forest [23]. All this research shows good model performance, as measured in terms of "standard" metrics such as accuracy (ACC) and area under the ROC curve (AUC), which are above 0.7.

Nevertheless, in this contribution, we will compare obtained results to those reported in [24, 25]. This is related to the fact that they use similar datasets. Moreover, they follow the same general methodology (see, Sect. 5). Nevertheless, we recognize the fact that a more broad comparative study (similar to that reported in [22], but aligned with our investigation) could be of value.

3.2 Classifiers Used in Our Work

Let us now summarize the specific machine learning approaches that have been used in the context of credit risk assessment.

3.2.1 K-Nearest Neighbours

K-Nearest Neighbours (KNN) is a classifier, which is very often used in pattern recognition [26], and to build credit scoring models [27]. It is a simple algorithm, based on adding new elements to the class that they are "best fitting", on the basis of a process of comparing it to the nearest set of observations (k-nearest neighbours).

The typical distance function is an Euclidean distance. In our work, we used the KNN modules available in *sklearn.neighbors.KNeighborsClassifier*.[1]

3.2.2 Support Vector Machines

Support vector machines (SVMs) is an ML method proposed by Vapnik [28]. It is based on the search for a hyperplanes separating classes. Established hyperplanes should be such that the closest observations from each class are as far away from each other as possible. In our work, we have experimented with the SVM in two versions: with (a) linear and (b) radial kernel functions. Here, we used *sklearn.svm.LinearSVC*[2] and *sklearn.svm.SVC*[3] modules.

3.2.3 Random Forest (RF)

The Random Forest (RF) algorithm was proposed by Breiman [29, 30]. It is based on the decision tree, and the classification committee. This means that the final response of the algorithm is selected as the average response of each decision tree (seen as an "independent classifier"). For the random forest, we used *sklearn.ensemble.RandomForestClassifier*[4] modules.

3.2.4 Gradient Boosted Decision Trees

Gradient Boosted Decision Trees (GBDT) is a generalization of boosting [31] to an arbitrary differential loss functions [32]. GBDT is an accurate and effective off-the-shelf procedure that can be used for both the regression and the classification problems. It can be applied in a variety of areas. To experiment with the GBDT, we used *sklearn.ensemble.GradientBoostingClassifier*[5] modules.

3.2.5 AdaBoost

An AdaBoost is an adaptive boosting algorithm, proposed by Freund and Schapire [33]. This is a meta-estimator that begins by fitting a classifier into the original

---

[1] https://scikit-learn.org/stable/modules/generated/sklearn.neighbors.KNeighborsClassifier.html#sklearn.neighbors.KNeighborsClassifier.

[2] https://scikit-learn.org/stable/modules/generated/sklearn.svm.LinearSVC.html?highlight=svm#sklearn.svm.LinearSVC.

[3] https://scikit-learn.org/stable/modules/generated/sklearn.svm.SVC.html#sklearn.svm.SVC.

[4] https://scikit-learn.org/stable/modules/generated/sklearn.ensemble.RandomForestClassifier.html?highlight=random%20forest#sklearn.ensemble.RandomForestClassifier.

[5] https://scikit-learn.org/stable/modules/generated/sklearn.ensemble.GradientBoostingClassifier.html#sklearn.ensemble.GradientBoostingClassifier.

dataset and then fits additional copies of the classifier on the same dataset. However, in this case, the weights of incorrectly classified instances are adjusted, such that the subsequent classifiers focus more on difficult cases. In case of AdaBoost, we used *sklearn.ensemble.AdaBoostClassifier*[6] modules.

3.2.6 Extreme Gradient Boosting

The Extreme Gradient Boosting (XGBoost) algorithm is one of the most popular, and most effectively implemented algorithms, in the family of algorithms based on the gradient boosting method [34]. For the XGBoost, we used *xgboost.XGBClassifier*[7] modules.

3.2.7 Meta-Classifiers

While the above listed approaches can be (and have been) applied individually to credit risk assessment (see, Sect. 3.1), recently it has been realized that each one of them captures "separate aspects" of the data. The question has thus been posed: Is it possible to combine results from multiple classifiers to develop a meta-classifier, which will improve the overall quality of prediction? Such meta-classifiers have been tried, with moderate success, in non-financial domains (see, for instance, [35, 36]). However, they involve relatively complex methods that require a lot of attention to be correctly implemented.

Searching in the same direction, recently, a different class of simple meta-classifiers has been proposed. Specifically, Weighted Average Recall Error (WARE) and Weighted Average Type-1 Error (WA-T1E) algorithms are based on an easy to implement consensus models. These classifiers are similar to the Weighted Average Prediction Error (WA-PE) method used in [37]. Here, the data sample is divided into two parts, one for training the classifiers and one to calculate the Recall Recall_m, and the Type I error $T1err_m$ (where m is the index of the classifier). After normalizing the weights, for the Recall_m and the $T1err_m$, posteriori probabilistic classes of observations are computed as follows:

$$\text{WARE}_1(x) = \sum_{m=1}^{M} \text{Recall}_m \, p_1^{h_m}(x) \tag{1}$$

$$\text{WA} - T1E_1(x) = \sum_{m=1}^{M} T1err_m \, p_1^{h_m}(x) \tag{2}$$

---

[6]https://scikit-learn.org/stable/modules/generated/sklearn.ensemble.AdaBoostClassifier.html?highlight=adaboost#sklearn.ensemble.AdaBoostClassifier.

[7]https://xgboost.readthedocs.io/en/latest/python/python_api.html#module-xgboost.sklearn.

where p_1 is the posteriori probability of belonging to class 1 and h_m is a classifier. Here, observation x is assigned to class 1 if $\mathrm{WARE}_1(x) > \mathrm{WARE}_0(x)$, or to class 0 otherwise. The same applies to the WA-T1E method, where the observation x is assigned to class 1 if $WA - T1E_1(x) > WA - T1E_0(x)$, or to class 0 otherwise.

Therefore, taking into account their simplicity, we have decided to experiment also with these two meta-classifiers, to see if their application can improve the overall performance of the credit risk estimators.

4 Datasets Used in Experiments

Let us now move our attention to the data used in our experiments. As mentioned above, there exist multiple tagged datasets that can be used to evaluate performance of approaches to credit risk prediction. In our work, we have used two of them: (a) German Credit,[8] and (b) Give Me Some Credit.[9] Both datasets contain information about the demographic characteristics and the financial situation, of potential borrowers. Let us now describe each one of them in more detail.

4.1 German Credit Dataset

The German redit data contains 1,000,000 observations, representing two classes: good creditor (699,774 observations) and bad creditor (300,226 observations). The dataset is described by 20 features, including 7 continuous features.

- *duration*—Duration in month
- *credit_amount*—Credit amount
- *installment_commitment*—Installment rate in percentage of disposable income
- *residence_since*—Present residence since
- *age*—Age in years
- *existing_credits*—Number of existing credits at this bank
- *num_dependents*—Number of people being liable to provide maintenance for

and 13 categorical features

- *checking_status*—Status of existing checking account
- *credit_history*—Credit history
- *purpose*—Purpose of loan
- *savings_status*—Status of savings account/bonds
- *employment*—Present employment since
- *personal_status*—Personal status and sex

[8]https://www.openml.org/d/260.

[9]https://www.kaggle.com/c/GiveMeSomeCredit/data?select=cs-training.csv.

Table 1 Information about missing values in categorical features of the German credit dataset

Feature name	Number of missing values	Representations of missing values	Ratio of missing values (%)
checking_status	394,051	'no checking'	39.40
credit_history	40,856	'no credits/all paid'	4.08
savings_status	184,016	'no known savings'	18.40
other_parties	905,898	none	90.58
property_magnitude	151,898	'no known property'	15.18
other_payment_plans	808,505	none	80.85
own_telephone	594,649	none	59.46

- *other_parties*—Other debtors/guarantors
- *property_magnitude*—Magnitude of personal property
- *other_payment_plans*—Other installment plans
- *housing*—Housing (own or rent)
- *job*—Status of present job
- *own_telephone*—Telephone
- *foreign_worker*—Foreign worker.

The data has been curated, resulting in "no data missing". However, some categorical variables contain information that they do not represent "real data". For example, in some cases, the variable *saving_status* contains the value *no known savings*, indicating that the actual data is missing. All such cases are summarized in Table 1.

To apply machine learning, all categorical variables have been manually converted, from string variables, to numerical variables. In addition, strings listed in Table 1 were transformed into label 0. Nevertheless, no data was removed from the dataset. This decision may be questioned, and we plan to investigate this aspect further in the future.

4.2 Give Me Some Credit Dataset

This dataset is smaller and uses 10 features, to describe 150,000 borrowers. On the basis of the data provided, it should be anticipated whether the person described there will experience financial difficulties, in the next two years. This feature is tagged as *SeriousDlqin2yrs*. This information allows one to determine the credibility of the borrower and can be used to make a decision about approving loan application. Below, we present features available in the dataset, with their description.

- *Age*—Age in years
- *NumberOfDependents*—Number of the borrower's dependants
- *MonthlyIncome*—Amount of monthly income

- *DebtRatio*—The ratio of the sum of monthly debt. alimony and maintenance costs to monthly income expressed as a percentage
- *RevolvingUtilizationOfUnsecuredLines*—The ratio of the total balance on cards and personal lines of credit to the sum of credit limits expressed as a percentage
- *NumberOfOpenCreditLinesAndLoans*—Number of loans and open credit lines
- *NumberRealEstateLoansOrLines*—Number of housing loans and credits and household credit lines
- *NumberOfTime30–59DaysPastDueNotWorse*—Number of cases where the borrower was 30–59 days late in making payments in the last 2 years
- *NumberOfTime60-89DaysPastDueNotWorse*—Number of cases where the borrower was 60–89 days late in making payments in the last 2 years
- *NumberOfTimes90DaysLate*—Number of cases where the borrower was above 90 days late in making payments in the last 2 years

Data available in this dataset is incomplete. Specifically, there are two features that have considerable data gaps. For the *MonthlyIncome* attribute, data is missing for 29,731 observations, while for the *NumberOfDependents* attribute, data is missing for 3924 observations.

However, most of the tested algorithms (mentioned in Sect. 3.1) require that the dataset should *not* contain any missing data. To avoid "simple imputing", which means: to fill in the missing values with 0 or -1, we used the KNNImputer algorithm from the sklearn package. This algorithm calculates the k-th nearest neighbours of a given observation, based on the Euclidean distance. Next, it selects the value of the feature to be completed from the k-neighbors. This data augmentation step was applied after the division of the dataset into the teaching and the testing parts.

5 Experimental Setup

Let us now summarize the key technical aspects of the performed experiments.

- We have separately experimented (including applied preprocessing), with the two datasets described in Sect. 4.
- In each case, we have applied all classifiers, described in Sect. 3.
- Only open-source classifiers were used (as specified in Sect. 3.1).
- In each case, available data was randomly divided into the standard 7:3 ratio, between the training and testing subsets.
- During the training, the models were fitted, and their hyperparameters were optimized using special software, described in what follows.
- During the testing, all metrics described in Sect. 5.1 were calculated.
- Meta-classifiers were used to combine "suggestions" of the individual classifiers.

It should be noted that all algorithms were automatically optimized. For the scientific context of tuning of hyperparameters of classifiers, see [38]. Overall, the main goal of optimization was to minimize the function:

$$\text{loss}(f(X)) = 1 - \text{AUC}(f(X), y). \tag{3}$$

where X is the feature, y is the target, and f is the decision function. To achieve the needed optimization, the *hyperopt*[10] package has been used. Additionally, all algorithms had a parameter set to unbalanced classes (matching the characteristics of the datasets), to further improve the performance.

5.1 Performance Evaluation—Methodological Considerations

In the works summarized in Sect. 3, to assess performance of the proposed approaches, the following standard performance metrics have been used: accuracy (ACC), Matthews correlation coefficient (MCC), the area under the ROC curve (AUC), Recall, Precision, Type I error, and Type II error. The formulae, representing these metrics, have been summarized in the following equations:

$$\text{ACC} = \frac{\text{TP} + \text{TN}}{\text{TP} + \text{TN} + \text{FP} + \text{FN}} \tag{4}$$

$$\text{MCC} = \frac{\text{TP} \times \text{TN} - \text{FP} \times \text{FN}}{\sqrt{(\text{TP} + \text{FP})(\text{TP} + \text{FN})(\text{TN} + \text{FP})(\text{TN} + \text{FN})}} \tag{5}$$

$$\text{Recall} = \frac{\text{TP}}{\text{TP} + \text{FN}} \tag{6}$$

$$\text{Precision} = \frac{\text{TP}}{\text{TP} + \text{FP}} \tag{7}$$

$$\text{Type I error} = \frac{\text{FP}}{\text{TN} + \text{FP}} \tag{8}$$

$$\text{Type II error} = \frac{\text{FN}}{\text{TP} + \text{FN}} \tag{9}$$

All variables, found in equations above, are calculated on the basis of the error matrix, represented in Table 2.

Table 2 Confusion matrix used in this study

Predicted values			
		Positive[a]	Negative[b]
True values	Positive	True Positive (TP)	False Negative (FN)
	Negative	False Positive (FP)	True Negative (TN)

[a]Positive (True) means bad debtors
[b]Negative (False) means good debtors

[10]https://github.com/hyperopt/hyperopt.

It is important to make the following meta-level observation. Regardless of the academic discussions concerning value of each performance metrics (e.g., the famous "tug-of-war" between the Precision and the Recall), our approach is rooted in the real-world practices, followed by the banking industry. As noted, one of authors has worked in a large bank and, therefore, we can reasonably claim that from the bank perspective it are the Recall and the Type I error that are the most important. In other words, bank's focus is on making sure that bad loans will not happen. Hence, these two metrics are crucial from the point of view of credit risk, as they indicate how much the financial institution could potentially lose, by giving funding to a person who will stop paying the debt. Type I error indicates the ratio of how many people the model predicted as good[11] and they turned out to be bad[12] to the total number of all "good people". Recall, on the other hand, indicates the fraction of how many people the model correct predicted as bad, to all bad creditors. In an ideal world, the financial institution would like the Recall to be close to 1. This fact provided us with the guideline, which results should be considered "the best". Therefore, in our work, *we have focused on obtaining the best Recall score.*

This fact has two important consequences. (1) Since our goal was to achieve the best Recall score, it could have happened that the fact that our results are better than that reported in the literature/obtained using other classifiers can be seen as slightly unfair (if they were focused on improving other performance metrics). (2) At the same time, our results are *are real-world grounded* as the performance metrics of our choice are the ones that are actually pursued by the banks (the Recall, in particular). Obviously, in this way our approach differs from the academic research found in the literature.

6 Experimental Results

Let us now summarize the obtained results (keeping in mind the methodological considerations described in Sect. 5). For performance comparison, the results from [24, 25] were used as a "baseline". Let us now describe individual results for the two datasets.

6.1 Results for the German Credit Data

In Table 3, we summarize the results obtained for the German Credit dataset. Moreover, as a comparison, first, we show results from [24, 25], where its author used the same classifiers, but did not optimize the hyperparameters. Second, we report results from [25], where authors, firstly, used unsupervised algorithms to cluster

---

[11] good—debtors not delayed in repayment.

[12] bad—debtors delayed in repayment.

Table 3 Results for the German credit data

	AUC	ACC	MCC	Precision	Type I error	Type II error	Recall	Recall [24]	Recall [25]
XGBoost	**0.890**	**0.832**	**0.587**	**0.761**	0.143	0.239	0.645	0.393	–
Adaboost	0.779	0.831	0.584	0.754	0.091	0.351	0.649	0.44	–
GBDT	0.779	0.830	0.583	0.752	0.092	0.350	0.650	0.393	0.418
Random Forest	0.648	0.750	0.350	0.639	0.095	0.610	0.390	0.343	0.754
SVM_rbf	0.530	0.672	0.082	0.398	0.112	0.828	0.172	0.097	–
LinearSVM	0.516	0.328	0.081	0.308	0.957	**0.011**	**0.989**	0.477	0.459
KNN	0.548	0.695	0.139	0.481	**0.083**	0.821	0.179	0.373	0.443

observations, and then they build supervised individual models. We believe that both comparison fairly illustrate the comparative quality of our results.

The first conclusion is that there is no single model that has the highest score across all results. However, the model that "leads the way" is the XGBoost (highest AUC, ACC, MCC and Precision). However, as mentioned above, we focus on the Recall value. Hence, the "winner" seems to be the LinearSVM algorithm that has the highest score (0.989). Unfortunately, it also has the highest Type I error (0.957). This means that, in the banking practice, this approach should be avoided (as discussed in Sect. 5). Therefore, we cannot say that this is the best algorithm.

This being the case, the AdaBoost and the GBDT are the "best overall algorithms", to apply to our problem, from the bank's point of view. The Recall score is high (0.65), and they have relatively low Type I error (0.09; although, again, not the lowest), so these are the algorithms that banks may want to be most interested in.

It is also worth mentioning that, compared to [24], algorithms XGBoost, AdaBoost, GBDT, LinearSVM, SVM_rbf and RF have a higher Recall score. Finally, compared to [25], GBDT and LinearSVM also have a higher Recall score. However, the Random Forest and the KNN algorithm have worse scores. This may be due to differences in data preparation for modeling.

6.2 *Results for the Give Me Some Credit Dataset*

In order to verify the effectiveness and stability of our strategy, we have also applied the same methodology to the Give Me Some Credit dataset. Comparison between our classifiers and these from [24] is shown in Table 4. Recall, that there, authors build the same classifiers, but did not optimize hyperparameters.

Note that, here, the XGBoost algorithm did not have the highest values of the four performance metrics, as in above Section. In this case, only the MCC metric (0.348) had the highest value, but this value is much lower than the satisfactory one (above 0.5). The best algorithm in terms of the number of "best" values of metrics was

Table 4 Results for the give me some credit data

	AUC	ACC	MCC	Precision	Type I error	Type II error	Recall	Recall [24]
XGBoost	0.869	0.81	**0.348**	0.228	0.02	0.772	0.769	0.17
Adaboost	**0.873**	0.801	0.345	0.222	0.019	0.778	0.784	0.178
GBDT	0.872	0.792	0.341	0.215	**0.018**	0.785	**0.799**	0.17
Random Forest	0.867	0.778	0.326	0.204	**0.018**	0.796	0.796	0.152
SVM_rbf	0.656	0.643	0.12	0.106	0.044	0.894	0.584	0.0
LinearSVM	0.513	**0.934**	0.12	**0.621**	0.065	**0.379**	0.027	0.006
KNN	0.54	0.931	0.055	0.265	0.066	0.735	0.019	0.013

the SVM with the linear kernel, because it had the highest ACC (0.934), Precision (0.621) and the lowest Type II error (0.379).

However, note that the most important metric for us is Recall. Here, the algorithm that had the highest Recall score (0.799) was the GBDT. It also had the second highest AUC (0.872), and the lowest Type I error (0.018, including Random Forest), which (as noted above) is the second equally important metric in credit risk control. Note also that AdaBoost had the highest AUC (0.873). The worst results were achieved by the SVM_rbf and the KNN models.

Compared to the [24], all algorithms calculated with our methodology achieve higher Recall scores, especially in boosting and bagging models. For instance, for XGBoost, AdaBoost, GBDT and Random Forest, the difference is more than 50% percentage points. This is an improvement in the quality of the classifiers of over 300%. This may be due to the fact that the data was very unbalanced and our algorithms took this fact into account.

6.3 Results for the Meta-Classifiers

Let us now discuss the results for the two consensus models: the **Weighted Average Recall Error (WARE)** and the **Weighted Average Type-1 Error (WA-T1E)**, described in Sect. 3.1. Both were applied to the two datasets. They were tested because they had promising results reported in [37]. In Table 5, results for the German Credit dataset are presented. Next, in the Table 5, results for the Give Me Some Credit dataset are summarized.

In Table 5, we can notice that the meta-classifiers have not reached a single higher metric, over the best metric from the individual models, shown in Table 3. However, the MCC metric is high and is comparable to the results of the XGBoost model. The Recall score was lower than expected, and the Type I error was one of the highest.

Table 6 shows more promising results. The MCC metric achieved a better result (0.395) than the best result from individual models (0.348, for the XGBoost model).

Table 5 Results for the German credit data for consensus models

	WARE	WA-T1E
AUC	0.884	0.869
ACC	0.827	0.806
MCC	0.568	0.507
Recall	0.585	0.458
Precision	0.786	0.817
Type I error	0.161	0.196
Type II error	0.214	0.183

Table 6 Results for the give me some credit data for consensus models

	WARE	WA-T1E
AUC	0.873	0.867
ACC	0.876	0.924
MCC	0.395	0.391
Recall	0.667	0.432
Precision	0.305	0.432
Type I error	0.026	0.041
Type II error	0.695	0.568

The AUC achieved the same result (0.873), as the best result from the individual models (for the AdaBoost model). Moreover, ACC, Precision and Type II error have the second best result, compared to the individual models shows in Table 4. However, for our most important metrics, i.e., Recall and Type I error, the results are one of the worst.

7 Concluding Remarks

With the expanding and rapidly growing credit market, and the popularity of machine learning methods, ML-based credit scoring approaches are becoming an increasingly important aspect of differentiating between good and bad borrowers. Due to the area in which banks operate, even a small change in the predictive power of the models results in large income growth. In this work, we propose to change the popular (in scientific literature) approach to credit scoring models so that the main goal is the best possible Recall score and a methodology to achieve this.

In this context we have discussed: optimization of hyperparameters of the models, in terms of the best Recall score, and taking into account class imbalance. We have conducted experiments on two publicly available benchmarks (Sect. 4). In both cases, the best Recall was achieved by the SVM model with the linear kernel, but it had the highest Type I error, which is the second most important parameter from the bank's point of view. Therefore, among the examined algorithms, the Gradient Boosting Decision Trees turned out to be "the best" in both cases. They had high Recall score and, at the same time, very low Type I error. Besides, for both datasets, the models with the worst results were the SVM with radial kernel and the K-nearest neighbors.

Then, we used two related methods, to create meta-classifiers. However, it turned out that this did not bring the expected (positive) results. The Recall score and the Type I error turned out to be worse than in the case of individual models. However, we observed improvement of results for the Give Me Some Data (which is very unbalanced) for the MCC and the AUC metrics. This means that these two simple meta-classifiers are an interesting solution, but for our problem it is not a good solutions. In our future work, we focus on more complicated meta-classificators, such as these presented in [36] and we try to establish if this methodology can result in a better Recall score. We will report on our work in subsequent publications.

References

1. M. Machowiak, Personal communication
2. M. Paprzycki, M. Machowiak, Development of e-banking in poland, pp. 1–8 (2000)
3. Raport Kredyt Trendy, pp. 1–45 (2019)
4. N. Chen, B. Ribeiro, A. Chen, Financial credit risk assessment: a recent review. Artif. Intell. Rev. **45** (2015)
5. E.I. Altman, Financial ratios, discriminant analysis and the prediction of corporate bankruptcy. J. Fin. pp. 589–609 (1968)
6. J.C. Wiginton, A note on the comparison of logit and discriminant models of consumer credit behavior. J. Fin. Quant. Anal. pp. 757–770 (1980)
7. R. Florez, J. Ramon, Enhancing accuracy and interpretability of ensemble strategies in credit risk assessment. A correlated-adjusted decision forest proposal. Expert Syst. Appl. **42** (2015)
8. P. Wysiński, The use of credit scoring in credit risk management. Int. Bus. Global Econ. pp. 253–268 (2013)
9. L. Thomas, R. Oliver, D. Hand, A survey of the issues in consumer credit modelling research. J. Oper. Res. Soc. **56**, 1006–1015 (2005)
10. A. Hanic, E. Žunić Dželihodžić, A. Dzelihodzic, Scoring models of bank credit policy management. Econ. Anal. **46**, 12–27 (2013)
11. J. Abellán, F. Castellano, A comparative study on base classifiers in ensemble methods for credit scoring. Expert Syst. Appl. **73** (2016)
12. M. Alaraj, M. Abbod, Classifiers consensus system approach for credit scoring. Knowl.-Based Syst. **104** (2016)
13. A. Bequé, S. Lessmann, Extreme learning machines for credit scoring: an empirical evaluation. Expert Syst. Appl. **86** (2017)
14. S.H. Shashi Dahiya, N. Singh, A feature selection enabled hybrid-bagging algorithm for credit risk evaluation. Expert Syst. **34**(6) (2017)
15. Y. Xia, C. Liu, B. Da, F. Xie, A novel heterogeneous ensemble credit scoring model based on bstacking approach. Expert Syst. Appl. **93** (2017)
16. A. Ben-David, E. Frank, Accuracy of machine learning models versus "hand crafted" expert systems–a credit scoring case study. Expert Syst. Appl. **36**, 5264–5271 (2009)
17. S. Sohn, D.-H. Kim, J. Yoon, Technology credit scoring model with fuzzy logistic regression. Appl. Soft Comput. **43**, 150–158 (2016)

18. B. Twala, Multiple classifier application to credit risk assessment. Expert Syst. Appl. **37**, 3326–3336 (2010)
19. M. Alaraj, M. Abbod, A systematic credit scoring model based on heterogeneous classifier ensembles (2015)
20. M. Alaraj, M. Abbod, A new hybrid ensemble credit scoring model based on classifiers consensus system approach. Expert Syst. Appl. **64** (2016)
21. L. Cleofas, V. García, A. Marqués, J. Sánchez, Financial distress prediction using the hybrid associative memory with translation. Appl. Soft Comput. **44**, 144–152 (2016)
22. S. Lessmann, B. Baesens, H.-V. Seow, L. Thomas, Benchmarking state-of-the-art classification algorithms for credit scoring: an update of research. Euro. J. Oper. Res. (2015). https://doi.org/10.1016/j.ejor.2015.05.030
23. G. Wang, J. Ma, L. Huang, K. Xu, Two credit scoring models based on dual strategy ensemble trees. Knowl.-Based Syst. **26**, 61–68 (2012)
24. A. Hernik, J. Balicki, Metodyka porównania algorytmów klasyfikacji z pomoca benchmarków do oceny wiarygodności kredytowej (2018)
25. B. Wang, Y. Kong, Y. Zhang, D. Liu, L. Ning, Integration of unsupervised and supervised machine learning algorithms for credit risk assessment. Expert Syst. Appl. **128** (2019)
26. T.M. Cover, P.E. Hart, Nearest neighbor pattern classification. IEEE Trans. Inf. Theory **13**, 21–27 (1967)
27. W. Henley, D.J. Hand, A k-nearest-neighbour classifier for assessing consumer credit risk (1996)
28. C. Cortes, V. Vapnik, Support vector machine. Mach. Learn. pp. 1303–1308 (1995)
29. L. Breiman, Some infinity theory for predictor ensembles. Technical Report **577** (2000)
30. L. Breiman, Consistency for a simple model of random forest. Technical Report **670** (2004)
31. R.E. Schapire, K. Yoav Freund, Boosting: foundations and algorithms. Kybernetes **42**, 164–166 (2013)
32. J.H. Friedman, Greedy function approximation: a gradient boosting machine. IMS 1999 Reitz Lectures (1999)
33. Y. Freund, R.E. Schapire, A decision-theoretic generalization of on-line learning and an application to boosting. Comput. Syst. Sci. pp. 119–139 (1996)
34. J. Friedman, Greedy function approximation: a gradient boosting machine. Ann. Stat. **29**, 1189–1232 (2001)
35. A. Byczynska, M. Ganzha, M. Paprzycki, M. Kutka, Evidence quality estimation using selected machine learning approaches, pp. 1–8 (2020)
36. M. Ganzha, M. Paprzycki, S. Jakub, Combining information from multiple search engines–preliminary comparison. Inform. Sci. **180**, 1908–1923 (2010)
37. M. Bourel, C. Crisci, A. Martínez, Consensus methods based on machine learning techniques for marine phytoplankton presence-absence prediction. Ecol. Inform. **42** (2017)
38. J. Bergstra, D. Yamins, D.D. Cox, Hyperopt: a python library for optimizing the hyperparameters of machine learning algorithms (2013)

Analytical Study on Algorithms for Content-Based Mobile Phone Recommendation System

P. V. S. M. S. Kartik, B. Abhilash, Durga Naga Sai Sravan Nekkanti, and G. Jeyakumar

Abstract Recommendation of anything helps in filtering out unwanted and irrelevant products from the entire work set which are not of any use or do not add any importance or value to the current task set to be accomplished. Mobile phone recommendation systems could be a probable revolution in the near future, as the telecommunication and mobile phone production industries are increasing exponentially in the market. Such systems would extract description of similar mobile phones which are most correlated with the mobile phone of user interest. The proposed system carries out this task taking the dataset which comprises of mobile phones and its features extracted from the e-commerce Web site '*Flipkart*'. The dataset comprises of 6917 records in total each one corresponding to a unique mobile phone in the Web site. The proposed model finds the similarities and does the predictions based on the input features in the dataset.

Keywords Natural language processing · Tokenizing · Cosine similarity · Corpus · *LSI* model · Similarity index

1 Introduction

These days every customer is confronted with various choices. For example, if a customer searching for a book to examine with no particular thought of what he

P. V. S. M. S. Kartik (✉) · B. Abhilash · D. N. S. S. Nekkanti · G. Jeyakumar
Department of Computer Science and Engineering, Amrita School of Engineering, Amrita Vishwa Vidyapeetham, Coimbatore, India
e-mail: cb.en.u4cse17033@cb.amrita.students.edu

B. Abhilash
e-mail: cb.en.u4cse17002@cb.amrita.students.edu

D. N. S. S. Nekkanti
e-mail: 3cb.en.u4cse17018@cb.amrita.students.edu

G. Jeyakumar
e-mail: g_jeyakumar@cb.amrita.edu

M. Paprzycki et al. (eds.), *Intelligent Systems, Technologies and Applications*, Advances in Intelligent Systems and Computing 1353,
https://doi.org/10.1007/978-981-16-0730-1_2

needs, there is a wide extent of chances for how his interest may work out. He may consume lot of time examining around the Internet content and trawling through. In such cases, he tends to take suggestions/recommendations from other people. This has led to the idea of developing recommendation systems instead of getting it from other people. The force of recommendation engines is being benefited by most associations these days. These systems wipe out data over-burden issues, i.e., the normal state of a Web customer, of having an exorbitant measure of information to choose a decision or remain instructed about a subject. A recommendation system is a kind of data sifting framework which endeavors to anticipate the inclinations of a client, and make recommendations based on these inclinations. There are a wide variety of applications for these systems. They have become progressively notable in the course of the most recent years and are presently used in most online stages are used. Frequently, these systems can gather data about a client's decisions and can utilize this data to improve their recommendations later on.

One of the two primary methodologies that are broadly utilized in recommendation systems is the content-based filtering, where it is attempted to profile the clients' interests utilizing data gathered and suggest things dependent on that profile. Content-based recommendation is the approach that has been applied in this work and is experimented for mobile phone recommendation.

The rest of the paper is organized as follows—Sect. 2 to present the works related to this work, the Sect. 3 to discuss the design of experiments used for this study, the Sect. 4 details the architecture of the proposed system, the Sect. 5 to discuss the results, and the Sect. 6 to conclude the work.

2 Related Works

The proposed architecture involves computing the similarity index using both Count Vectorizer and *TF-IDF* to recommend mobile phones to the user. As there are no research works reported on content-based recommendation of mobile phones, the research works related to recommending products from e-commerce sites like *Flipkart* and *Amazon* are discussed here. Also, the research works on content-based filtering that have been done over the past and where some prominent exploration is needed are highlighted.

Techniques involved in content-based filtering (*Count Vectorizer* and *TF-ID*) for calculating the similarity index were presented in [1]. The author of [2] suggested combining recommender systems innovations with mobile phone search can become beneficial for mobile users, as the mobile phones are becoming an essential device for accessing the information. According to [3], mobile recommenders form a new exploration field that has not been efficiently studied and further analysis is needed to concentrate on the implementation of recommendation systems in mobile phones. The authors of [4] explained the recommendation process's multiple phases, namely data collection, learning, and recommendation. According to [5], use of recommender systems is crucial for the service providers as well as for the users. The reasons that the providers use such systems are—increase in sales, employ a decent variety of items, increase in client satisfaction, and increase in faithfulness. Using specifications of an item to recommend additional items with similar properties by the model in content-based filtering was proposed in [6]. Various methods involved in preprocessing the dataset like tokenization, stop word removal, stemming, and dictionary were explained in [7]. The authors of [8] proposed an approach to build a recommender system that can be adapted to customizable products such as computers and home theater system. The authors expanded the means of recommendation systems, so that they recommend the products alongside their customizable parts to customers and provide them a more significant alternative in looking. Various challenges of recommender systems like cold start, sparsity, and scalability were explained in [9]. The authors of [10] described how the implementation of recommender system has enlarged on the Internet, that demonstrated its use in various areas like recommending movies, applications, gadgets, songs, Web sites and wherever the user will provide ratings to an item of his choice. According to [11], for improving the retrieval method of latent semantic indexing (*LSI*) stop words are enforced to spot normally used words (which can be neglected before or when the processing of a natural language) and removing them plays a crucial role. The authors of [12] explained why it is vital to reduce the dimension of the vectors used to represent the search space for retrieving textual information. Representation of vectors in high dimension will cause scattered vectors, which will affect the efficiency of retrieval of information.

The proposed system used *LSI* in order to reduce the dimension of the term space into a lower-dimensional *LSI* space, which is more compact. Thus, it could identify similarities that go beyond term similarity [13]. The authors of [14] described how personalized recommendation helps in improving the satisfaction of the user and identified the gaps and the issues in traditional methods of recommendation systems. The authors of [15] provided a comprehensive study on the RS covering the different recommendation approaches, associated issues, and techniques used for information retrieval. The main purpose of this paper is to spot the research trend in RS. Furthermore, this paper also envisions the future of RS. Considering the massive gap between the users and service providers of mobile phone recommendation systems, this work focuses on exploring various algorithms for mobile phone recommendation systems and thus filling the gap between users and service providers.

3 Design of Experiments

With late progressions in innovation, a mobile phone can do all the tasks a personal computer is prepared to do and furthermore the greater part of the Web traffic is originating from mobile phone. It is a clear indication that the interest and use of mobile phones are expanding at a tremendous measure. Prior, the quantity of mobile phone user in a family was restricted to few but these days every single individual have a mobile phone, because of which demand of mobile phones is going up and the mobile phone industries are enhancing the specification and configuration of them to pull in an ever-increasing number of clients. Hence, mobile phone variations are additionally expanding at an exponential rate. This makes a great deal of disarray for the purchasers and they may wind up purchasing an unseemly item. To beat these issues of data over-burden, a recommender system is suggested. Because of quick development in the varieties of mobile phone dependent on their structure and functionalities, it is getting hard for the client to choose the best mobile phone as per their prerequisites and inclinations. The primary issue in mobile phone determination is the examination between the various things, which is very complicated and physically it is impossible. So, an efficient methodology is necessary for the complicated examination between the things and to recommend the client most proper thing as per their inclinations. Therefore, recommender system helps in doing an examination and assessment over enormous assortments of mobile phone functionalities, highlights, structure, and brand to give a productive recommendation to the user.

3.1 Dataset

The proposed system used a dataset which consists of over 6900 mobile phone of various brands with their specifications, descriptions, user ratings on a 5-point scale and reviews. This information was scraped from the *Flipkart* Web site for the corresponding mobiles as *Flipkart* is one of the most popular sites and the data here is more organized when compared to other Web sites. Sample dataset is shown in Fig. 1.

<table>
<tr><th>index</th><th>Product Name</th><th>RamRom</th><th>Price</th><th>Ratings</th><th>RatingReview</th><th>Features</th><th>Smalldesc</th><th>TotalDesc</th><th>Specification</th><th>DescriptionHeadings</th><th>EachRate</th><th>SpecialFeatures</th></tr>
<tr><td>0</td><td>Acer Z630S (Black & Gold/Black - G, 32 GB)</td><td>3 GB RAM | 32 GB ROM | Expandable Upto 32 GB</td><td>7190</td><td>3.8</td><td>5609 Ratings ,1131 Reviews,</td><td>3 GB RAM | 32 GB ROM | Expandable Upto 32 GB13...</td><td>This Acer phone has an HD screen and a octa co...</td><td>Thanks to the innovative hybrid camera UI, you...</td><td>Handset, Headset, Charger, AC Adaptor, USB Cab...</td><td>Innovative Hybrid UI,Voice Control,Bright Magi...</td><td>[('5', '2550'), ('4', '1209'), ('3', '704'), (...</td><td>[('Camera', '3.1'), ('Battery', '2.5'), ('Valu...</td></tr>
<tr><td>1</td><td>Acer Liquid Z530 (Black, 16 GB)</td><td>2 GB RAM | 16 GB ROM | Expandable Upto 32 GB</td><td>5750</td><td>3.8</td><td>9356 Ratings ,1361 Reviews,</td><td>2 GB RAM | 16 GB ROM | Expandable Upto 32 GB12...</td><td>Everything's better and simpler with the Acer ...</td><td>The Acer Z530 features an 8 MP rear camera as ...</td><td>Handset, AC Adaptor, USB Cable, Headset, Quick...</td><td>Camera,Voice Camera,Bright Magic,Photo With So...</td><td>[('5', '4090'), ('4', '2201'), ('3', '1403'), ...</td><td>[('None', 'None')]</td></tr>
<tr><td>2</td><td>Adcom J1</td><td>32 MB RAM | 32 MB ROM | Expandable Upto 8 GB</td><td>699</td><td>3</td><td>9 Ratings ,2 Reviews,</td><td>32 MB RAM | 32 MB ROM | Expandable Upto 8 GB4...</td><td>None</td><td>None,</td><td>Handset, Charger, Battery, Headset,User Manual...</td><td>None,</td><td>[('5', '3'), ('4', '1'), ('3', '1'), ('2', '1' ...</td><td>[('None', 'None')]</td></tr>
<tr><td>3</td><td>Adcom AJ2</td><td>32 MB RAM | 32 MB ROM | Expandable Upto 8 GB</td><td>699</td><td>3.8</td><td>233 Ratings ,38 Reviews,</td><td>32 MB RAM | 32 MB ROM | Expandable Upto 8 GB4...</td><td>None</td><td>None,</td><td>1 Handset, 1 Battery, 1 Charger, Warranty Card...</td><td>None,</td><td>[('5', '112'), ('4', '38'), ('3', '42'), ('2',...</td><td>[('None', 'None')]</td></tr>
<tr><td>4</td><td>Adcom A-J1</td><td>32 MB RAM | 32 MB ROM | Expandable Upto 8 GB</td><td>699</td><td>3.7</td><td>568 Ratings ,76 Reviews,</td><td>32 MB RAM | 32 MB ROM | Expandable Upto 8 GB4...</td><td>None</td><td>None,</td><td>Handset, Charger, Battery, Headset,User Manual...</td><td>None,</td><td>[('5', '262'), ('4', '98'), ('3', '90'), ('2',...</td><td>[('None', 'None')]</td></tr>
</table>

Fig. 1 Sample dataset

3.2 Content-Based Implementation

Data Scraping—To obtain data consisting of mobile phones and their specifications, the *Flipkart* Web site page content was accessed using the *Beautiful Soup* Python library and *Requests* module.

Beautiful Soup is a tool for parsing *HTML* code and grabbing exactly the information you need. *Requests* is a module in Python which can be used to send *HTTP requests*. Once the structure of the Web site was identified, a script to automatically extract the relevant information about a mobile phone was developed.

Way of data collection is described in Fig. 2, the class name of the element to be extracted is highlighted in red color, using this name, the data required is extracted and this process is repeated for extracting data for each mobile of different brands.

The list of columns present in the dataset is shown in Fig. 3.

In our dataset used, we have 13 columns as shown in figure (col names fig). For further analysis and modeling, we have taken all the columns into consideration, i.e., a combined column is created where each row of the new column has all the data combined from the corresponding rows of the other 13 columns. This is done because

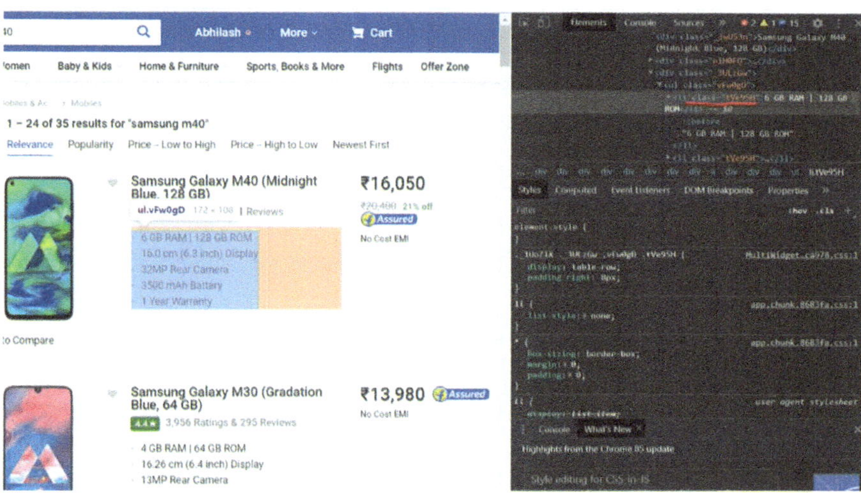

Fig. 2 Way of data collection

```
df.columns

Index(['index', 'Product Name', 'RamRom', 'Price', 'Ratings', 'RatingReview',
       'Features', 'Smalldesc', 'TotalDesc', 'Specification',
       'DescriptionHeadings', 'EachRate', 'SpecialFeatures', 'HREFs'],
      dtype='object')
```

Fig. 3 List of columns in the dataset

as we analyze the overall semantics of the textual data, each keyword contributes to the model and has its own weight of contribution. So, to get an overall contribution from all the data, we take in the combined features.

The details extracted are:

Product Name. *Samsung Galaxy A50 (Blue, 64 GB).*

RAM & ROM-*6 GB RAM | 64 GB ROM | expandable up to 512 GB.*

Price—*18,999,* **Rating**—*4.4,*

Number of Ratings and Reviews—*10,235 ratings and 920 reviews.*

Features—*6 GB RAM | 64 GB ROM | expandable up to 512 GB,16.26 cm (6.4 inch) full HD + Display 25 MP + 5 MP + 8 MP | 25 MP front camera, 4000 mAh lithium-ion battery, Exynos 9610 processor, super-AMOLED display, brand warranty of 1 year available for mobile and 6 months for accessories.*

Description—Do a lot more than just text and call your friends with the *Samsung Galaxy A50* smartphone. The *Exynos 9610 octa-core Processor* makes multitasking a breeze. Take photography a notch higher with its revolutionary *Triple camera system* that comprises a *25 MP low-light camera, 8 MP ultra-wide* Camera and *5 MP live.*

Focus Camera.

[(‘***5Star***’, ‘6430′), (‘***4Star***’, ‘2371′), (‘***3Star***’, ‘675′), (‘***2Star***’, ‘215′), (‘***1Star***’, ‘544′)], i.e.,6430/10,235 users rated 5/5 for this product.

[(‘***Camera***’, ‘3.6′), (‘***Battery***’, ‘3.0′), (‘***Display***’, ‘3.7′), (‘***Value for money***’, ‘4.2′)] –Ratings out of 5.

All the above details were considered while recommending the mobile phone to the users. A training model is proposed to use all the details extracted and to recommend the mobiles phone.

4 Proposed System

The architecture of the proposed mobile phone recommendation system is presented in Fig. 4.

4.1 Prediction Using LSI Model

After extracting the combined features for each mobile phone in the dataset as mentioned above, tokenizing the combined features column was carried out using *RegExpTokenizer*. After tokenization, extracted the words which are not ‘stop words’. Stop words are the common words in a language which do not add value to the learning process and they do not add to the uniqueness factor of the feature set. Then these extracted unique words, i.e., non-stop words were combined to form a dictionary. The dictionary consists of a combination of distinctive words.

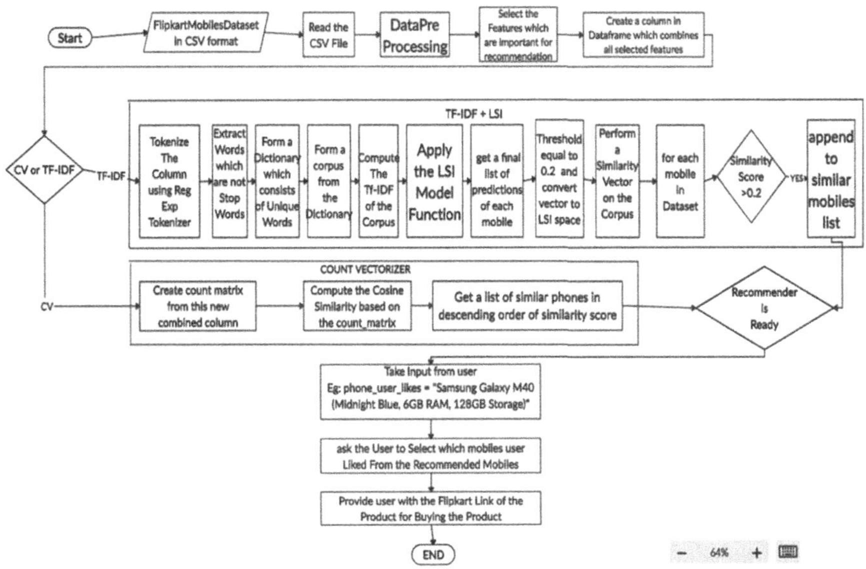

Fig. 4 Architecture diagram—proposed recommendation system

In the next step, a corpus is formed from the dictionary. The corpus is the group of gathered information of all synopses preprocessed and transformed using the dictionary. Next, the term frequency-inverse document frequency (*tf-idf*) transformation was performed. The *tf-idf* transformation gives an indication of weights or importance of words in the corpus. The *tf-idf* value is on the count of number of words in a document and also on the number of times it appears in the corpus. The '*tf-idf*' function transforms the corpus from a grouped integer counts format into *tf-idf* real values weight. Then, computed the *tf-idf* of the corpus which was extracted earlier (corpus-*tf-idf*). Next, the '*LSI* Model' was applied to transform *tf-idf* corpus into a latent n-dimensional space. The *LSI* model is present in the *Gensim* library of Python. After transforming the corpus to *LSI* space, it was indexed. This was done using *MatrixSimilarity* function in similarities library. To get a final list of predictions of each mobile phone, a threshold was set to 0.2 initially, then iterated over each element of the list which stored the *tf-idf* of the corpus. During the iteration, the current element of the corpus *tf-idf* was converted to LSI space, then performed a similarity vector construction against the corpus to find the index of the *LSI* space transformed vector. Now, for each mobile phone in the dataset, it was checked that the corresponding index in the similarity vector (similarity score) is greater than the threshold 0.2. If yes, it is appended to the similar mobile phones list of the current mobile in the iteration. So, at the end, a list was generated where for each mobile phone in the dataset its corresponding row showed all the similar mobile phones (the mobiles with similarity index >0.2). Thus, a list of recommended mobile phones based on the content-based prediction was obtained.

4.2 Prediction Using Cosine Similarity

Cosine similarity is the measure of the similarity between two documents by counting the frequencies of similar terms and considering them as vectors. It is calculated by the cosine of the angle between two vectors and it ranges from 0 to +1 in the case of documents, because the frequencies of the terms cannot be negative. This measure suffers less from the curse of dimensionality than Euclidean distance. Document vectors tend to get long because there will be many words in a given document. So *cosine similarity* is a way to avoid the curse of dimensionality, which is measured by Eq. (1).

$$\text{Cosine Similarity}(|x.y|) = \frac{x.y}{\|x\|.\|y\|} \tag{1}$$

where $\|x\|$ is the norm or magnitude of vector $x = (x_1, x_2, x_3, \ldots, x_n)$ *is* defined as $\sqrt{x_1^2 + x_2^2 + x_3^2 + \cdots, x_n^2}$.

Basically, it is the Euclidean Distance of the vector from origin. Similarly, $\|y\|$ is the magnitude of vector y. If the *cosine similarity* value is closer to 1 then angle between the vectors is smaller and similarity between the vectors is greater.

After extracting the combined features for each mobile phone in the dataset, all the selected features are combined together. A count matrix was created from these new combined features. Now, *cosine similarity* was computed based on the count matrix. After this process, input was taken from the user. Based on the input given by the user, index of the mobile phone is retrieved from its name. A list of similar mobile phones was generated in descending order of similarity score, then names and details of first 20 similar mobile phones corresponding to the mobile phone the user had chosen were displayed. Then, the user can choose any of the mobile phone from the given top 20 recommendations and then a link to buy that mobile phone is given to the user.

5 Results and Discussions

The final result obtained based on the content-based filtering is shown as an image in Fig. 5. In Fig. 5, the column—'*Mobile Name*' is the mobile phone which the user likes, and the columns *Mob1,Mob2,...,Mob20* are the first 20 similar mobile phones recommended to the user and *Score1,Score2, ..., Score20* are the similarity indices of the top 20 recommendation, respectively. The set of recommendations for additional set of user inputs is depicted in Fig. 6, for reference.

The comparison of *LSI* and *cosine similarity* technique was carried out for the mobile models 'OnePlus 7T Froster Silver, 256 GB' and 'Realme XT Pearl Blue, 64 GB'. The methods are compared based on the similarity score. The results obtained for 'OnePlus 7T Froster Silver, 256 GB' and 'Realme XT Pearl Blue, 64 GB' are

	Unnamed: 0	Mobile Name	Mob1	Score1	Mob2	Score2	Mob3	Score3	Mob4	Score4	...	Score16	Mob17	Score17	Mob18	Score1
0	0	Acer Z630S (Black & Gold/Black - G, 32 GB)	Acer Liquid Z530 (Black, 16 GB)	0.949191	Panasonic P90 (Black, 16 GB)	0.742600	Yu Yunicorn (Gold Rush, 32 GB)	0.738004	Lava X50 4G (Blue Silver, 8 GB)	0.731686	...	0.675502	Samsung Galaxy A8 (Black, 32 GB)	0.664698	Honor 4X (White, 8 GB)	0.66127
1	1	Acer Liquid Z530 (Black, 16 GB)	Acer Z630S (Black & Gold/Black - G, 32 GB)	0.949191	Lava X81 4G with VoLTE (Space Grey, 16 GB)	0.772154	Panasonic P95 (Blue, 16 GB)	0.736302	Panasonic P95 (Grey, 16 GB)	0.733992	...	0.678175	Yu Yunicorn (Rush Silver, 32 GB)	0.675824	Alcatel Idol 4 (Dark Grey, 16 GB)	0.67337
2	2	Adcom J1	Adcom A-J1	0.985032	Gfive Classic	0.663438	Gfive Audi	0.635590	Mesee M3	0.632486	...	0.527197	Diamond D09	0.525580	Snexian Splender	0.52392
3	3	Adcom AJ2	Adcom AJ2	0.715455	InFocus F110	0.703642	Adcom A101 Voice Changer	0.688862	M-tech G24	0.672982	...	0.536631	Gfive Classic	0.533683	Karbonn K88	0.52246

Fig. 5 Results for content-based filtering

Mobile Name	Mob1	Score1	Mob2	Score2	Mob3	Score3	Mob4	Score4	Mob5	Score5	Mob6	Score6	Mob7	Score7	Mob8	Score8	Mob9
OPPO F15 (Light	OPPO F15 (Unicorn	0.99930245	Realme X2 (Pearl Wh	0.866089	OPPO A52 (Stre	0.84243	OPPO Reno2 (	0.839641	Realme 5 Pro	0.828523	Realme 5 Pro (S	0.82676	Realme 5 Pro (Cr	0.826084	Realme X50	0.814911	OPPO Ren
OPPO K1 (Piano	OPPO K1 (Piano Bla	1	OPPO K1 (Astral Blue	0.898959	OPPO A1K (Red,	0.767842	OPPO F11 (Mi	0.72053	OPPO F11 (Je	0.71967214	OPPO F11 Pro (	0.709833	OPPO A5s (Gold,	0.704337	OPPO A5s ((	0.704301	OPPO A5s
OPPO F11 (Marb	OPPO F11 (Marble (	1	OPPO F11 (Marble G	0.968495	OPPO F11 (Fluo	0.955917	OPPO F11 Pro	0.923688	OPPO F11 Pro	0.90648144	OPPO A1K (Red	0.897017	OPPO F9 Pro (Su	0.77572	OPPO A5s ((	0.765102	OPPO A5s
OPPO A7 (Glarin	OPPO A7 (Glaring G	0.99999994	OPPO A7 (Glaze Blue	0.998804	OPPO A7 (Glarir	0.988488	OPPO A7 (Gla	0.987396	OPPO A5s (Bl	0.9077667	OPPO A5s (Blue	0.906894	OPPO A5s (Black,	0.906089	OPPO A5s ((	0.906055	OPPO A5s
OPPO F11 Pro (A	OPPO F11 Pro (Aur	1	OPPO F11 Pro (Wate	0.971895	OPPO F11 (Mari	0.950457	OPPO F11 (Mi	0.923688	OPPO F11 (Je	0.92232174	OPPO A1K (Blac	0.87172	OPPO F9 Pro (Su	0.871	OPPO F15 ((	0.779923	OPPO R17
OPPO A7 (Glarin	OPPO A7 (Glaring G	1	OPPO A7 (Glaze Blue	0.998811	OPPO A7 (Glarir	0.988488	OPPO A7 (Gla	0.987203	OPPO A5s (Bl	0.8854592	OPPO A5s (Blue	0.884524	OPPO A5s (Black,	0.883472	OPPO A5s ((	0.883181	OPPO A5s
OPPO A7 (Glaze	OPPO A7 (Glaze Blu	1	OPPO A7 (Glaring Gr	0.998811	OPPO A7 (Glaze	0.988494	OPPO A7 (Gla	0.987396	OPPO A5s (Bl	0.8871393	OPPO A5s (Blac	0.887013	OPPO A5s (Blue,	0.88513	OPPO A5s ((	0.884974	OPPO A5s
OPPO A7 (Glaze	OPPO A7 (Glaze Blu	0.9999999	OPPO A7 (Glaring Gr	0.998804	OPPO A7 (Glaze	0.988494	OPPO A7 (Gla	0.987203	OPPO A5s (Bl	0.9093351	OPPO A5s (Blac	0.909142	OPPO A5s (Blue,	0.907612	OPPO A5s ((	0.907412	OPPO A5s
OPPO Reno2 Z (	OPPO Reno2 Z (Sky	1	OPPO Reno2 (Ocean	0.900154	OPPO F15 (Blazi	0.818827	OPPO F15 (Lig	0.805521	OPPO A52 (Tw	0.7490188	OPPO F9 Pro (St	0.74024	OPPO F9 Pro (Sta	0.740028	Realme XT (	0.732774	OPPO F11
OPPO A5 2020 (	OPPO A5 2020 (Mir	1	OPPO A5 2020 (Dazz	0.999471	OPPO A9 2020 (	0.980609	OPPO A9 2020	0.979974	OPPO A9 202	0.91733825	OPPO A52 (Twi	0.702163	OPPO Reno2 Z (S	0.646654	Realme 5s (	0.633565	Realme 5s
OPPO F9 Pro (Su	OPPO F9 Pro (Sunris	1	OPPO F11 Pro (Auro	0.871	OPPO F11 Pro (	0.863965	OPPO F11 (Mi	0.826601	OPPO F11 Pro	0.8133728	Realme 3 (Black	0.781326	OPPO F11 (Jewel	0.777772	OPPO F11 (	0.77572	Realme 3 (
OPPO A5s (Gree	OPPO A5s (Green, 6	1	OPPO A5s (Black, 32	0.997845	OPPO A5s (Blac	0.997739	OPPO A5s (Bl	0.997084	OPPO A5s (Bl	0.99692434	OPPO A5s (Red,	0.996904	OPPO A7 (Glaze (	0.907271	OPPO A7 (G	0.906055	OPPO A1K
OPPO A5s (Blue,	OPPO A5s (Blue, 32	1	OPPO A5s (Blue, 32 (	0.999893	OPPO A5s (Blac	0.999348	OPPO A5s (Re	0.998123	OPPO A5s (Gi	0.99692434	OPPO A5s (Gold	0.996342	OPPO A7 (Glaze (	0.909335	OPPO A7 (G	0.906894	OPPO A1K
OPPO A5s (Red,	OPPO A5s (Red, 32	1	OPPO A5s (Black, 32	0.99922	OPPO A5s (Blue	0.998287	OPPO A5s (Bk	0.998123	OPPO A5s (Gi	0.996904	OPPO A5s (Gold	0.996442	OPPO A7 (Glaze (	0.906609	OPPO A7 (G	0.905528	OPPO A1K
OPPO A5s (Blue,	OPPO A5s (Blue, 32	0.9999999	OPPO A5s (Blue, 32 (	0.999893	OPPO A5s (Blac	0.999294	OPPO A5s (Re	0.998286	OPPO A5s (Gi	0.99708414	OPPO A5s (Gold	0.99651	OPPO A7 (Glaze (	0.907612	OPPO A7 (G	0.905229	OPPO A1K
OPPO A5s (Black	OPPO A5s (Black, 3	1	OPPO A5s (Blue, 32 (	0.999348	OPPO A5s (Blue	0.999294	OPPO A5s (Re	0.99922	OPPO A5s (Gi	0.99773854	OPPO A5s (Gold	0.997162	OPPO A7 (Glaze (	0.909142	OPPO A7 (G	0.907767	OPPO A1K
OPPO R17 (Ambi	OPPO R17 (Neon Pe	0.9585908	Samsung Galaxy M3	0.690574	Samsung Galaxy	0.672268	Huawei Y9 (Se	0.650127	Huawei Y9 (B	0.6077547	Samsung Galax	0.587142	Samsung Galaxy	0.566542	Vivo U20 (B	0.536896	Samsung G
OPPO R17 (Neon	OPPO R17 (Ambien	0.9585908	Honor View 20 (sapp	0.67726	Samsung Galaxy	0.643873	Samsung Gala	0.619999	Huawei Y9 (S	0.60010815	Samsung Galax	0.553696	LG Q60 (New Mo	0.551761	Samsung Ga	0.510756	OPPO K3 (
OPPO Reno2 (Oc	OPPO Reno2 (Ocea	1	OPPO Reno2 Z (Sky (	0.900154	OPPO F15 (Blazi	0.848071	OPPO F15 (Lig	0.839641	Realme X2 (P	0.79591024	OPPO A52 (Twi	0.778786	Realme 5 Pro (Sp	0.776446	Realme 5 Pr	0.775315	Realme XT
OPPO K1 (Piano	OPPO K1 (Piano Bla	1	OPPO K1 (Astral Blue	0.898959	OPPO A1K (Red,	0.767842	OPPO F11 (Mi	0.72053	OPPO F11 (Je	0.71967214	OPPO F11 Pro (	0.709833	OPPO A5s (Gold,	0.704337	OPPO A5s ((	0.704301	OPPO A5s
OPPO F11 (Marb	OPPO F11 (Marble (	1	OPPO F11 (Marble G	0.968495	OPPO F11 (Fluo	0.955917	OPPO F11 Pro	0.923688	OPPO F11 Pro	0.90648144	OPPO A1K (Red	0.897017	OPPO F9 Pro (Su	0.77572	OPPO A5s ((	0.765102	OPPO A5s
OPPO F11 Pro (A	OPPO F11 Pro (Aur	1	OPPO F11 Pro (Wate	0.971895	OPPO F11 (Mari	0.950457	OPPO F11 (Mi	0.923688	OPPO F11 (Je	0.92232174	OPPO A1K (Blac	0.87172	OPPO F9 Pro (Su	0.871	OPPO F15 ((	0.779923	OPPO R17
OPPO A7 (Glarin	OPPO A7 (Glaring G	0.99999994	OPPO A7 (Glaze Blue	0.998804	OPPO A7 (Glarir	0.988488	OPPO A7 (Gla	0.987396	OPPO A5s (Bl	0.9077667	OPPO A5s (Blue	0.906894	OPPO A5s (Black,	0.906089	OPPO A5s ((	0.906055	OPPO A5s
OPPO A7 (Glarin	OPPO A7 (Glaring G	1	OPPO A7 (Glaze Blue	0.998811	OPPO A7 (Glarir	0.988488	OPPO A7 (Gla	0.987203	OPPO A5s (Bl	0.8854592	OPPO A5s (Blue	0.884524	OPPO A5s (Black,	0.883472	OPPO A5s ((	0.883181	OPPO A5s
OPPO A7 (Glaze	OPPO A7 (Glaze Blu	1	OPPO A7 (Glaring Gr	0.998811	OPPO A7 (Glaze	0.988494	OPPO A7 (Gla	0.987396	OPPO A5s (Bl	0.8871393	OPPO A5s (Blac	0.887013	OPPO A5s (Blue,	0.88513	OPPO A5s ((	0.884974	OPPO A5s
OPPO A7 (Glaze	OPPO A7 (Glaze Blu	0.9999999	OPPO A7 (Glaring Gr	0.998804	OPPO A7 (Glaze	0.988494	OPPO A7 (Gla	0.987203	OPPO A5s (Bl	0.9093351	OPPO A5s (Blac	0.909142	OPPO A5s (Blue,	0.907612	OPPO A5s ((	0.907412	OPPO A5s

Fig. 6 Recommendation for additional set of user inputs

presented in Figs. 7 and 8, respectively. The figures display the mobile phones recommended by the two techniques for a given mobile phone in x-axis. The similarity score is used as reference for y-axis. It was found from the results that the similarity scores of many of the recommended mobiles phones are similar by the cosine similarity technique. However, the similarity scores are varying for each of the mobile phone recommended by the *LSI* technique.

6 Conclusion

A system for mobile phone recommendation was proposed and its architecture, implementation details, and the results were discussed in this paper. The experimental analysis of the proposed system was done based on the *LSI* and *cosine similarity* methods. The similarity indices were taken as reference for the recommendation. The proposed system could map each of the mobile phone to its similar mobile phones in the dataset based on the filtering done by keeping a threshold for relevance. Proper and nearly accurate prediction of similar mobile phones based on all the input features and descriptions was achieved.

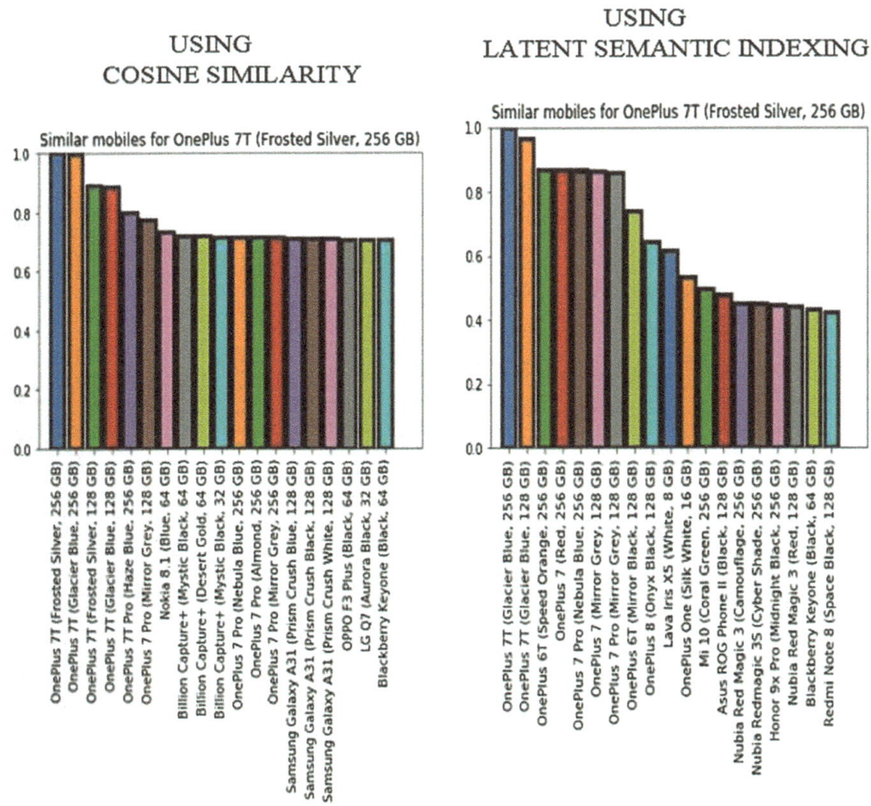

Fig. 7 Comparison of LSI and cosine similarity techniques for 'OnePlus 7T Froster Silver, 256 GB

The work can be further extended to be applied in real life where it can be implemented in the form of an application which does comparisons between multiple Web sites selling the same product and gives an in depth analysis of all the similar mobile phone to the input/searched mobile phone.

Also the work can be further extended to be applied to give an audio-based output/recommendations and later when the final target is chosen from the similar choices, the text from the audio can be extracted and in depth analysis of the final choice can be provided.

This work can further be expanded to several recommendation systems of multiple different applications and also this work is confined to Flipkart's Mobile data and recommendations are done on this data. This algorithm can also be extended to work on an application which uses a different dataset as the analysis of semantics of the textual data is done in the preprocessing of the model irrespective of the dataset used.

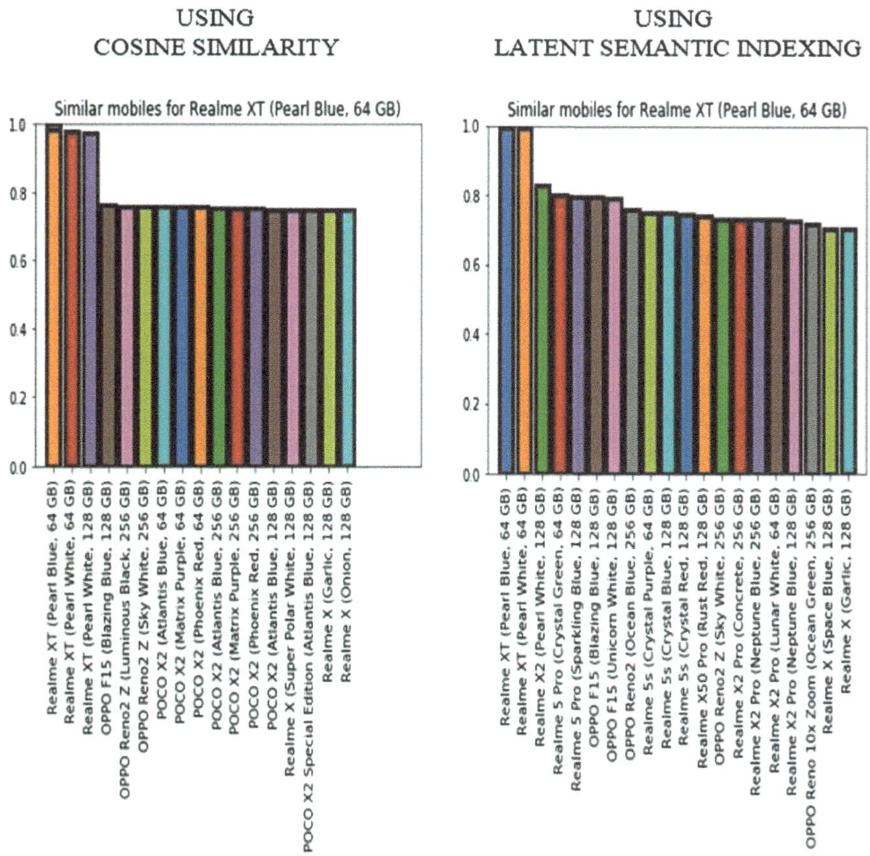

Fig. 8 Comparison of LSI and cosine similarity techniques for 'Realme XT Pearl Blue, 64 GB'

References

1. S. Sivapalan, A. Sadeghian, H. Rahnama, A.M. Madni, Recommender systems in E-commerce, in *Proceedings of 2014 World Automation Congress (WAC)* (2014). https://doi.org/10.13140/2.1.3235.5847
2. F. Ricci, Mobile recommender systems. J. Inform. Technol. Tourism **12**(2), 205–231 (2011)
3. Y. Sun, W. Chong, Y.-S. Han, S. Rho, K.L. Man, Key factors affecting user experience of mobile recommendation systems, in *Lecture Notes in Engineering and Computer Science*, vol. 2, pp. 724–728 (2015)
4. F. Isinkaye, Y. Folajimi, B. Ojokoh, Recommendation systems: principles, methods and evaluation. Egypt. Inform. J. **16** (2015)
5. N. Polatidis, C.K. Georgiadis, Mobile recommender systems: an overview of technologies and challenges, in *Proceedings of 2nd International Conference on Informatics and Applications* (2013)

6. M.H. Mohamed, M.H. Khafagy, M.H. Ibrahim, Recommender systems challenges and solutions survey, in*Proceeding of International Conference on Innovative Trends in Computer Engineering (ITCE)* (2019)
7. N.V. Phadnis, J. Gadge, Framework for document retrieval using latent semantic indexing. Int. J. Comput. Appl. **94**, 37–41 (2014)
8. A. Sivaramakrishnan, M. Krishnamachari, V. Balasubramanian, Recommending customizable products: a multiple choice Knapsack solution, in *Proceedings of the 5th International Conference on Web Intelligence, Mining and Semantics*, New York, NY, USA (2015)
9. J.-M. Low, I.K.T. Tan, C.-Y. Ting, Recent developments in recommender systems, in*Book Multi-disciplinary Trends in Artificial Intelligence, 13th International Conference MIWAI 2019*, pp. 38–51 (2019)
10. K.B. Yedugiri, S. Chandni, P. Sini Raj, S. Souparnika, Recommender systems—a deeper insight. Int. J. Appl. Eng. Res. **9**, 28,521–28,531 (2014)
11. A. Aquino, E. Chavez, Analysis on the use of Latent Semantic Indexing (LSI) for document classification and retrieval system of PNP files, in*Proceedings of 2nd International Conference on Material Engineering and Advanced Manufacturing Technology (MATEC)*, vol. 189, No. 4 (2018)
12. Z. Samia, R. Khaled, Ontology-based approach for semantic indexing of Arabic documents, in*Proceedings of 1st International Conference on Artificial Intelligence and its Applications (AIAP' 2018)* (2018)
13. S. Santhiappan, V. Gopalan, Finding the optimal rank for LSI models, in *Proceedings of ICAET* (2010)
14. P.M. Alamdari, N.J. Navimipour, M. Hosseinzadeh, A.A. Safaei, A. Darwesh, A systematic study on the recommender systems in the E-commerce. IEEE Access **8**, 115694–115716 (2020). https://doi.org/10.1109/ACCESS.2020.3002803
15. P.K. Singh, P. Dutta Pramanik, A. Dey, P. Choudhury, Recommender systems: an overview, research trends and future directions. Int. J. Bus. Syst. Res. (2020)

ShExMap and IPSM-AF—Comparison of RDF Transformation Technologies

Paweł Szmeja and Eric Prud'hommeaux

Abstract RDF is growing in popularity for enterprise data harmonization software (e.g. Virtuoso, or Corporate Memory). Interoperability techniques, e.g. query rewriting, and language transformation rely on the expressivity and flexibility of RDF. As a meta-model capable of expressing information with the use of a variety of vocabularies or ontologies, RDF enables a range of solutions for the problem of data interoperability. Exploiting RDF for interoperability surfaces the need for transformations between homologous RDF structures. This paper examines two independently developed technologies that enable mapping of RDF graphs: Shape Expressions Mapping Language (*ShExMap*) and *IPSM* (Inter Platform Semantic Mediator). We investigate the formal languages used by both, examining their theoretical foundation, peculiarities, and design. We consider two practical examples of transformations and compare effectively equivalent constructs that can be expressed in either approach. We also describe both technologies in broader context of practical applicability.

Keywords Semantics · RDF · alignment · graph rewriting

1 Introduction

RDF[1] is widely recognized for its expressivity, i.e. the ability to convey a wide range of meanings, enabling users to write down data with few limitations. The straightforward structure of RDF triples provides simple representation of graph-oriented data, structuring information in accordance with various schemas, vocabularies, and

[1] http://www.w3.org/TR/rdf11-concepts/.

P. Szmeja (✉)
Systems Research Institute, Polish Academy of Sciences, Warsaw, Poland
e-mail: pawel.szmeja@ibspan.waw.pl

E. Prud'hommeaux
W3C/Massachusetts Institute of Technology, Cambridge, MA, USA
e-mail: eric@w3.org

M. Paprzycki et al. (eds.), *Intelligent Systems, Technologies and Applications*,
Advances in Intelligent Systems and Computing 1353,
https://doi.org/10.1007/978-981-16-0730-1_3

ontologies, as well as offering many formats to present RDF itself (e.g. N-Quads, JSON-LD). The Web Architecture upon which it is based is designed to provide unambiguous term identification and encourage the re-use of existing vocabularies, which should in turn promote data interoperability.

In practice, this landscape reduces the diffusion of information across redundant vocabularies, but does not eliminate it. There are many sources of diffusion:

- overlapping but not identical semantics and use cases,
- parallel development before competing vocabularies are mature and deployed,
- legacy data structures from dictated by non-RDF representations,
- language barriers increasing the cost of and reducing the utility of re-use,
- institutional requirements for in-house vocabularies or self-promotion,
- initial cost of learning exceeds the cost of inventing,
- expanding use cases encroaching upon other vocabularies ("mission creep")
- good faith efforts to create unifying vocabularies https://xkcd.com/927/.

All of these give rise to the need for mapping between schemas, to be used at a conceptual level for documentation, or for translation of instance data [3, 4, 11], i.e. (RDF) graph rewriting [5, 10].

This paper focuses on two platforms used for RDF translation, *ShEx* (Sect. 3.1) and *IPSM* (Sect. 3.2), and the languages they use to express transformations—*ShExMap* (Sect. 4.1) and *IPSM-AF* (Sect. 4.2). Section 2 discusses related work and puts *ShEx* and *IPSM* in the context of other technologies and software. Section 5 explores two case studies, the first a core *IPSM-AF* use case translated to *ShExMap* and the second a *ShExMap* use case translated to *IPSM-AF*. We conclude in Sect. 6 with the summary of our comparison of *ShEx* and *IPSM*, their place in RDF transformation ecosystem, and their applicability to various RDF-related problems and domains.

2 Related Work

While there are many ad hoc tools for structure transformation, some systems are based on more widely standardized and adopted languages. For instance, SPARQL Update[2] is widely used for in-place manipulation of datasets and is generally considered quite approachable by RDF experts. Languages, and vocabularies such as SPIN,[3] SWRL,[4] or SKOS[5] are also used as means of interoperability, although, in a strict sense, they do not transform data, but annotate it. In particular, SPIN expresses SPARQL functions in RDF, allowing them to be embedded in the data and enabling meta analysis of the functions with SPARQL (or SPIN) itself. "raw" OWL can also be used with either simple `owl:sameAs` assertions or more complex restriction classes.

[2] https://www.w3.org/Submission/SPARQL-Update/.

[3] https://www.w3.org/Submission/spin-overview/.

[4] https://www.w3.org/Submission/SWRL/.

[5] https://www.w3.org/2004/02/skos/.

On the application side, tools like Pentaho[6] and Silk[7] can be used to produce, map, and transform RDF data. Wikidata's use of OpenRefine[8] produces Wikibase assertions, which are then available at the Wikidata SPARQL interface. Other tools, such as Virtuoso,[9] or Corporate Memory[10] are more focused on provision of interoperability by mapping data sources (e.g. structured or flat files, relational or NoSQL databases) and transforming them to an RDF representation.

3 ShEx and IPSM

3.1 Shape Expressions

Shape Expressions [2] (*ShEx*) is language for modeling and disseminating RDF graphs and for validating instances against those models. It was designed as a general purpose schema language for RDF, with attention paid to creating reusable components, especially for clinical informatics use cases [12]. We introduce *ShEx* with the case study describing a `sosa:Observation` (Listing 1).

```
s:Observation {                   # An observation has:
  a [sosa:Observation]? ;         #   an optional type arc,
  sosa:hasResult @s:Result+;      #   one or more results,
  sosa:madeBySensor {             #   a sensor object, which has:
    a [iiot:IoTDevice]? ;         #     an optional IoT type,
    a [sosa:Sensor]               #     a required Sensor type
  }
}
s:Result {                        # A result has:
  a [sosa:Result]? ;              #   an optional type arc,
  iiot:hasUnit IRI ;              #   an IRI identifying the units,
  iiot:hasValue xsd:float         #   a value as an XML Schema float.
}
```

Listing 1: *ShExMap* sample schema

A *ShEx* schema is essentially a regular expression executed over a bag of triples, rather than a string. This implies a validation processing model (though others are possible) analogous to a typical regex engine with a list of threads representing different paths through cardinality ranges and disjunctions. A similar processing model can be used in *ShExMap* to synthesize valid documents.

[6] https://www.hitachivantara.com/go/pentaho.html.

[7] http://silkframework.org/.

[8] http://openrefine.org/.

[9] https://virtuoso.openlinksw.com/.

[10] https://www.eccenca.com/en/products/eccenca-corporate-memory.html/.

3.2 *IPSM*

Inter-Platform Semantic Mediator [8] (*IPSM*) is a software component for semantic translation, framed in terms of the layered interoperability model [1]. It was developed by Systems Research Institute, Polish Academy of Sciences (SRIPAS[11]) as part of the INTER-IoT project,[12] that aimed to solve the problem of interoperability between Internet of Things artifacts (platforms, devices, services, etc). *IPSM* was designed to work together with other software from the project to provide a multi-layer IoT suite, which offered syntactic (e.g. between JSON, JSON-LD, TTL, XML, or other formats used by IoT platforms) and semantic interoperability. The latest version of *IPSM* can also work as a standalone translator with RDF as data exchange format. It offers a scalable architecture for semantic translation available as a service, or as a reactive stream processor. *IPSM* translation is configurable with alignment files written in Inter-Platform Semantic Mediator Alignment Format (*IPSM-AF*; described in Sect. 4.2). *IPSM* code[13] and documentation[14] are available online.

4 RDF transformation languages

4.1 *ShExMap*

ShExMap was designed to map instance documents from one schema to another with a minimum of effort and expertise. The source and target schemas are annotated with semantic identifiers (URLs, e.g. `x:patientName`) for the shared concepts. By labeling a constraint in both schemas with the same identifier, the annotator is asserting that instance data matching one of those constraints can be re-written to match the other. This goes well beyond the expressivity of `owl:sameAs` as the annotations are context sensitive, and complements the OWL best practice of using generic properties wherever possible.

Procedurally, instance mapping takes as input a source schema and target schema, and a source instance document conformant to the source schema. The process can be broken down into a parsing and a materialization step. Parsing is simply a validation, where the RDF terms matching semantic identifiers are recorded in a binding tree. This binding tree acts as an intermediate representation, which can then be passed to a materializer. The materializer can be thought of as a regular expression engine running in reverse; each time a property is called for, the value is extracted from the binding tree and that element in the binding tree is marked as used. If the binding tree doesn't have sufficient information to instantiate a property or shape, any regular expression engine threads dependent on that property or shape are eliminated. If the

---

[11] http://www.ibspan.waw.pl/.

[12] http://inter-iot.eu.

[13] https://github.com/INTER-IoT.

[14] https://inter-iot-ipsm.readthedocs.io/en/latest/.

materializer consumes all the data and has at least one thread reaching an endpoint regular expression state machine, the materialization was successful and an instance graph conformant to the target schema is emitted.

A small library of ETL functions for transmuting lexical form during the parsing and materialization steps was developed for project porting clinical data from the US Department of Defense to the US Department of Veterans Affairs. This library follows *ShExMap*'s precedent for bidirectional transforms, where possible. For instance, regular expressions with named capture groups `(/(?<x:familyName>[a-zA-Z]+),(?<x:givenName>[a-zA-Z]+)/)` can both parse a name field into two name field, an materialize a single name field from binds for given and family names.

4.2 IPSM-AF

IPSM-AF was designed around the idea of using ontology alignments to configure and perform semantic translation (in particular on streaming data). It is based on, and compatible with, the Alignment API [6]—an RDF/XML format used in Ontology Alignment Evaluation Initiative (OAEI[15]). An *IPSM-AF* alignment is, in essence, a sequence of translation **steps** (driven by rules), applied in a given **order**. Every **step** is described by an alignment **cell** that consist primarily of input and output **RDF patterns**, called **entity1** and **entity2** respectively, with optional **transformations** between them. Formally, a single *IPSM-AF* file defines a one-way mapping from an input ontology `onto1` to an output ontology `onto2`. We refer interested readers to [7, 13], where more information about *IPSM-AF* and concrete translation examples can be found.

```
<align:Cell rdf:about="[cell id]">
<align:entity1 rdf:parseType="Literal">
        { source RDF pattern }
</align:entity1>
<align:entity2 rdf:parseType="Literal">
        { target RDF pattern }
</align:entity2>
<align:relation>{relation}</align:relation>
<align:measure rdf:datatype="&xsd;float">
        { confidence measure }
</align:measure>
<sripas:filters rdf:parseType="Literal">
        { filters }
</sripas:filters>
<sripas:typings rdf:parseType="Literal">
        { typings }
</sripas:typings>
<sripas:transformation rdf:parseType="Literal">
        { functional constraints }
</sripas:transformation>
</align:Cell>
```

Listing 2: *IPSM-AF* – alignment cell structure

[15] http://oaei.ontologymatching.org/.

The core of any alignment file are the **cells** The order of application, and which cells are actually applied (and how many times) is configurable. Cells are atomic and transactional, but do not need to be idempotent, so the order of application is important. Each cell (Listing 2) describes a parameterized mapping that can be applied to an RDF graph. This is done using so-called RDF patterns, which can be any valid RDF graph, with the optional addition of alignment variables, which work in a way reminiscent of SPARQL variables. `<entity1>` defines an RDF pattern that needs to match with an input RDF graph, i.e. the input RDF needs to contain a sub-graph isomorphic to the RDF pattern (where variables are wildcard nodes). When matched, the variables are bound to corresponding nodes in the input RDF, and the rest of the alignment cell is applied. `<entity2>` defines an output RDF graph, i.e. a graph that is added to the input RDF data as the cell is applied, with inclusion of the bound variables. By default, the sub-graph that matched`<entity1>` is removed from the result. In effect, `<entity1>` pattern is substituted with `<entity2>` pattern.

Transformations are (optional) functions applied to variables. Results of such functions may be bound to new variables and used in `<entity2>`. Use of any function may be declared in an *IPSM-AF*, and *IPSM* supports a range of functions available in SPARQL, as well as user-defined custom function libraries. A common use case for transformations is conversion between a string and an URI, or string operations, such as concatenation, or tokenization with regular expressions. The optional `<filters>` and `<typings>` add information about data types.

5 Case studies

In this section, we examine case study RDF mappings, expressed in *ShExMap* and *IPSM-AF*. The two following subsections describe two pairs of mapping definitions—one where a pre-existing *IPSM-AF* alignment was re-written as a *ShExMap* schema, and the other, where the opposite was done. The goal of this exercise was to represent equivalent RDF transformation in the two technologies, in order to learn their practical similarities and differences. The original mappings used in case studies were developed independently.

5.1 From IPSM-AF into ShExMap

The first case study describes a mapping of a message coming from an eHealth platform BodyCloud,[16] used in INTER-IoT pilots. Because *IPSM* is a streaming processor dedicated to translation of messages, any *IPSM* input is called a "message",

[16]http://bodycloud.dimes.unical.it/.

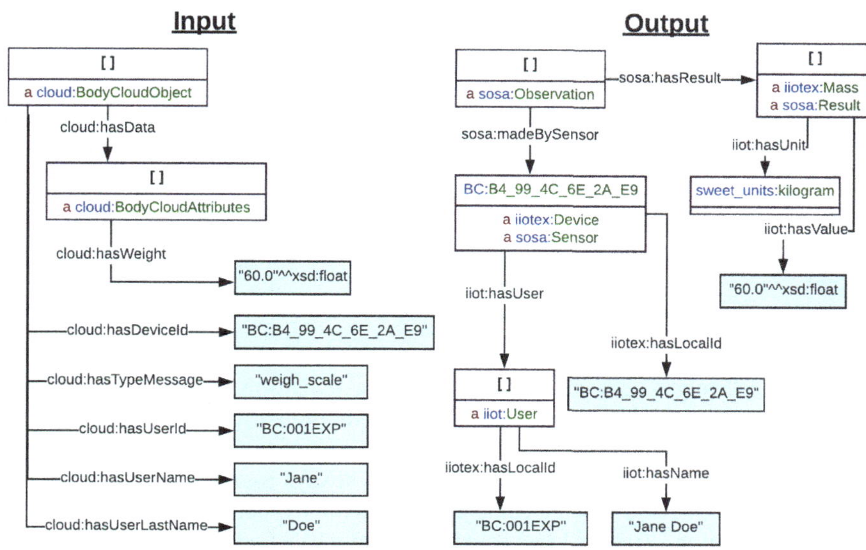

Fig. 1 *IPSM-AF* to *ShExMap* case study—input and output data

although technically, it can be any RDF graph. The task of *IPSM* was to present this information using the GOIoTP[17] ontology, prefixed "iiot".

The input data (Listing 3) describes a simple weight measurement coming from a weighing scale. It informs about a so-called BodyCloudObject (i.e. a device making an observation) with a type of "weight_scale" and other metadata that observed a weight of "60.0" (note the lack of explicit unit declaration).

The input message is re-written into a new RDF structure compliant with the output ontology (GOIoTP). Additional information implicit in BodyCloud, e.g. the unit of weight measurement, is made explicit in the output (Fig. 1).

input

```
[] a cloud:BodyCloudObject ;
 cloud:hasData [
  a cloud:BodyCloudAttributes  ;
  cloud:hasWeight "60.0"^^xsd:float] ;
 cloud:hasDeviceId
"http://inter-iot.eu/bodycloud/B4_99_4C_6E_2A_E9";
 cloud:hasTypeMessage "weight_scale" ;
 cloud:hasUserId
 "http://inter-iot.eu/bodycloud/001EXP";
 cloud:hasUserLastName "Doe" ;
 cloud:hasUserName "Jane" .
```

output

```
[] a sosa:Observation ;
 sosa:madeBySensor BC:B4_99_4C_6E_2A_E9 ;
 sosa:hasResult [
  a sosa:Result ;
  iiot:hasUnit sweet_units:kilogram ;
  iiot:hasValue "60.0"^^xsd:float] .
BC:B4_99_4C_6E_2A_E9
 a iiot:IoTDevice , sosa:Sensor ;
 iiot:hasUser [
  a iiot:User ;
  iiot:hasName "Jane Doe" ;
  iiotex:hasLocalId
"http://inter-iot.eu/bodycloud/001EXP"] .
```

Listing 3: *IPSM-AF* to *ShExMap* case study – input and output data

[17]http://inter-iot.eu/GOIoTP#.

5.1.1 IPSM-AF transformation

The *IPSM-AF* file that defines the transformation described above has 4 *IPSM-AF* translation cells, whose excerpts are presented in Listings 4, 5, 6 and 7. The first cell (Listing 4) is a simple substitution, where any node (variable `VAR:X`) of type `cloud:BodyCloudObject` is annotated with type `sosa:Observation`, with the original type removed (i.e. substituted).

Entity 1	Entity 2
`VAR:X a cloud:BodyCloudObject .`	`VAR:X a sosa:Observation .`

Listing 4: *IPSM-AF* to *ShExMap* case study – observation cell

Next, the "weight" cell (Listing 5) defines a transformation of the subgraph that stores the actual observed weight value. Additionally, an annotation about `sweet_units:kilogram` is added to the output. This information is implicit in the source data model, i.e. the BodyCloud platform always uses kilograms as weight units, which is described in its documentation, but not explicitly stated in messages. Such "specification" (i.e. making something explicit) often appears in *IPSM-AF* alignments that are used in ecosystems with many platforms, because otherwise such details would be unavailable to platforms that make different implicit assumptions and may even lead to misunderstandings and errors.

Note that the property `cloud:hasTypeMessage` is repeated for variable `VAR:X` in both input and output RDF patterns. This is to ensure that on one hand only nodes that have this property assertion are translated, and on the other, that this assertion is preserved (i.e. copied) to the output.

Entity 1

```
VAR:X
 cloud:hasTypeMessage "weight_scale" ;
 cloud:hasData [
  a VAR:Y ,
  cloud:BodyCloudAttributes ;
  cloud:hasWeight VAR:weight
 ] .
```

Entity 2

```
VAR:X
 cloud:hasTypeMessage "weight_scale" ;
 sosa:hasResult [
  a VAR:Y, sosa:Result, iiotex:Mass ;
  iiot:hasUnit sweet_units:kilogram ;
  iiot:hasValue VAR:weight
 ] .
```

Listing 5: *IPSM-AF* to *ShExMap* case study – weight cell

Next, the "device" cell (Listing 6) "splits" the device identifier (a string) bound to `VAR:deviceId` and puts it in two different places in the output RDF graph. First, the variable is copied as is under a different graph. Second, a SPARQL function IRI[18] is applied to it, resulting in another variable `VAR:deviceUri`, which is used in the output graph as a named node, and connected to `VAR:X` with the `sosa:madeBySensor` property. Variables are cell-local, so `VAR:X` here has no relation to `VAR:X` in other cells.

[18]https://www.w3.org/TR/sparql11-query/#func-iri.

Finally, the "user" cell (Listing 7) copies user data, in particular first and last name. Here, a SPARQL STR[19] and CONCAT[20] functions are applied to concatenate first and last name into a single string.

ShExMap and IPSM-AF Comparison 9

Entity 1

```
VAR:X
 cloud:hasTypeMessage "weight_scale" ;
 cloud:hasDeviceId VAR:deviceId .
```

Entity 2

```
VAR:X
 sosa:madeBySensor VAR:deviceUri ;
 iiotex:hasLocalId VAR:deviceId .
 VAR:deviceUri a sosa:Sensor, iiot:IoTDevice .
```

Transformation

```
<sripas:function sripas:about="IRI">
        <sripas:param sripas:order="1" sripas:about="&VAR;deviceId"/>
        <sripas:return sripas:about="&VAR;deviceUri"/>
</sripas:function>
```

Listing 6: *IPSM-AF* to *ShExMap* case study – device cell

Entity 1

```
VAR:X
 sosa:madeBySensor VAR:sensor ;
 cloud:hasUserId VAR:userId ;
 cloud:hasUserName VAR:fName ;
 cloud:hasUserLastName VAR:lName .
```

Entity 2

```
VAR:X
 sosa:madeBySensor [
  a VAR:sensor ;
  iiot:hasUser [
   a iiot:User ;
   iiotex:hasLocalId VAR:userId ;
   iiot:hasName VAR:name
  ]
 ] .
```

Transformation

```
<sripas:function sripas:about="str">
  <sripas:param sripas:order="1" sripas:about="&VAR;fName"/>
  <sripas:return sripas:about="&VAR;str_fName"/>
</sripas:function>
<sripas:function sripas:about="str">
  <sripas:param sripas:order="1" sripas:about="&VAR;lName"/>
  <sripas:return sripas:about="&VAR;str_lName"/>
</sripas:function>
<sripas:function sripas:about="concat">
  <sripas:param sripas:order="1" sripas:about="&VAR;str_fName"/>
  <sripas:param sripas:order="2" sripas:val=" "/>
  <sripas:param sripas:order="3" sripas:about="&VAR;str_lName"/>
  <sripas:return sripas:about="&VAR;name"/>
</sripas:function>
```

Listing 7: *IPSM-AF* to *ShExMap* case study – user cell

Application of the cells in the order in which they were described results in translation, as in Fig. 1. The cells are atomic and transactional, so even if the input message is in some way "incomplete" the alignment may still be applied, and some cells will be matched. This is the default behavior of *IPSM-AF* that allows for definition of "optional" cells that may or may not match.

[19]https://www.w3.org/TR/sparql11-query/#func-str.

[20]https://www.w3.org/TR/sparql11-query/#func-concat.

5.1.2 ShExMap transformation

To reiterate, the goal of the exercise was to achieve the same transformation (depicted in Fig. 1) with a different tool. The input and output *ShExMap* schemas are presented in Listing 8 and correspond to all 4 *IPSM-AF* cells written under one definition.

5.2 *From ShExMap into IPSM-AF*

The second case study transformation comes from HL7 demonstration of mapping a blood pressure observation from HL7 FHIR ontology[21] into an another medical ontology. As opposed to the first case study from Sect. 5.1, this mapping performs a "reduction" of the RDF graph by presenting the same information in a simpler structure.

The most significant information in the input data (Listing 9) are the observed values and measurement units of systolic and diastolic blood pressure, encoded under a tree with named and anonymous individuals. The same information is represented in the output (Listing 9) under a much simpler tree with no explicit types.

input schema

```
{a [cloud:BodyCloudObject] ;
 cloud:hasData {
 a [cloud:BodyCloudAttributes] ;
 cloud:hasWeight
  xsd:float %Map:{ cloud:hasWeight %} ;
} ;
cloud:hasDeviceId
 IRI %Map:{ cloud:deviceId %} ;
cloud:hasTypeMessage
 xsd:string %Map:{ cloud:typeMessage %} ;
cloud:hasUserId
 xsd:string %Map:{ cloud:userId %} ;
cloud:hasUserLastName
 xsd:string %Map:{ cloud:userLastName %} ;
cloud:hasUserName
 xsd:string %Map:{ cloud:userName %}
}
```

output schema

```
{<X> a cloud:BodyCloudObject ;
 cloud:hasData [
 a cloud:BodyCloudAttributes ;
 cloud:hasWeight "60.0"^^xsd:float ;
] ;
 cloud:hasDeviceId
<http://inter-iot.eu/bodycloud/B4_99_4C_6E_2A_E9>;
 cloud:hasTypeMessage "weight_scale" ;
 cloud:hasUserId
"http://inter-iot.eu/bodycloud/001EXP";
 cloud:hasUserLastName "Doe" ;
 cloud:hasUserName "Jane" .
}
```

Listing 8: *IPSM-AF* to *ShExMap* case study – *ShExMap* schemas

21 https://www.hl7.org/fhir/.

input

```
<tag:BPfhir123>
 a fhir:Observation;
 fhir:coding [
  fhir:code sct:Blood_Pressure ];
 fhir:related [
  fhir:type "has-component";
 fhir:target _:sysBP123 ];
 fhir:related [ fhir:type "has-component";
 fhir:target _:diaBP123 ] .

_:sysBP123
 a fhir:Observation;
 fhir:coding [
 fhir:code sct:Systolic_Blood_Pressure ];
 fhir:valueQuantity [
 a fhir:Quantity;
 fhir:value "110"^^xsd:float;
 fhir:units "mmHg" ].

_:diaBP123
 a fhir:Observation;
 fhir:coding [
 fhir:code sct:Diastolic_Blood_Pressure ];
 fhir:valueQuantity [
 a fhir:Quantity;
 fhir:value "70"^^xsd:float;
 fhir:units "mmHg" ].
```

output

```
tag:BPfhir123
 damMed:diastolic [
 damMed:units "mmHg" ;
 damMed:value "70"^^xsd:float;
] ;
damMed:systolic [
 damMed:units "70"^^xsd:float; ;
 damMed:value "mmHg"
] .
```

Listing 9: *ShExMap* to *IPSM-AF* case study – input and output data

5.2.1 ShExMap transformation

The *ShExMap* schemas for the transformation are defined in Listing 9. Additionally, to change of data structure, a constant value is asserted in the output schema, as a demonstrative case study.

input schema

```
start = @<BPfhir>

<collector> {
fhir:item @<BPfhir>*
}

<BPfhir> {
 a [fhir:Observation]?;
 # fhir:subject @<Patient>;
 fhir:coding { fhir:code [
  sct:Blood_Pressure] };
 fhir:related {
  fhir:type ["has-component"];
  fhir:target @<sysBP> };
 fhir:related {
  fhir:type ["has-component"];
  fhir:target @<diaBP> }
}
<sysBP> {
 a [fhir:Observation]?;
 fhir:coding { fhir:code [
  sct:Systolic_Blood_Pressure] };
 fhir:valueQuantity {
 a [fhir:Quantity]?;
 fhir:value
  xsd:float %Map:{ bp:sysVal %};
 fhir:units
  xsd:string %Map:{ bp:sysUnits %}
}
}
<diaBP> {
 a [fhir:Observation]?;
 fhir:coding {
  fhir:code [sct:Diastolic_Blood_Pressure] };
 fhir:valueQuantity {
 a [fhir:Quantity]?;
 fhir:value
  xsd:float %Map:{ bp:diaVal %};
 fhir:units
  xsd:string %Map:{ bp:diaUnits %}
}
}
```

output schema

```
start = @<BPunitsDAM>

<BPunitsDAM> {
 em:systolic {
 em:value xsd:float %Map:{ bp:sysVal %};
 em:units xsd:string %Map:{ bp:sysUnits %}
};
em:diastolic {
 em:value xsd:float %Map:{ bp:diaVal %};
 em:units xsd:string %Map:{ bp:diaUnits %}
};
em:someConstProp
  xsd:string?%Map:{
  <http://abc.example/someConstant> %}
}
```

Listing 10: ShEx to *IPSM-AF* case study – *ShExMap* schemas

5.2.2 IPSM-AF transformation

An equivalent mapping that defines the transformation depicted in Fig. 2 can be achieved in *IPSM-AF* as a single-cell alignment. The input RDF pattern (Listing 11) binds the three most important individuals to variables. Those are `VAR:observation` for the "root" of the input graph, and `VAR:systolic`, and `VAR:diastolic` for the `fhir:Observations`. The values of interest (blood pressure value and unit) are bound to literal variables "deeper" in – `VAR:systolicValue`, `VAR:systolicUnits` for systolic blood pressure, and correspondent variables for diastolic blood pressure. Additionally, datatypes for literal variables are restricted to `xsd:float` (Listing 11).

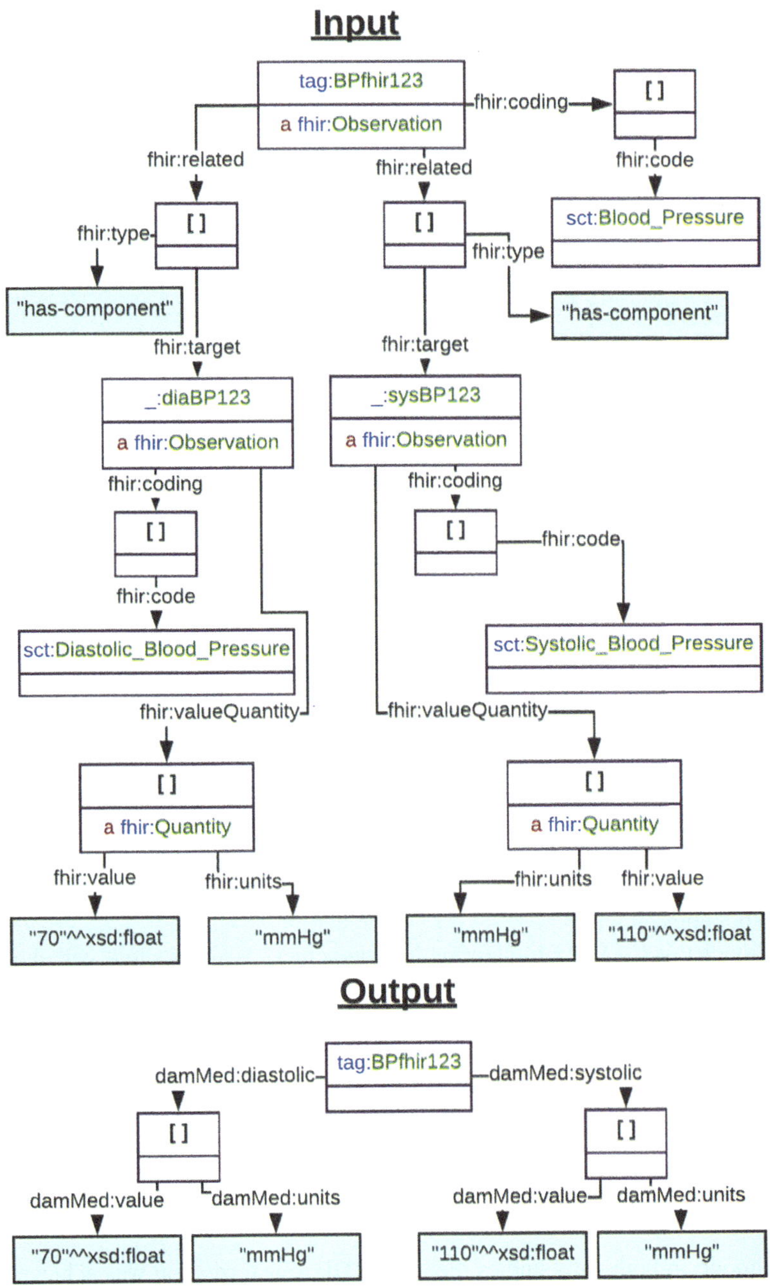

Fig. 2 *ShExMap* to *IPSM-AF* case study—input and output data

Entity 1

```
1  [] a VAR:observation,
2   fhir:Observation ;
3   fhir:coding [
4    fhir:code sct:Blood_Pressure ] ;
5   fhir:related
6    [ a VAR:diastolic ],
7    [ a VAR:systolic ] .
8
9  [] a VAR:systolic ;
10  fhir:type "has-component" ;
11  fhir:target [
12   a fhir:Observation ;
13   fhir:coding [
14    fhir:code sct:Systolic_Blood_Pressure ] ;
15   fhir:valueQuantity [
16    a fhir:Quantity ;
17    fhir:units [ a VAR:systolicUnits ] ;
18    fhir:value [ a VAR:systolicValue ]
19   ]
20 ] .
21
22 [] a VAR:diastolic ;
23  fhir:target [ a fhir:Observation ;
24   fhir:coding [
25    fhir:code sct:Diastolic_Blood_Pressure ] ;
26   fhir:valueQuantity [ a fhir:Quantity ;
27   fhir:units [ a VAR:diastolicUnits ] ;
28   fhir:value [ a VAR:diastolicValue ]
29   ] ] ;
30  fhir:type "has-component" .
```

Entity 2

```
1  [] a VAR:observation ;
2   damMed:diastolic [
3    damMed:units [ a VAR:diastolicUnits ] ;
4    damMed:value [ a VAR:diastolicValue ]
5   ] ;
6   damMed:systolic [
7    damMed:units [ a VAR:systolicUnits ] ;
8    damMed:value [ a VAR:systolicValue ]
9  ] .
```

Filters and Typings

```
<filter about="&VAR;systolicValue" datatype="&xsd;float"/>
<filter about="&VAR;diastolicValue" datatype="&xsd;float"/>

<typing about="&VAR;systolicValue" datatype="&xsd;float"/>
<typing about="&VAR;diastolicValue" datatype="&xsd;float"/>
```

Listing 11: ShEx to *IPSM-AF* case study – IPSM-AF alignment cell

Note that input pattern in Listing 11 will match `VAR:observation` with any named or anonymous node that fits the described subgraph, while in the original *ShExMap* transformation, the name of the node (`tag:BPfhir123`) is explicitly specified. To simulate such "focus node", line 1 in **`entity1`** and **`entity2`** of the *IPSM-AF* cell should be replaced with `tag:BPfhir123 a VAR:observation.`.

Although this *IPSM-AF* transformation was achieved using a single cell, the usage of "top-level" variables `VAR:observation`, `VAR:systolic`, `VAR:diastolic` suggests an alternative, where the diastolic and systolic mapping are split into two cells. In fact, there are many ways in which this alignment may be written, and whether a single, or multiple cells should be used, depends on the use case.

6 Concluding remarks

Even though in both case studies identical RDF transformation was achieved, this exercise exposed properties and advanced features specific to either language. Generally, the properties of both technologies stem from their origins and intended usage. *IPSM* is primarily a stream processor focused on throughput and scalability. *IPSM-AF* alignments are linear descriptions of translation steps that may or may not match (and transform) the data independently of each other. In fact, even if no translation cell is matched, by default no error is thrown, and input message remains unchanged. Moreover, *IPSM-AF* cells can match any RDF pattern, including those with anonymous root, or many roots. A focus node approach is available, but not required. Consequently, *IPSM* can use partial *IPSM-AF* alignments, that can be extended, or combined at any time.

By default *IPSM* operates in a replacement semantics mode, which means that input RDF patterns are removed and substituted with output patterns. This behavior can be changed with skillful construction of alignments, albeit not by configuration option, or a flag. This is documented in a catalog of *IPSM-AF* usage patterns[22]. It is also important to note that the transactional cells change the data at the moment they are applied, so the next cell is applied to an already modified "intermediate" graph—the changes are incremental. The atomic, transactional, nonrestrictive and "pass-through" nature of *IPSM-AF* cells make *IPSM* a good technology for a distributed messaging environment, in which message contents may vary, and new annotations may appear often.

ShExMap, on the other hand, is directed toward translation of whole schemas, where data items have a formally defined shape. Where *IPSM-AF* has a replacement semantics, *ShExMap* has "constructive" semantics, in that nothing from the input graph appears in the output graph with being prescribed by semantic identifiers shared between the input and output schema. Having library implementations in multiple languages, *ShExMap* offers much more support for programmers that want to build their own translation systems for specific vocabularies, or use cases. For workflows that have (or will need) formal schema description, the incremental effort in creating and maintaining *ShExMap* annotations is relatively small. *ShExMap* itself is an extension to *ShEx*, which offers intuitive, human-facing schemas with well-defined and portable parsing (validating) semantics.

A feature that is not immediately apparent in *ShExMap*, is that *ShExMap* builds an intermediate representation during translation, that can be then used for verification[9]. This allows not just for debugging, but for quick analysis of the differences between a shape and some input data.

The conspicuous contrast between *IPSM-AF*'s replacement semantics and *ShExMap*'s constructive semantics reflects their respective use cases rather than limitations in the technologies. If *IPSM-AF* generated a new graph with novel blank nodes, it could have a constructive semantics, and if *ShExMap* reused the input graph, it could have a replacement semantics.

Since both technologies offer a library of functions/transformations, and as both are periodically updated, we can expect more additions and extensions. In general, *ShExMap* and *IPSM-AF* are significantly different, but very similarly capable. The biggest practical difference lies in their software implementations. More about *IPSM* and *IPSM-AF*, including installation, configuration, alignment patterns and case stud-

[22]https://inter-iot-cookbook.readthedocs.io/en/latest/inter-layer/ds2ds/recipes/alignment/.

ies can be found online[23] [24]. *ShEx* and its library implementations are available on the web page[25].

Acknowledgements This research was partially supported by the European Union's "Horizon 2020" research and innovation programme as part of the "Interoperability of Heterogeneous IoT Platforms" (INTER-IoT) project under Grant Agreement No. 687283.

Appendix: Prefixes

Declarations of prefixes used in the paper:

```
BC: <http://inter-iot.eu/bodycloud/>
bp: <http://shex.io/extensions/Map/#BPDAM->
cloud: <http://inter-iot.eu/syntax/BodyCloud#>
damMed: <http://dam.example/med#>
em: <http://dam.example/med#>
fhir: <http://hl7.org/fhir-rdf/>
iiot: <http://inter-iot.eu/GOIoTP#>
iiotex: <http://inter-iot.eu/GOIoTPex#>
Map: <http://shex.io/extensions/Map/#>
owl: <http://www.w3.org/2002/07/owl#>
rdf: <http://www.w3.org/1999/02/22-rdf-syntax-ns#>
rdfs: <http://www.w3.org/2000/01/rdf-schema#>
sct: <http://snomed.info/sct/>
sosa: <http://www.w3.org/ns/sosa/>
sripas: <http://www.inter-iot.eu/sripas#>
sweet_units: <http://sweet.jpl.nasa.gov/2.3/reprSciUnits.owl>
VAR: <http://www.inter-iot.eu/sripas#node_>
xsd: <http://www.w3.org/2001/XMLSchema#>
```

References

1. *Achieving Technical Tnteroperability—the ETSI approach*. ETSI White Paper No. 3
2. I. Boneva, J.E. Labra Gayo, E. Prud'hommeaux, *Semantics and Validation of shapes schemas for rdf* (2017)
3. O. Corby, C. Faron-Zucker, F. Gandon, A generic rdf transformation software and its application to an online translation service for common languages of linked data, in *International Semantic Web Conference* (2015), pp. 150–165
4. O. Corby, C.F. Zucker, *STTL: A SPARQL-based Transformation Language for RDF*
5. G. Correndo, M. Salvadores, I. Millard, H. Glaser, N. Shadbolt, Sparql query rewriting for implementing data integration over linked data. in *1st International Workshop on Data Semantics (DataSem 2010)* (2010)
6. J. David, J. Euzenat, F. Scharffe, C. Trojahn dos Santos, The Alignment API 4.0. Semantic Web **2**(1), 3–10 (2011)
7. M. Ganzha, M. Paprzycki, W. Pawłowski, P. Szmeja, K. Wasielewska, Alignment-based semantic translation of geospatial data, in *3rd International Conference on Advances in Computing, Communication & Automation (ICACCA), Proceedings* (2017)
8. M. Ganzha, M. Paprzycki, W. Pawłowski, P. Szmeja, K. Wasielewska, Streaming semantic translations, in *21st International Conference on System Theory, Control and Computing ICSTCC, Proceedings* (2017)
9. J.E. Labra Gayo, E. Prud'Hommeaux, I. Boneva, D. Kontokostas, *Validating RDF Data*. Morgan & Claypool Publishers (2017)

[23] https://inter-iot-ipsm.readthedocs.io/en/latest/.

[24] https://inter-iot-cookbook.readthedocs.io/en/latest/.

[25] http://shex.io/.

10. K. Makris, N. Bikakis, N. Gioldasis, S. Christodoulakis, Sparql-rw: transparent query access over mapped rdf data sources, in *15th International Conference on Extending Database Technology* (2012), pp. 610–613
11. F. Michel, J. Montagnat, C.F. Zucker, A survey of RDB to RDF translation approaches and tools. Tech. rep
12. H.R. Solbrig, E. Prud'hommeaux, G. Grieve, L. McKenzie, J.C. Mandel, D.K. Sharma, G. Jiang, Modeling and validating HL7 FHIR profiles using semantic web Shape Expressions (ShEx). J Biomed Inform **67**, 90–100 (2017)
13. P. Szmeja, M. Ganzha, M. Paprzycki, W. Pawłowski, K. Wasielewska, Declarative ontology alignment format for semantic translation, in *3rd International Conference On Internet of Things: Smart Innovation and Usages (IoT-SIU'18), Proceedings* (2018)

AI Approaches for IoT Security Analysis

Mohamed Abou Messaad, Chadlia Jerad, and Axel Sikora

Abstract IoT networks are increasingly used as entry points for cyberattacks, as often they offer low-security levels, as they may allow the control of physical systems and as they potentially also open the access to other IT networks and infrastructures. Existing intrusion detection systems (IDS) and intrusion prevention systems (IPS) mostly concentrate on legacy IT networks. Nowadays, they come with a high degree of complexity and adaptivity, including the use of artificial intelligence. It is only recently that these techniques are also applied to IoT networks. In this paper, we present a survey of machine learning and deep learning methods for intrusion detection, and we investigate how previous works used federated learning for IoT cybersecurity. For this, we present an overview of IoT protocols and potential security risks. We also report the techniques and the datasets used in the studied works, discuss the challenges of using ML, DL and FL for IoT cybersecurity and provide future insights.

1 Overview and Introduction

The thought of Internet of Things first became popular in the late nineties of the last century. The British entrepreneur Kevin Ashton first used the term Internet of Things in 1999 while working at Auto-ID laboratories [1]. IoT is an emerging technology,

M. A. Messaad (✉) · C. Jerad
University of Manouba, La Manouba, Tunisia
e-mail: mohamed.abou@hs-offenburg.de

C. Jerad
e-mail: chadlia.jerad@ensi-uma.tn

M. A. Messaad · A. Sikora
Institute of Reliable Embedded Systems and Communication Electronics (ivESK), Offenburg University, 77652 Offenburg, Germany
e-mail: axel.sikora@hs-offenburg.de

C. Jerad
University of Carthage, Carthage, Tunisia

M. Paprzycki et al. (eds.), *Intelligent Systems, Technologies and Applications*,
Advances in Intelligent Systems and Computing 1353,
https://doi.org/10.1007/978-981-16-0730-1_4

and it presents a network in which all physical objects are connected to the Internet through network devices or routers, or gateways to exchange information. IoT allows objects to be monitored and controlled remotely across network infrastructure. IoT environment is dynamic and interrelated. According to [2], IoT connected 22 billion things in 2018, and estimations predict that it will connect 38.6 billion things by 2025, reaching 50 billion by 2030. Sharing data between such huge number of devices around the globe enlarges the issue of IoT security and cyberattacks. Cybercrime has become on top-reported crimes in the world. Accordingly, IoT attacks were up by 600% in 2017, and the global cost of online crime is expected to reach 6 trillion dollars by 2021 [3]. Cybersecurity is a critical area, in which artificial intelligence [4] is more and more investigated. AI refers to the simulation of human intelligence in machines in order to rationalize and easily execute tasks to achieve a specific goal. The goals of AI include learning, reasoning and perception. Machine learning (ML) [5] and deep learning (DL) [6] are two subsets of AI. ML algorithms can modify themselves without human intervention in order to make predictions or decisions without being explicitly programmed to do so. It could be possible by feeding them through structured data known as "training data." However, DL is a subset of ML and its algorithms behave similar to ML algorithms but they involve different layers that behave like a network in order to imitate the function of the human's brain neural networks. Federated learning (FL) [7] is a ML technique that involves network edge devices to train a global model locally with private fragmented data. The advantage of FL is to blend ML and edge computing to improve a global model with the users data while preserving data privacy.

In this paper, we give a survey of the existing works on ML, DL, as well as FL use for IoT security analysis, which is the analysis of potential threats and vulnerabilities. Such analysis helps to identify risks and potentially add actions aiming to secure IoT devices, as well as the networks they are connected to against possible cyberattacks. To the best of our knowledge, none of the existing surveys covered FL techniques for IoT security analysis. In addition, only one previous survey [8] did target IoT applications, but with a focus on ML techniques only. For this, we start by positioning our work with regard to existing surveys (c.f. Sect. 2), followed by a general overview of IoT protocols, where we set a taxonomy for IoT attacks (c.f. Sect. 3). The study that we conduct covers a larger number of papers. We classify them according to the used techniques: ML (c.f. Sect. 5), DL (c.f. Sect. 6) and FL (c.f. Sect. 7) and analyze the general trend with regard to three perspectives: used algorithms, hybrid approaches (algorithms that combine multiple techniques) and the used dataset. Additionally, several discussions and possible future research directions are reported (c.f. Sect. 8) based on the aforementioned classification criteria. Thus, this work can be seen as a contribution to the determination of the AI approaches and IoT security co-exploration. Finally, Sect. 9 concludes the article.

2 Related Work

In the literature, different related studies present an overview on the use of AI techniques for cybersecurity and intrusion detection systems. They are illustrated in Table 1. The year in which the survey has been conducted clearly shows that the field is getting more attention over the last 3 years. We mention which AI approaches have been used (i.e., ML, DL or FL). We mention as well if the survey details the used datasets and if there is a focus on IoT application in particular.

Five surveys ([8–11, 14]) among the seven we identified studied the use of machine learning techniques for intrusion detection systems. The least recent one is [9], where Buzack et al. presented a survey of machine learning and data mining methods used by the intrusion detection systems for cybersecurity. This study classified the datasets into three main types: packet-level data, netflow data and public datasets. Furthermore, Boutaba et al. [10] presented a survey on the application of ML techniques in various areas of networking. The authors provided a comprehensive discussion on the different learning paradigms with a focus on ML techniques applied to solve fundamental problems in networking. They summarized different ML/DL methods by algorithms categories and intrusion detection types. Meanwhile, Verma et al. [8] investigated the use of ML classification algorithms for securing IoT against DoS attacks. A comprehensive study is carried on the classifiers, and they evaluated the performance assessment of seven machine learning classification algorithms. The authors performed a benchmarking of all the classifiers on the popular datasets on the matter: CIDDS-001, UNSW-NB15 and NSL-KDD. In addition, they evaluated the average response time of all classifiers on Raspberry Pi hardware device.

Two of the surveys dealt with the use of DL [11, 14], as well as ML. Xin et al. [11], for example, studied these methods for network analysis of intrusion detection and provided a brief tutorial description of each ML/DL method. The authors introduced some of the security datasets that are accessible on the Internet. Similarly, Ferrag et al. [14] presented a survey that focuses on deep learning approaches for cybersecurity intrusion detection and provided a detailed study on 35 well-known cyber datasets

Table 1 Related surveys on AI application for cybersecurity intrusion detection

References	Year	ML	DL	FL	Datasets	IoT
Buczak et al. [9]	2015	Yes	No	No	Yes	No
Boutaba et al. [10]	2018	Yes	Partial	No	Yes	No
Xin et al. [11]	2018	Yes	Yes	No	Partial	No
Ring et al. [12]	2019	No	No	No	Yes	No
Lim et al. [13]	2020	No	No	Yes	No	Partial
Verma et al. [8]	2020	Yes	No	No	Yes	Yes
Ferrag et al. [14]	2020	Partial	Yes	No	Yes	No
Our study	2020	Yes	Yes	Yes	Yes	Yes

and provided a classification of these datasets into seven categories. They analyzed as well seven deep learning approaches. Furthermore, the authors compared the methods using two new datasets: CSE-CIC- IDS2018 dataset and the Bot-IoT dataset. We report that one among the seven mentioned studies targeted IoT.

FL techniques use, however, has been explored in only one of the surveys, that is [13]. In this paper, Lim et al. presented a tutorial of FL as an enabling technology for collaborative model training at mobile edge networks and its implementation in several applications such as cybersecurity. The authors provided detailed reviews, analyses and comparisons of approaches for emerging implementation challenges in FL. The survey focuses also on the issues regarding FL implementation such as communication cost, resource allocation, data privacy and data security.

Furthermore, and as datasets are central in AI techniques, Ring et al. focused in [12] on the datasets used for intrusion detection systems. They identified 34 network-based intrusion detection datasets. Specifically, the paper presents the peculiarities of each one. An investigation of the attack scenarios within the datasets and their interrelationships were presented.

3 Overview of IoT protocols and Security Risks

The IoT environment is heterogeneous with hundreds of communication technologies and protocols. One of the biggest challenges is the standardization of the IoT protocols stack. There are hundreds of proprietary protocols, nevertheless they are supported by an alliance of product vendors. In addition, the IoT environment is exposed to several risks and cyberattacks. This section presents an overview of different IoT protocols and the potential security risks.

3.1 IoT layers and protocols

We present in Fig. 1 a number of IoT protocols. This is not an exhaustive list, as they are numerous and new ones are to appear. For a detailed presentation of most used IoT protocols and standards, readers may refer to [15]. The physical and data link layers combined together present the first layer of the model that defines a machine's network interface, where the physical layer describes the way data is actually transmitted on the network medium and the data link layer defines the format in the network and defines also the algorithm of how the medium (the physical) is accessed in case of shared medium. For instance, the first includes short-range wireless protocols [16, 17] such as Wi-Fi, BLE, RFID and NFC. Home automation protocols include ZigBee [18] (an IEEE 802.15.4-based specification), Z-Wave [19] and KNX RF [20]. Low-power wide area networks [21] include LoRaWAN, Sigfox, NB-IoT and MIOTY. Smart metering protocols include DLMS/COSEM [22] and wireless M-BUS [23]. Besides, industrial automation protocols include Ethernet/IP [24], Eth-

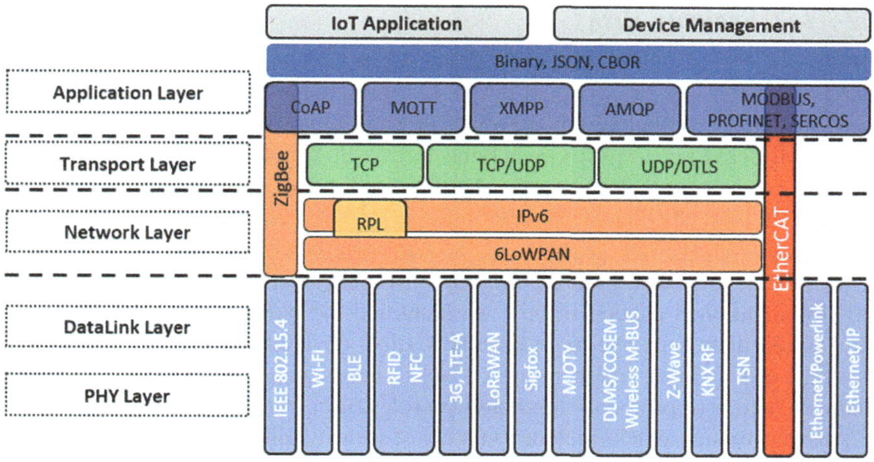

Fig. 1 Possible elements of IoT protocol stack

ernet/Powerlink [25] and EtherCat [26]. It should be noted that most of the previous industrial protocols may cover more than just the lower layers. TSN-Ethernet [27] (time-sensitive networking) for real-time audio/video streaming and real-time control streams are used in industrial facilities, which is defined in the data link layer. Meanwhile, Sercos [28], Profinet [29] and Modbus [30] are popular protocols in industrial environments for data communication in the application layer. Finally, cellular protocols [31] include 3G and LTE-advanced. The network layer is responsible for packet forwarding, which is divided into two sub-layers (Routing and Encapsulation), the network layer protocols [15] include RPL the routing protocol for low-power and lossy networks and its extension CORPL (cognitive RPL, designed for cognitive networks). Besides 6LoWPAN (standing for IPv6 over low-power wireless personal area networks) makes the encapsulation and header compression. The transport layer is responsible for end-to-end delivery of data from the source host to the destination host. The protocols of this layer are transmission control protocol (TCP) is used for connection-oriented transmissions, [32] user datagram protocol (UDP) is used for simpler messaging transmissions. Another point is communication security, [33] transport layer security (TLS) is a cryptographic protocol designed to provide communications security, and [34] datagram transport layer security (DTLS) provides security for datagram-based applications. The application layer specifies shared communications protocols [15] and interface methods used by hosts such as constrained application protocol (CoAP) which is designed to provide low overhead RESTful (HTTP) interface, message queue telemetry transport (MQTT) provides the connectivity between applications and users also between network and communications, and extensible messaging and presence protocol (XMPP) is designed for chats and messages exchange applications. DDS, AMQTT and AMQP are further examples of application layer protocols.

3.2 IoT Security Risks

Due to the heterogeneity of smart devices, communication protocols, applications, and services, the IoT network is exposed to various kinds of cyberattacks. For example, the Mirai IoT botnet [35], Green lights forever in Ann Arbor, Michigan state [36] are attack examples that show case the threat related to wide deployment of IoT devices. Such attacks against resource-constrained IoT devices have continued to increase over the last few years [37]. Recently, several research papers [38–40] surveyed IoT cyberattacks. Figure 2 present a taxonomy that we elaborated, based on the aforementioned papers. In this, we identified the potential vulnerabilities and categorized them into five main classes: (i) IoT devices attacks (which could be hardware or software), (ii) encryption attacks, (iii) data attacks, (iv) communication channels (active or protocol-based) and (v) roles (user or device identification).

Answers to such attacks include a number of security solutions, ranging from next-generation firewalls, intrusion detection systems, malware protection, to antivirus software. IDSs monitor a network or systems for malicious activity or policy violations by identifying unauthorized system behavior such as use, copying, modification and destruction [41]. In [42], authors review traditional intrusion detection techniques that are used in the IoT. Sikora et al. [43] highlighted in their work the ongoing and open research activities on Industrial Internet of Things (IIoT), from

Fig. 2 IoT attacks taxonomy adopted from [38–40].

novel cryptographic algorithms over security protocol integration and testing to security architectures for the full lifetime of devices and systems. Along with the effort of detecting and preventing IoT cyberattacks, the call for more *intelligent* security systems let to the use of AI techniques to provide new solutions. We discuss in the next section the structure of such techniques.

4 Structure of ML, DL and FL algorithms

This section introduces popular ML techniques used for cybersecurity intrusion detection. Within each of the AI techniques, several algorithms have been developed, investigated and assessed.

Popular ML approaches include (i) decision trees (DT) [44], a predictive modeling approach, which represent classifications and branches, (ii) random forest (RF) [45], an ensemble learning method for classification and regression, (iii) support vector machine (SVM) [46], a supervised learning model with associated learning algorithms that analyze data used for classification and regression analysis, (iv) K-nearest neighbors (KNN) [47], a nonparametric method used for classification and regression that assumes the similarity between the new case/data and available cases and put the new case into the category that is most similar to the available categories, (v) K-means [48], which partitions n observations into k clusters in which each observation belongs to the cluster with the nearest mean and (vi) genetic algorithms (GA) [49], a stochastic search algorithms which act on a population of possible solutions that is inspired by the process of natural selection and belongs to the larger class of evolutionary algorithms.

Popular DL algorithms, however, include: (i) recurrent neural network (RNN) [50], which utilizes the sequential information in the neural network, (ii) convolutional neural network (CNN) [51], a class of deep neural networks that is similar to the traditional neural networks with a structure inspired by the neurons in human brains, (iii) deep belief network (DBN), a class of deep neural network that is composed of multiple layers of latent variables (hidden units), with connections between the layers but not between units within each layer, (iv) deep neural network (DNN), an artificial neural network (ANN) is with multiple layers between the input and output layers and is typically feed-forward networks in which data flows from the input layer to the output layer and (v) long short-term memory (LSTM) [52], which is an RNN that has feedback connections that can process entire sequences of data by providing memory blocks in its recurrent connections.

Federated learning [53] is a machine learning technique introduced by Google in 2016, also known as collaborative learning because its training process involves multiple nodes. FL trains an algorithm across multiple decentralized edge devices while preserving data privacy. The approach is to train the algorithm on multiple local datasets contained in local nodes without explicitly exchanging their data samples. In the second round, the server aggregates the updates coming from the nodes and regenerates an updated global model. Motivated by privacy concerns among data owners, FL can serve as an enabling technology for ML model training.

5 Machine Learning for Cybersecurity

5.1 State of the Art

Over the last years, ML has been successfully applied in various areas of networking. We report in Table 2 the 52 research works we identified, where the authors investigated the use of ML techniques for cybersecurity.

We classify these works with regard to (i) the used algorithm(s), (ii) the dataset for training and (iii) the achieved results in terms of accuracy and precision. Accuracy is the proportion of correct predictions among the total number of predictions. However, it is not reliable for skewed class-wise data. For this, precision, which is the proportion of positive predictions that are correctly predicted, is added to asses further the results quality. For the used dataset, the symbol (*) means that its use was partial, while the symbol (+) means that it was augmented with a custom one.

5.2 General Trend Analysis

In Fig. 3, we report statistics based on Table 2. The chart of Fig. 3a counts the number of times a given algorithm (x-axes) is used.

Among those we detailed in sub-section 4, SVM and DT were clearly the most used ones, 15 times each, against six algorithms (RF, K-means, GA, C4.5, SBF and SOM) below 4 times. This disparity is likely due to the ease of use SVM and DT algorithms. It is important to mention that the sum of algorithm's use exceeds by far the total number of referenced papers, as 55.8% of the research works use several ML techniques. This is highlighted in Fig. 3b. As for the used datasets, KDD Cup 99 [102] comes first with a usage rate of 65.5% against only 7.7% of custom datasets. Indeed, it is one of the most used datasets for intrusion detection. It contains basic attributes about TCP connections and high-level attributes such as number of failed logins, but no IP addresses. It contains 20 different types of attacks such as DoS and buffer overflow.

6 Deep Learning use for Cybersecurity

6.1 State of the Art

As in Table 2, we report in Table 3 the 66 identified papers that perform intrusion detection using deep learning techniques.

Table 2 Mapping of ML techniques to cybersecurity intrusion detection

References	Year	Machine Learning Techniques	Dataset	Results	
				Accuracy (%)	Precision (%)
Pfahringer et al. [54]	2000	DT, C5	KDD Cup		
Mukkamala et al. [55]	2002	SVM, RBF	KDD Cup	99.50	
Pan et al. [56]	2003	DT, c4.5, Sup. NN	KDD Cup	93.28	
Kayacik et al. [57]	2003	SOM	KDD Cup		
Moradi et al. [58]	2004	Sup. NN	KDD Cup	91	
Amor et al. [59]	2004	Sup. NB	KDD Cup		
Zanero et al. [60]	2004	SOM	KDD Cup		
Wang et al. [61]	2004	SVM, Unsup. Centroid model	KDD Cup + CUCS	100	
Chebrolu et al. [62]	2005	Sup. BN, CART	KDD Cup	100	
Stein et al. [68]	2005	DT, C4.5	KDD Cup		
Kim et al. [63]	2005	SVM, GA	KDD Cup		
Zhang et al. [64]	2005	RBF	KDD Cup		
Depren et al. [65]	2005	SOM	KDD Cup		
Zhang et al. [66]	2006	RF	KDD Cup		
Jiang et al. [67]	2006	Unsup NN.	KDD Cup		
Zhang et al. [66]	2006	RF	KDD Cup		
Paddabachigari et al. [68]	2007	SVM, DT	KDD Cup		
Shon et al. [69]	2007	SVM, GA, OCSVM	KDD Cup (+)		
Paddabachigari et al. [68]	2007	SVM, DT	KDD Cup	100	
Li et al. [70]	2007	KNN, TCM	KDD Cup	99.7	
Ahmed et al. [71]	2007	Sup. Kernel Function	Abilence Network		
Giacinto et al. [72]	2008	k-means, Multiple Classifiers	KDD Cup		
Giacinto et al. [72]	2008	Multiple Classifiers	KDD Cup		
Hu et al. [73]	2008	Sup. Decision stumps	KDD Cup		
Perdisci et al. [74]	2009	SVM	KDD Cup + GATECH		
Gornitz et al. [75]	2009	SVDD	- -		
Sangkatsanee et al. [76]	2011	DT, C4.5	RLD09		
Panda et al. [77]	2012	RF, ND	NSL-KDD		99.90
Muniyandi et al. [78]	2012	k-means, DT	KDD Cup	95.80	95.60
Saxena et al. [79]	2014	SVM, PSO	KDD Cup	99	84.20
Chandrasekhar et al. [80]	2014	SVM, C	KDD Cup (*)	98.90	99.50
Yan et al. [81]	2010	SVM	DARPA 98 (*)	80.10	81.20
Pervez et al. [82]	2014	SVM	NSL-KDD	82.37	74
Kokila et al. [83]	2014	SVM	DARPA 98	95.11	
Kotpalliwar et al. [84]	2015	SVM, RBF	KDD Cup (*)	99.90	
Relan et al. [85]	2015	DT, C4.5	KDD Cup		98.45
Modinat et al. [86]	2015	DT, CFS	KDD Cup	94.60	
Balogun et al. [87]	2015	KNN DT	KDD Cup	100	
Sharifi et al. [88]	2015	k-means, KNN	NSL-KDD	90	
Azad et al. [89]	2015	GA, C4.5	KDD Cup	99.89	

(continued)

Table 2 (continued)

References	Year	Machine Learning Techniques	Dataset	Results	
				Accuracy (%)	Precision (%)
Ammar et al. [90]	2015	DT	Netflow	99	
Meng et al. [91]	2015	KNN	DARPA 99	85.30	
Vishwakarma et al. [92]	2017	KNN, ACO	KDD Cup (*)	94.70	
Rao et al. [93]	2017	KNN, IPDS	NSL-KDD (*)	99.60	
Shapoorifard et al. [94]	2017	KNN, kFN	NSL-KDD	99	98
Ingre et al. [95]	2017	DT, CFS	NSL-KDD	90.30	
Anderson and McGrew. [96]	2017	SVM, RF, DT, LogReg	Custom dataset	~ 99	
Moo et al. [97]	2017	DT	Netflow	84.70	
Dada [98]	2017	KNN	KDD Cup	98.55	
Miller et al. [99]	2017	Sup. MPML	NSL-KDD		
Boero et al. [100]	2017	SVM, RBF	U. of Genoa Malwares		
Malik et al. [101]	2018	DT	KDD Cup	91.94%	99.8895%

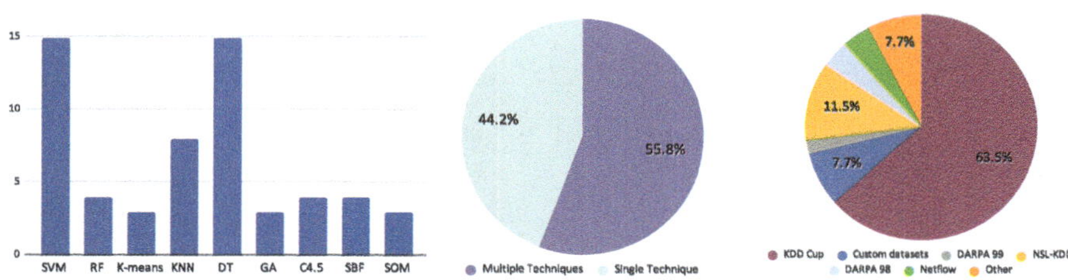

(a) Most used ML algorithms (b) ML algorithms multiplicity (c) Datasets

Fig. 3 Analysis of the use of ML in cybersecurity

Here, we use the same classification criteria as in the previous section, recalling the DL technique, the dataset, as well as the achieved results in terms of accuracy and precision.

6.2 General Trend Analysis

We also report in Fig. 4 statistics based on Table 3. The chart of Fig. 4a shows that DBN, CNN and DNN are the most used algorithms. However, and unlike ML, there is no notable disparity with the remaining algorithms. In addition, and unlike ML also, only 7.6% of the research works use several DL techniques, which is presented in the pie chart of Fig. 4b. As for the used datasets, KDD Cup 99 comes first, but with a much smaller rate of 25.8%, while custom datasets rate 15.2%.

Table 3 Mapping of DL techniques to cybersecurity

References	Year	Deep Learning techniques	Dataset	Results	
				Accuracy (%)	Precision (%)
Cannady et al. [103]	2000	RL, CNN, CMAC-NN	Prototype Application		
Servin et al. [104]	2008	RL, Q-Learning	Generated using NS-2	70–99	
Salama et al. [105]	2011	RBM	NSL-KDD		
Fiore et al. [106]	2013	RBM	KDD Cup		
Gao et al. [107]	2014	DBN	KDD Cup	93.49	92.33
Staudemeyer [108]	2015	LSTM	KDD Cup (*)	93.85	
Li et al. [109]	2015	DBN	KDD Cup	88.95–92.10	
Alom et al. [110]	2015	DBN	NSL-KDD	97.5	
Alrawashdeh et al. [111]	2016	DBN	KDD Cup (*)		97.90
Tan et al. [112]	2016	DBN	Netflow	97.60	
Krishnan et al. [113]	2016	RNN	KDD Cup	77.55	84.60
Kim et al. [114]	2016	LSTM	KDD Cup	99.80	
Kolosnjaji et al. [115]	2016	CNN	Netflow		93
Tang et al. [116]	2016	DNN	NSL-KDD	72.0–75.75	79–83
Kim et al. [117]	2016	RNN, LSTM	KDD Cup	96.93	
Javaid et al. [118]	2016	Self-taught Learning	NSL-KDD		85.44
Ding et al. [119]	2016	DBN	Netflow	96.10	
Nadeem et al. [120]	2016	DBN	KDD Cup	99.18	
Potluri et al. [121]	2016	DNN	NSL-KDD		
Kang et al. [122]	2016	DNN	Vehicular Net. Com.		
Taylor et al. [123]	2016	RNN	Attacks against Vehicles		
Fu et al. [124]	2016	CNN	Credit Card transaction data		
Wang et al. [125]	2016	Denoising Auto-Encoder	Heritrix	95	
Cordero et al. [126]	2016	Replicator Neural Network	MAWI		
Zhao et al. [127]	2017	DBN, PNN	DARPA 99	99.14	93.25
Yin et al. [128]	2017	RNN	NSL-KDD	97.09	
Yin et al. [128]	2017	RNN	NSL-KDD	83.28	
Le et al. [129]	2017	LSTM	KDD Cup	97.54	98.95
Yu et al. [130]	2017	CNN	CTU-UNB		98.44
Saxe and Berlin [131]	2017	CNN	Netflow	92	
Wang et al. [132]	2017	CNN	ISCX dataset		97.30
Wang et al. [133]	2017	CNN	Netflow	99.41	
Loukas et al. [134]	2017	RNN	Attacks against Vehicles		
yang et al. [135]	2017	RBM	Netflow		
Zhao et al. [127]	2017	DBN	KDD Cup	99.14	
He et al. [136]	2017	DBN	IEEE-118 bus, IEEE-300 bus		

(continued)

Table 3 (continued)

References	Year	Deep Learning techniques	Dataset	Results	
				Accuracy (%)	Precision (%)
Roy et al. [137]	2017	DNN	KDD Cup		
Kim et al. [138]	2017	DNN	KDD Cup	99	
Zhang et al. [139]	2018	DNN	Attacks against Vehicles		
Azmoodeh et al. [140]	2018	Deep Eigenspace Learning	Custom Dataset	99.68	98.59
Agarap [141]	2018	GRU	Netflow	84.15	
Tang et al. [142]	2018	RNN	NSL-KDD	89	91
Jiang et al. [143]	2018	RNN	NSL-KDD		
Shone et al. [144]	2018	Deep Auto-Encoder	NSL-KDD	85.42	100
Zhang et al. [145]	2018	CNN	Online Transaction data		
Nasr et al. [146]	2018	CNN	UMASS	96	
Aldwairi et al. [147]	2018	RBM	ISCX dataset		
Zhou et al. [148]	2018	DNN	Custom Dataset	96	
Feng et al. [149]	2019	DNN	KDD Cup		
Zhang et al. [150]	2019	DNN	KDD Cup		
Holbrook et al. [151]	2019	DNN	Mirai, Gafgyt	99	
Ferrag et al. [152]	2019	RNN	CICIDS2017		
Basumallik et al. [153]	2019	CNN	IEEE-30 bus + IEEE-118 bus		
Otoum et al. [154]	2019	RBM	KDD Cup	99.91	
Karimipour et al. [155]	2019	RBM	IEEE 39, 118, 2848 bus systems	99	
Thamilarasu et al. [156]	2019	DBN	IoT simulation dataset		98.47
Zhang et al. [157]	2019	CNN	CICIDS2017		
Zeng et al. [158]	2019	CNN	ISCX 2012		
Khan et al. [159]	2019	Deep Auto-Encoder	UNSW-NB15	89.13	
Papamartzivanos et al. [160]	2019	Deep Auto-Encoder	NSL-KDD		
yang et al. [161]	2019	Deep Auto-Encoder	NSL-KDD + UNSW-NB15	89.08	
Abusitta et al. [162]	2019	Denoising Auto-Encoder	KDD Cup	95	
Zhang et al. [163]	2019	DBN	NSL-KDD		
Aloqaily et al. [164]	2019	DBN	NSL-KDD + NS-3 traffic	99.43	
Kasongo et al. [165]	2019	DNN	NSL-KDD	98.66	
Thamilarasu et al. [156]	2019	DBN, DNN	–		95

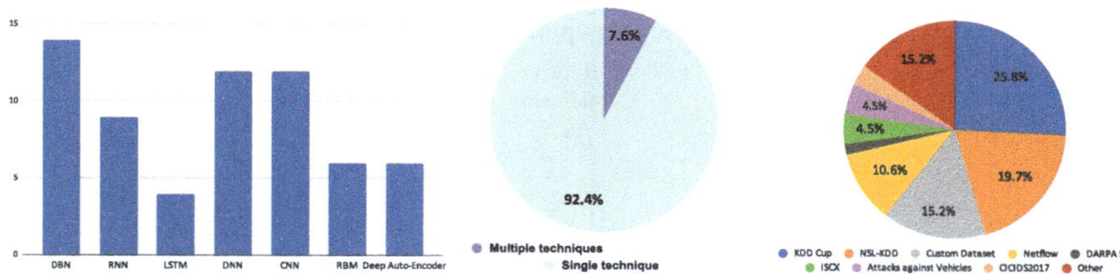

(a) Most used DL algorithms (b) DL algorithms multiplicity (c) Datasets

Fig. 4 Analysis of the use of DL techniques in cybersecurity

7 Federated Learning use for IoT Cybersecurity

7.1 State of the Art

A rule of thumb in ML and DL techniques is "the larger is the dataset, the better is the detection". To address data collection problems, especially data privacy, new attack detection modes have been introduced based on FL technique. This section presents the state of the art of FL-based implementations for cybersecurity, with a focus on IoT field and intrusion detection and communication optimization.

Table 4 presents the nine identified research works. The classification criteria include those from used in previous sections (algorithm dataset and results), while adding the cybersecurity system type, the target domain and the protocol.

In [166], the authors proposed a cyberattack detection model with edge nodes as participants that owns a set of local data. Empowered by FL, each edge node can learn from the other participants without sharing its data. This technique is used to increase detection accuracy and guarantee data privacy. Similarly, authors in [167] presented the same idea where the gateways operate as FL participants. In this model, the server that aggregates the trained model is an IoT security provider. In [168], the authors evaluate their model using a real-world smart home deployment, and the model has successfully reached 95.6% of attacks detection accuracy without raising any false alarm. The authors in [169] presented a study of FL application to URLLC in vehicular networks. It shows that FL solution reached good accuracy compared to centralized solution while reducing the average of power consumption. In [170], however, the authors presented a FL-based implementation of an anomaly detection DNN model deployed on the edge using autoencoder learning technique. The aim of this latter is to learn a representation (encoding) for a set of data while ignoring noise. The proposed solution shows similar results compared to centralized one, and it reduces the amount of transferred data. Authors in [171] implemented a data-sharing scheme empowered by blockchain and FL for industrial IoT applications to

Table 4 FL-based approaches for IoT security applications and communication efficiency

References	Algorithm	Dataset	Sys type	Target domain	Protocols	Results
[166]	DL	NSL-KDD	IDS	IoT	TCP	Accuracy = 99.20 %
					UDP	Detection Rate = 99.27 %
					ICMP	False Alarm Rate = 0.85 %
[167]	DL	Mirai attack	IDS	IoT	Wi-Fi	Detection Rate = 95.6 %
					Ethernet	False Alarm Rate = 0 %
						Detection Speed = 257 ms
[168]	DL	CICIDS2017	IDS	IoT	PoW	Performance overhead = 5-15%
[169]	n.s	Nnn-IID	URLLC	V2V Com.	PDCP 5G NR	Reductions of VUEs = 60 %
						Accuracy = 79 %
[170]	DL	KDD Cup	IDS	IoT	TCP	Accuracy = 99.05%
					ICMP	
					UDP	
[171]	ML	Reuters	Data Privacy	Industrial IoT	PoW	Resource optimization
					PoQ	Increased efficiency
[172]	DRL	n.s	Com.	IoT	n.s	
[173]	n.s	NSL-KDD	IDS	IoT	TCP	Accuracy = 83.09%
					UDP	
					ICMP	
[174]	DL	non-IID	Com.	Industrial IoT	5G	

address data-sharing problem. According to them, the proposed architecture achieves high efficiency, good accuracy and enhanced security. In [172], the authors deployed DRL agents on multiple edge nodes for IoT dynamic system. The FL-based approach has shown the effectiveness of the decision scheme. In the context of FL-based solution for IoT cybersecurity, the authors in [173] proposed intrusion detection model empowered by FL to maintain data privacy, the model's performance and accuracy are close to non-distributed approach. In [174], the authors introduced new FL-based solution integrated within 5G for wireless IoT networks. They validated the approach on an industrial IoT scenario without relying on the support of the server coordinator.

7.2 General Trend Analysis

As the number of identified papers is already small (nine), it is hard to build reliable analysis about the used algorithms or dataset use. However, the field being at its beginning, the disparity of the used datasets as well as the reached results make FL a real promising technique for IoT cybersecurity.

8 Discussion and Future Insights

Although more research works used DL techniques for cybersecurity compared to ML, a fewer rate did combine different algorithms in order to achieve better detection results. Since evaluation metrics of algorithms performance differ for each study, it is hard to compare the achieved results with one another. Still, the tendency of DL use is undeniable. Moreover, they come with more datasets and have been more investigated in FL.

With regard to the datasets, KDD Cup 99 is used in the two-thirds of ML-based implementations and the quarter of the DL ones, and this shows the limited number of existing representative datasets. However, the community is aware of the importance of realistic network-based data so they tried to use novel datasets or to train their model on netflow data. Added to that, DL gives the possibility to handle large amounts of data without human involvement. It is clear that most of the works using ML and DL did not focus on IoT security nor using datasets based on IoT data in the training set of their models. The works that give importance to IoT-based IDSs are still rare. For a more insightful analysis, we attempted to find out a correlation relationship between the achieved accuracy, the dataset and the used algorithms. No significant results were found. From all the aforementioned analysis and discussion, we draw the following conclusions and insights about future works:

1. ***Lack of real-world data***: Few public datasets are available, and the same dataset is used. Added to that, no perfect network-based dataset exists. Thus, the problems with datasets have widely limited the development of research in this area. Furthermore, most of the algorithms were trained offline. Online learning gives the opportunity to adapt new real-world data to the trained model. Therefore, FL technique is one of the emerging solutions to deal with issues related to data in order to build a robust ML model, thus allowing to address critical data issues.
2. ***IoT protocols diversity***: The increasing number of cyberattacks and their diversity reflects the lack of security in cyberspace. Additionally, it should be taken into consideration that IoT keeps expanding every day and more devices become connected through IoT networks. Hence, researchers should further investigate protocols issues and give deeper importance to the application of machine learning for IoT cybersecurity challenges.

3. ***Effective method for intrusion detection***: Despite the large application of ML and DL approaches in the field of intrusion detection, the most effective method has not been established yet, because each method has its own advantages and disadvantages. The evaluation metrics of algorithms performance differ for each study, thus results cannot be compared with one another. Thus, a part of the presented studies combined two or more techniques in order to achieve better intrusion detection performance of the algorithm.

9 Conclusion

The paper has presented the literature review of ML and DL methods used for cybersecurity applications in general and in intrusion detection specifically. Special emphasis was placed on finding example papers that describe the implementation of FL technique in IoT cybersecurity for intrusion detection. Unfortunately, there is no one ML or DL algorithm capable of effectively detecting all the types of attacks. Researches tend to combine

several algorithms and focus on a number of attacks. This makes an empty room for a complete, yet adaptive ML or DL method for cybersecurity to be established. Still, in such applications, data is another important aspect especially for algorithms training. Only a few public datasets are available, thus the lack of data is still holding the field back.

FL seems a promising alternative though. If generalized to support the large spectrum of IoT attacks, further enhancements in efficiency of IoT cybersecurity systems can be reached. A lot of work needs to be done toward this end.

References

1. S. Madakam, V. Lake, V. Lake, V. Lake et al., Internet of things (iot): a literature review. J. Comput. Commun. **3**(05), 164 (2015)
2. H. Tankovska, Iot connected devices worldwide 2030. https://www.statista.com/statistics/802690/worldwide-connected-devices-by-access-technology/ (2020)
3. I. Newsflash, Cyber security statistics for 2019. https://www.cyberdefensemagazine.com/cyber-security-statistics-for-2019/ (2019)
4. M.R. Genesereth, N.J. Nilsson, *Logical Foundations of Artificial Intelligence* (Morgan Kaufmann, Burlington, MA, 2012)
5. E. Alpaydin, *Introduction to Machine Learning*. MIT Press (2020)
6. Y. LeCun, Y. Bengio, G. Hinton, Deep learning. Nature **521**(7553), 436–444 (2015)
7. J. Konečnỳ, H.B. McMahan, F.X. Yu, P. Richtárik, A.T. Suresh, D. Bacon, Federated learning: strategies for improving communication efficiency.' *arXiv preprint* arXiv:1610.05492 (2016)
8. A. Verma, V. Ranga, Machine learning based intrusion detection systems for iot applications. Wirel. Personal Commun. **111**(4), 2287–2310 (2020)
9. A.L. Buczak, E. Guven, A survey of data mining and machine learning methods for cyber security intrusion detection. IEEE Commun. Surv. Tutori. **18**(2), 1153–1176 (2015)

10. R. Boutaba, M.A. Salahuddin, N. Limam, S. Ayoubi, N. Shahriar, F. Estrada-Solano, O.M. Caicedo, A comprehensive survey on machine learning for networking: evolution, applications and research opportunities. J. Internet Serv. Appl. **9**(1), 16 (2018)
11. Y. Xin, L. Kong, Z. Liu, Y. Chen, Y. Li, H. Zhu, M. Gao, H. Hou, C. Wang, Machine learning and deep learning methods for cybersecurity. IEEE Access **6**, 35365–35381 (2018)
12. M. Ring, S. Wunderlich, D. Scheuring, D. Landes, A. Hotho, A survey of network-based intrusion detection data sets. Comput. Sec. **86**, 147–167 (2019)
13. W.Y.B. Lim, N.C. Luong, D.T. Hoang, Y. Jiao, Y.-C. Liang, Q. Yang, D. Niyato, C. Miao, Federated learning in mobile edge networks: a comprehensive survey. IEEE Commun. Surv. Tutorials **22**(3), 2031–2063 (2020)
14. M.A. Ferrag, L. Maglaras, S. Moschoyiannis, H. Janicke, Deep learning for cyber security intrusion detection: approaches, datasets, and comparative study. J. Inf. Sec. Appl. **50**, 102419 (2020)
15. T. Salman, R. Jain, A survey of protocols and standards for internet of things. *arXiv preprint* arXiv:1903.11549 (2019)
16. J.-S. Lee, Y.-W. Su, C.-C. Shen, A comparative study of wireless protocols: bluetooth, uwb, zigbee, and wi-fi," in *IECON 2007-33rd Annual Conference of the IEEE Industrial Electronics Society* (IEEE, 2007), pp. 46–51
17. A. Lahtela, M. Hassinen, V. Jylha, "Rfid and nfc in healthcare: safety of hospitals medication care," in *2008 Second International Conference on Pervasive Computing Technologies for Healthcare*. (IEEE, 2008), pp. 241–244
18. P. Baronti, P. Pillai, V.W. Chook, S. Chessa, A. Gotta, Y.F. Hu, Wireless sensor networks: a survey on the state of the art and the 802.15. 4 and zigbee standards. Comput. commun. **30**(7), 1655–1695 (2007)
19. M.B. Yassein, W. Mardini, A. Khalil, Smart homes automation using z-wave protocol, in *2016 International Conference on Engineering & MIS (ICEMIS)*. (IEEE, 2016), pp. 1–6
20. S. Marksteiner, V.J.E. Jiménez, H. Valiant, H. Zeiner, An overview of wireless iot protocol security in the smart home domain, in *2017 Internet of Things Business Models, Users, and Networks*. (IEEE, 2017), pp. 1–8
21. U. Raza, P. Kulkarni, M. Sooriyabandara, Low power wide area networks: an overview. IEEE Commun. Surv. Tutorials **19**(2), 855–873 (2017)
22. D. Vyas, H. Pandya, Advance metering infrastructure and dlms/cosem standards for smart grid.' Int. J. Eng. Res. **1**(1) (2012)
23. V. Mohan, *An Introduction to Wireless M-bus*. Silicon Labs (2015)
24. P. Brooks, Ethernet/ip-industrial protocol, in *ETFA 2001. 8th International Conference on Emerging Technologies and Factory Automation. Proceedings (Cat. No. 01TH8597)*, vol. 2 (IEEE, 2001), pp. 505–514
25. G. Cena, L. Seno, A. Valenzano, S. Vitturi, Performance analysis of ethernet powerlink networks for distributed control and automation systems. Comput. Stand. Interf. **31**(3), 566–572 (2009)
26. M. Rostan, J.E. Stubbs, D. Dzilno, Ethercat enabled advanced control architecture, in *2010 IEEE/SEMI Advanced Semiconductor Manufacturing Conference (ASMC)* (IEEE, 2010), pp. 39–44
27. S. Kehrer, O. Kleineberg, D. Heffernan, A comparison of fault-tolerance concepts for ieee 802.1 time sensitive networks (tsn), in *Proceedings of the 2014 IEEE Emerging Technology and Factory Automation (ETFA)* (IEEE, 2014), pp. 1–8
28. E. Schemm, Sercos to link with ethernet for its third generation. Comput. Control Eng. **15**(2), 30–33 (2004)
29. J. Feld, Profinet-scalable factory communication for all applications, in *IEEE International Workshop on Factory Communication Systems, 2004. Proceedings*. (IEEE, 2004), pp. 33–38
30. S. Mohagheghi, J. Stoupis, Z. Wang, Communication protocols and networks for power systems-current status and future trends, in *2009 IEEE/PES Power Systems Conference and Exposition* (IEEE, 2009), pp. 1–9

31. E. Ezhilarasan, M. Dinakaran, A review on mobile technologies: 3g, 4g and 5g, in *2017 Second International Conference on Recent Trends and Challenges in Computational Models (ICRTCCM)*. (IEEE, 2017), pp. 369–373
32. J. Postel, *RFC0768: User Datagram Protocol* (1980)
33. E. Rescorla, T. Dierks, *The transport Layer Security (tls) Protocol Version 1.3* (2018)
34. T. Kothmayr, C. Schmitt, W. Hu, M. Brünig, G. Carle, Dtls based security and two-way authentication for the internet of things. Ad Hoc Netw. **11**(8), 2710–2723 (2013)
35. M. Antonakakis, T. April, M. Bailey, M. Bernhard, E. Bursztein, J. Cochran, Z. Durumeric, J.A. Halderman, L. Invernizzi, M. Kallitsis, *et al.*, Understanding the mirai botnet, in *26th* USENIX security symposium (USENIX *Security 17*) (2017), pp. 1093–1110
36. B. Ghena, W. Beyer, A. Hillaker, J. Pevarnek, J. A. Halderman, Green lights forever: Analyzing the security of traffic infrastructure, in *8th* USENIX Workshop on Offensive Technologies (WOOT **14**) (2014)
37. M. Nawir, A. Amir, N. Yaakob, O. B. Lynn, Internet of things (iot): taxonomy of security attacks," in *2016 3rd International Conference on Electronic Design (ICED)* (IEEE, 2016), pp. 321–326
38. J. Deogirikar, A. Vidhate, Security attacks in iot: a survey, in *2017 International Conference on I-SMAC (IoT in Social, Mobile, Analytics and Cloud)(I-SMAC)* (IEEE, 2017), pp. 32–37
39. H. A. Abdul-Ghani, D. Konstantas, M. Mahyoub, A comprehensive iot attacks survey based on a building-blocked reference model. IJACSA Int. J. Adv. Comput. Sci. Appl. **9**(3), 355–373, (2018)
40. M. Stoyanova, Y. Nikoloudakis, S. Panagiotakis, E. Pallis, E.K. Markakis, A survey on the internet of things (iot) forensics: challenges, approaches and open issues. IEEE Commun. Surv. Tutorials **22**(2), 1191–1221 (2020)
41. A. Milenkoski, M. Vieira, S. Kounev, A. Avritzer, B.D. Payne, Evaluating computer intrusion detection systems: A survey of common practices. ACM Comput. Surv. (CSUR) **48**(1), 1–41 (2015)
42. B.B. Zarpelão, R.S. Miani, C.T. Kawakani, S.C. de Alvarenga, A survey of intrusion detection in internet of things. J. Netw. Comput. Appl. **84**, 25–37 (2017)
43. A. Sikora, A. Walz, L. Zimmermann, Research aspects for secure communication in the industrial internet of things, in *2020 IEEE 11th International Conference on Dependable Systems, Services and Technologies (DESSERT)* (IEEE, 2020), pp. 284–289
44. S.R. Safavian, D. Landgrebe, A survey of decision tree classifier methodology. IEEE Trans. Syst. Man Cybern. **21**(3), 660–674 (1991)
45. T.K. Ho, Random decision forests, in *Proceedings of 3rd international conference on document analysis and recognition*, vol 1 (IEEE, 1995), pp. 278–282
46. J.A. Suykens, J. Vandewalle, Least squares support vector machine classifiers. Neural Process. Lett. **9**(3), 293–300 (1999)
47. T. Cover, P. Hart, Nearest neighbor pattern classification. IEEE Trans. Inf. Theory **13**(1), 21–27 (1967)
48. J. MacQueen et al., Some methods for classification and analysis of multivariate observations, in *Proceedings of the Fifth Berkeley Symposium on Mathematical Statistics and Probability*, vol 1 (Oakland, CA, USA, 1967), pp. 281–297
49. L. Davis, *Handbook of Genetic Algorithms* (CumInCAD, 1991)
50. D.E. Rumelhart, G.E. Hinton, R.J. Williams, *Learning Internal Representations by Error Propagation*, Technical Report (California University San Diego La Jolla Institution for Cognitive Science, 1985)
51. Y. LeCun, Y. Bengio et al., Convolutional networks for images, speech, and time series. Handbook brain Theory Neural Netw **3361**(10), 1995 (1995)
52. F.A. Gers, J. Schmidhuber, F. Cummins, *Learning to Forget: Continual Prediction with lstm*, (1999)
53. J. Konečnỳ, H. B. McMahan, D. Ramage, and P. Richtárik, "Federated optimization: Distributed machine learning for on-device intelligence," *arXiv preprint* arXiv:1610.02527, 2016

54. B. Pfahringer, Winning the kdd99 classification cup: bagged boosting. ACM SIGKDD Exp. Newslett. **1**(2), 65–66 (2000)
55. S. Mukkamala, G. Janoski, A. Sung, Intrusion detection using neural networks and support vector machines," in *Proceedings of the 2002 International Joint Conference on Neural Networks. IJCNN'02 (Cat. No. 02CH37290)*, vol 2 (IEEE, 2002), pp. 1702–1707
56. Z.-S. Pan, S.-C. Chen, G.-B. Hu, D.-Q. Zhang, Hybrid neural network and c4. 5 for misuse detection, in *Proceedings of the 2003 International Conference on Machine Learning and Cybernetics (IEEE Cat. No. 03EX693)*, vol 4 (IEEE, 2003), pp. 2463–2467
57. H.G. Kayacik, A.N. Zincir-Heywood, M.I. Heywood, On the capability of an som based intrusion detection system," in *Proceedings of the International Joint Conference on Neural Networks, 2003*, vol 3, (IEEE, 2003), pp. 1808–1813
58. M. Moradi, M. Zulkernine, A neural network based system for intrusion detection and classification of attacks, in *Proceedings of the IEEE International Conference on Advances in Intelligent Systems-theory and Applications* (IEEE, Lux-embourg-Kirchberg, Luxembourg, 2004), pp. 15–18
59. N.B. Amor, S. Benferhat, Z. Elouedi, Naive bayes vs decision trees in intrusion detection systems, in *Proceedings of the 2004 ACM symposium on Applied computing* (2004), pp. 420–424
60. S. Zanero, S. M. Savaresi, Unsupervised learning techniques for an intrusion detection system, in *Proceedings of the 2004 ACM Symposium on Applied Computing* (2004), pp. 412–419
61. K. Wang, S.J. Stolfo, Anomalous payload-based network intrusion detection, in *International Workshop on Recent Advances in Intrusion Detection* (Springer, 2004), pp. 203–222
62. S. Chebrolu, A. Abraham, J.P. Thomas, Feature deduction and ensemble design of intrusion detection systems. Comput. Sec. **24**(4), 295–307 (2005)
63. D.S. Kim, H.-N. Nguyen, J.S. Park, Genetic algorithm to improve svm based network intrusion detection system, in *19th International Conference on Advanced Information Networking and Applications (AINA'05) Volume 1 (AINA papers)*, vol 2 (IEEE, 2005), pp. 155–158
64. C. Zhang, J. Jiang, M. Kamel, Intrusion detection using hierarchical neural networks. Patt. Recogn. Lett. **26**(6), 779–791 (2005)
65. O. Depren, M. Topallar, E. Anarim, M.K. Ciliz, An intelligent intrusion detection system (ids) for anomaly and misuse detection in computer networks. Expert Syst. Appl. **29**(4), 713–722 (2005)
66. J. Zhang, M. Zulkernine, Anomaly based network intrusion detection with unsupervised outlier detection, in *2006 IEEE International Conference on Communications*, vol 5 (IEEE, 2006), pp. 2388–2393
67. S. Jiang, X. Song, H. Wang, J.-J. Han, Q.-H. Li, A clustering-based method for unsupervised intrusion detections. Patt. Recognition Lett. **27**(7), 802–810 (2006)
68. S. Peddabachigari, A. Abraham, C. Grosan, J. Thomas, Modeling intrusion detection system using hybrid intelligent systems. J. Netw. Comput. Appl. **30**(1), 114–132 (2007)
69. T. Shon, J. Moon, A hybrid machine learning approach to network anomaly detection. Inf. Sci. **177**(18), 3799–3821 (2007)
70. Y. Li, L. Guo, An active learning based tcm-knn algorithm for supervised network intrusion detection. Comput. Sec. **26**(7–8), 459–467 (2007)
71. T. Ahmed, M. Coates, A. Lakhina, Multivariate online anomaly detection using kernel recursive least squares, in *IEEE INFOCOM 2007-26th IEEE International Conference on Computer Communications* (IEEE, 2007), pp. 625–633
72. G. Giacinto, R. Perdisci, M. Del Rio, F. Roli, Intrusion detection in computer networks by a modular ensemble of one-class classifiers. Inf. Fusion **9**(1), 69–82 (2008)
73. W. Hu, W. Hu, S. Maybank, Adaboost-based algorithm for network intrusion detection. IEEE Trans. Syst. Man Cybern.Part B (Cybern.) **38**(2), 577–583 (2008)
74. R. Perdisci, D. Ariu, P. Fogla, G. Giacinto, W. Lee, Mcpad: A multiple classifier system for accurate payload-based anomaly detection. Comput. Netw. **53**(6), 864–881 (2009)
75. N.Görnitz, M. Kloft, K. Rieck, U. Brefeld, Active learning for network intrusion detection, in *Proceedings of the 2nd ACM Workshop on Security and Artificial Intelligence* (2009), pp. 47–54

76. P. Sangkatsanee, N. Wattanapongsakorn, C. Charnsripinyo, Practical real-time intrusion detection using machine learning approaches. Comput. Commun. **34**(18), 2227–2235 (2011)
77. M. Panda, A. Abraham, M.R. Patra, A hybrid intelligent approach for network intrusion detection. Procedia Eng. **30**, 1–9 (2012)
78. A.P. Muniyandi, R. Rajeswari, R. Rajaram, Network anomaly detection by cascading k-means clustering and c4. 5 decision tree algorithm. Procedia Eng. **30**, 174–182 (2012)
79. H. Saxena, V. Richariya, Intrusion detection in kdd99 dataset using svm-pso and feature reduction with information gain. Int. J. Comput. Appl. **98**(6) (2014)
80. A. Chandrasekhar, K. Raghuveer, Confederation of fcm clustering, ann and svm techniques to implement hybrid nids using corrected kdd cup 99 dataset, in *2014 International Conference on Communication and Signal Processing* (IEEE, 2014), pp. 672–676
81. M. Yan, Z. Liu, A new method of transductive svm-based network intrusion detection, in *International Conference on Computer and Computing Technologies in Agriculture* (Springer, 2010), pp. 87–95
82. M.S. Pervez, D.M. Farid, Feature selection and intrusion classification in nsl-kdd cup 99 dataset employing svms, in *The 8th International Conference on Software, Knowledge, Information Management and Applications (SKIMA 2014)* (IEEE, 2014), pp. 1–6
83. R. Kokila, S.T. Selvi, K. Govindarajan, Ddos detection and analysis in sdn-based environment using support vector machine classifier, in *2014 Sixth International Conference on Advanced Computing (ICoAC)* (IEEE, 2014), pp. 205–210
84. M.V. Kotpalliwar, R. Wajgi, Classification of attacks using support vector machine (svm) on kddcup'99 ids database, in *2015 Fifth International Conference on Communication Systems and Network Technologies* (IEEE, 2015), pp. 987–990
85. N.G. Relan, D.R. Patil, Implementation of network intrusion detection system using variant of decision tree algorithm, in *2015 International Conference on Nascent Technologies in the Engineering Field (ICNTE)* (IEEE, 2015), pp. 1–5
86. M.A.A. Akintola, A. Balogun, O. Ayilara, Gain ratio and decision tree classifier for intrusion detection. Int. J. Comput. Appl *126*(10), 975–8887 (2015)
87. A.O. Balogun, R.G. Jimoh, Anomaly intrusion detection using an hybrid of decision tree and k-nearest neighbor. J. Adv. Sci. Res. Appl. (JASRA) **2** (2015)
88. A.M. Sharifi, S.K. Amirgholipour, A. Pourebrahimi, Intrusion detection based on joint of k-means and knn. J. Converg. Inf. Technol. **10**(5), 42 (2015)
89. C. Azad, V.K. Jha, Genetic algorithm to solve the problem of small disjunct in the decision tree based intrusion detection system. Int. J. Comput. Netw. Inf. Sec. (IJCNIS) **7**(8), 56–71 (2015)
90. A. Ammar et al., A decision tree classifier for intrusion detection priority tagging. J. Comput. Commun. **3**(04), 52 (2015)
91. W. Meng, W. Li, L.-F. Kwok, Design of intelligent knn-based alarm filter using knowledge-based alert verification in intrusion detection. Sec. Commun. Netw. **8**(18), 3883–3895 (2015)
92. S. Vishwakarma, V. Sharma, A. Tiwari, An intrusion detection system using knn-aco algorithm. Int. J. Comput. Appl **171**(10), 18–23 (2017)
93. B.B. Rao, K. Swathi, Fast knn classifiers for network intrusion detection system. Indian J. Sci. Technol. **10**(14), 1–10 (2017)
94. H. Shapoorifard, P. Shamsinejad, Intrusion detection using a novel hybrid method incorporating an improved knn. Int. J. Comput. Appl **173**(1), 5–9 (2017)
95. B. Ingre, A. Yadav, A. K. Soni, Decision tree based intrusion detection system for nsl-kdd dataset, in *International Conference on Information and Communication Technology for Intelligent Systems* (Springer, 2017), pp. 207–218
96. B. Anderson, D. McGrew, Machine learning for encrypted malware traffic classification: accounting for noisy labels and non-stationarity, in *Proceedings of the 23rd ACM SIGKDD International Conference on knowledge discovery and data mining* (2017), pp. 1723–1732
97. D. Moon, H. Im, I. Kim, J.H. Park, Dtb-ids: an intrusion detection system based on decision tree using behavior analysis for preventing apt attacks. J. Supercomput. **73**(7), 2881–2895 (2017)

98. E. Dada, A hybridized svm-knn-pdapso approach to intrusion detection system, in *Proceedings of Faculty Seminar Series* (2017), pp. 14–21
99. S.T. Miller, C. Busby-Earle, Multi-perspective machine learning a classifier ensemble method for intrusion detection, in *Proceedings of the 2017 International Conference on Machine Learning and Soft Computing* (2017), pp. 7–12
100. L. Boero, M. Marchese, S. Zappatore, Support vector machine meets software defined networking in ids domain, in *2017 29th International Teletraffic Congress (ITC 29)*, vol 3 (IEEE, 2017), pp. 25–30
101. A.J. Malik, F.A. Khan, A hybrid technique using binary particle swarm optimization and decision tree pruning for network intrusion detection. Cluster Comput. **21**(1), 667–680 (2018)
102. M. Tavallaee, E. Bagheri, W. Lu, A.A. Ghorbani, A detailed analysis of the kdd cup 99 data set," in *2009 IEEE Symposium on Computational Intelligence for Security and Defense Applications* (IEEE, 2009), pp. 1–6
103. J. Cannady, Next generation intrusion detection: Autonomous reinforcement learning of network attacks, in *Proceedings of the 23rd National Information Systems Security Conference* (2000), pp. 1–12
104. A. Servin, D. Kudenko, Multi-agent reinforcement learning for intrusion detection: a case study and evaluation, in *German Conference on Multiagent System Technologies* (Springer, 2008), pp. 159–170
105. M.A. Salama, H.F. Eid, R.A. Ramadan, A. Darwish, A.E. Hassanien, Hybrid intelligent intrusion detection scheme, in *Soft Computing in Industrial Applications* (Springer, 2011), pp. 293–303
106. U. Fiore, F. Palmieri, A. Castiglione, A. De Santis, Network anomaly detection with the restricted boltzmann machine. Neurocomputing **122**, 13–23 (2013)
107. N. Gao, L. Gao, Q. Gao, H. Wang, An intrusion detection model based on deep belief networks, in *2014 Second International Conference on Advanced Cloud and Big Data* (IEEE, 2014), pp. 247–252
108. R.C. Staudemeyer, Applying long short-term memory recurrent neural networks to intrusion detection. South African Comput. J. **56**(1), 136–154 (2015)
109. Y. Li, R. Ma, R. Jiao, A hybrid malicious code detection method based on deep learning. Int. J. Sec. Appl. **9**(5), 205–216 (2015)
110. M.Z. Alom, V. Bontupalli, T.M. Taha, Intrusion detection using deep belief networks, in *2015 National Aerospace and Electronics Conference (NAECON)* (IEEE, 2015), pp. 339–344
111. K. Alrawashdeh, C. Purdy, Toward an online anomaly intrusion detection system based on deep learning, in *2016 15th IEEE International Conference on Machine Learning and Applications (ICMLA)* (IEEE, 2016), pp. 195–200
112. Q.-S. Tan, W. Huang, and Q. Li, An intrusion detection method based on dbn in ad hoc networks, in *Wireless Communication and Sensor Network: Proceedings of the International Conference on Wireless Communication and Sensor Network (WCSN 2015)* (World Scientific, 2016), pp. 477–485
113. R.B. Krishnan, N. Raajan, An intellectual intrusion detection system model for attacks classification using rnn. Int. J. Pharm. Technol. **8**(4), 23157–23164 (2016)
114. G. Kim, H. Yi, J. Lee, Y. Paek, S. Yoon, Lstm-based system-call language modeling and robust ensemble method for designing host-based intrusion detection systems. *arXiv preprint* arXiv:1611.01726 (2016)
115. B. Kolosnjaji, A. Zarras, G. Webster, C. Eckert, Deep learning for classification of malware system call sequences in *Australasian Joint Conference on Artificial Intelligence* (Springer, 2016), pp. 137–149
116. T.A. Tang, L. Mhamdi, D. McLernon, S.A.R. Zaidi, M. Ghogho, Deep learning approach for network intrusion detection in software defined networking, in *2016 International Conference on Wireless Networks and Mobile Communications (WINCOM)* (IEEE, 2016), pp. 258–263
117. J. Kim, J. Kim, H. L. T. Thu, H. Kim, Long short term memory recurrent neural network classifier for intrusion detection, in *2016 International Conference on Platform Technology and Service (PlatCon)* (IEEE, 2016), pp. 1–5

118. A. Javaid, Q. Niyaz, W. Sun, M. Alam, A deep learning approach for network intrusion detection system, in *Proceedings of the 9th EAI International Conference on Bio-inspired Information and Communications Technologies (formerly BIONETICS)* (2016), pp. 21–26
119. Y. Ding, S. Chen, J. Xu, Application of deep belief networks for opcode based malware detection, in *2016 International Joint Conference on Neural Networks (IJCNN)* (IEEE, 2016), pp. 3901–3908
120. M. Nadeem, O. Marshall, S. Singh, X. Fang, X. Yuan, Semi-supervised deep neural network for network intrusion detection, in *KSU Proceedings on Cybersecurity Education, Research and Practice*, 2 (2016)
121. S. Potluri, C. Diedrich, Accelerated deep neural networks for enhanced intrusion detection system, in *2016 IEEE 21st International Conference on Emerging Technologies and Factory Automation (ETFA)* (IEEE, 2016), pp. 1–8
122. M.-J. Kang, J.-W. Kang, Intrusion detection system using deep neural network for in-vehicle network security. PloS one **11**(6), e0155781 (2016)
123. A. Taylor, S. Leblanc, N. Japkowicz, Anomaly detection in automobile control network data with long short-term memory networks, in *2016 IEEE International Conference on Data Science and Advanced Analytics (DSAA)* (IEEE, 2016), pp. 130–139
124. K. Fu, D. Cheng, Y. Tu, L. Zhang, Credit card fraud detection using convolutional neural networks, in *International Conference on Neural Information Processing* (Springer, 2016), pp. 483–490
125. Y. Wang, W.-D. Cai, P.-C. Wei, A deep learning approach for detecting malicious javascript code. Sec. Commun. Netw. **9**(11), 1520–1534 (2016)
126. C.G. Cordero, S. Hauke, M. Mühlhäuser, M. Fischer, Analyzing flow-based anomaly intrusion detection using replicator neural networks, in *2016 14th Annual Conference on Privacy, Security and Trust (PST)* (IEEE, 2016), pp. 317–324
127. G. Zhao, C. Zhang, L. Zheng, Intrusion detection using deep belief network and probabilistic neural network, in *2017 IEEE International Conference on Computational Science and Engineering (CSE) and IEEE International Conference on Embedded and Ubiquitous Computing (EUC)*, vol 1 (IEEE, 2017), pp. 639–642
128. C. Yin, Y. Zhu, J. Fei, X. He, A deep learning approach for intrusion detection using recurrent neural networks. IEEE Access **5**, 21954–21961 (2017)
129. J. Kim, H. Kim, et al., An effective intrusion detection classifier using long short-term memory with gradient descent optimization, in *2017 International Conference on Platform Technology and Service (PlatCon)* (IEEE, 2017), pp. 1–6
130. Y. Yu, J. Long, Z. Cai, Network intrusion detection through stacking dilated convolutional autoencoders, *Security and Communication Networks* (2017)
131. J. Saxe, K. Berlin, Expose: A character-level convolutional neural network with embeddings for detecting malicious urls, file paths and registry keys, *arXiv preprint* arXiv:1702.08568 (2017)
132. W. Wang, M. Zhu, J. Wang, X. Zeng, Z. Yang, End-to-end encrypted traffic classification with one-dimensional convolution neural networks, in *2017 IEEE International Conference on Intelligence and Security Informatics (ISI)* (IEEE, 2017), pp. 43–48
133. W. Wang, M. Zhu, X. Zeng, X. Ye, Y. Sheng, Malware traffic classification using convolutional neural network for representation learning, in *2017 International Conference on Information Networking (ICOIN)* (IEEE, 2017), pp. 712–717
134. G. Loukas, T. Vuong, R. Heartfield, G. Sakellari, Y. Yoon, D. Gan, Cloud-based cyber-physical intrusion detection for vehicles using deep learning. IEEE Access **6**, 3491–3508 (2017)
135. J. Yang, J. Deng, S. Li, Y. Hao, Improved traffic detection with support vector machine based on restricted boltzmann machine. Soft Comput. **21**(11), 3101–3112 (2017)
136. Y. He, G.J. Mendis, J. Wei, Real-time detection of false data injection attacks in smart grid: A deep learning-based intelligent mechanism. IEEE Trans. Smart Grid **8**(5), 2505–2516 (2017)
137. S.S. Roy, A. Mallik, R. Gulati, M.S. Obaidat, P.V. Krishna, A deep learning based artificial neural network approach for intrusion detection, in *International Conference on Mathematics and Computing* (Springer, 2017), pp. 44–53

138. J. Kim, N. Shin, S.Y. Jo, S.H. Kim, Method of intrusion detection using deep neural network, in *2017 IEEE International Conference on Big Data and Smart Computing (BigComp)* (IEEE, 2017), pp. 313–316
139. L. Zhang, L. Shi, N. Kaja, D. Ma, A two-stage deep learning approach for can intrusion detection, in *Proceedings of Ground Vehicle Systems Engineering and Technology Symposium (GVSETS)* (2018), pp. 1–11
140. A. Azmoodeh, A. Dehghantanha, K.-K.R. Choo, Robust malware detection for internet of (battlefield) things devices using deep eigenspace learning. IEEE Trans. Sustainable Comput. **4**(1), 88–95 (2018)
141. A.F.M. Agarap, A neural network architecture combining gated recurrent unit (gru) and support vector machine (svm) for intrusion detection in network traffic data, in *Proceedings of the 2018 10th International Conference on Machine Learning and Computing* (2018), pp. 26–30
142. T.A. Tang, L. Mhamdi, D. McLernon, S.A.R. Zaidi, M. Ghogho, Deep recurrent neural network for intrusion detection in sdn-based networks, in *2018 4th IEEE Conference on Network Softwarization and Workshops (NetSoft)* (IEEE, 2018), pp. 202–206
143. F. Jiang, Y. Fu, B.B. Gupta, F. Lou, S. Rho, F. Meng, Z. Tian, Deep learning based multi-channel intelligent attack detection for data security, *IEEE transactions on Sustainable Computing* (2018)
144. N. Shone, T.N. Ngoc, V.D. Phai, Q. Shi, A deep learning approach to network intrusion detection. IEEE Trans. Emerging Topics Comput. Intell. **2**(1), 41–50 (2018)
145. Z. Zhang, X. Zhou, X. Zhang, L. Wang, P. Wang, A model based on convolutional neural network for online transaction fraud detection, *Security and Communication Networks* (2018)
146. M. Nasr, A. Bahramali, A. Houmansadr, Deepcorr: Strong flow correlation attacks on tor using deep learning, in *Proceedings of the 2018 ACM SIGSAC Conference on Computer and Communications Security* (2018), pp. 1962–1976
147. T. Aldwairi, D. Perera, M.A. Novotny, An evaluation of the performance of restricted boltzmann machines as a model for anomaly network intrusion detection. Comput. Netw. **144**, 111–119 (2018)
148. L. Zhou, X. Ouyang, H. Ying, L. Han, Y. Cheng, T. Zhang, Cyber-attack classification in smart grid via deep neural network, in *Proceedings of the 2nd International Conference on Computer Science and Application Engineering* (2018), pp. 1–5
149. F. Feng, X. Liu, B. Yong, R. Zhou, Q. Zhou, Anomaly detection in ad-hoc networks based on deep learning model: A plug and play device. Ad Hoc Netw. **84**, 82–89 (2019)
150. H. Zhang, X. Yu, P. Ren, C. Luo, G. Min, Deep adversarial learning in intrusion detection: A data augmentation enhanced framework, *arXiv preprint* arXiv:1901.07949 (2019)
151. L. Holbrook, M. Alamaniotis, Internet of things security analytics and solutions with deep learning, in *2019 IEEE 31st International Conference on Tools with Artificial Intelligence (ICTAI)* (IEEE, 2019), pp. 178–185
152. M.A. Ferrag, L. Maglaras, Deepcoin: a novel deep learning and blockchain-based energy exchange framework for smart grids. IEEE Trans. Eng. Manage. **67**(4), 1285–1297 (2019)
153. S. Basumallik, R. Ma, S. Eftekharnejad, Packet-data anomaly detection in pmu-based state estimator using convolutional neural network. Int. J. Electri. Power Energy Syst. **107**, 690–702 (2019)
154. S. Otoum, B. Kantarci, H.T. Mouftah, On the feasibility of deep learning in sensor network intrusion detection. IEEE Netw. Lett. **1**(2), 68–71 (2019)
155. H. Karimipour, A. Dehghantanha, R.M. Parizi, K.-K.R. Choo, H. Leung, A deep and scalable unsupervised machine learning system for cyber-attack detection in large-scale smart grids. IEEE Access **7**, 80778–80788 (2019)
156. G. Thamilarasu, S. Chawla, Towards deep-learning-driven intrusion detection for the internet of things. Sensors **19**(9), 1977 (2019)
157. Y. Zhang, X. Chen, L. Jin, X. Wang, D. Guo, Network intrusion detection: Based on deep hierarchical network and original flow data. IEEE Access **7**, 37004–37016 (2019)

158. Y. Zeng, H. Gu, W. Wei, Y. Guo, *deep − full − range*: A deep learning based network encrypted traffic classification and intrusion detection framework. IEEE Access **7**, 45182–45190 (2019)
159. F.A. Khan, A. Gumaei, A. Derhab, A. Hussain, A novel two-stage deep learning model for efficient network intrusion detection. IEEE Access **7**, 30373–30385 (2019)
160. D. Papamartzivanos, F.G. Mármol, G. Kambourakis, Introducing deep learning self-adaptive misuse network intrusion detection systems. IEEE Access **7**, 13546–13560 (2019)
161. Y. Yang, K. Zheng, C. Wu, Y. Yang, Improving the classification effectiveness of intrusion detection by using improved conditional variational autoencoder and deep neural network. Sensors **19**(11), 2528 (2019)
162. A. Abusitta, M. Bellaiche, M. Dagenais, T. Halabi, A deep learning approach for proactive multi-cloud cooperative intrusion detection system. Fut. Gener. Comput. Syst. **98**, 308–318 (2019)
163. Y. Zhang, P. Li, X. Wang, Intrusion detection for iot based on improved genetic algorithm and deep belief network. IEEE Access **7**, 31711–31722 (2019)
164. M. Aloqaily, S. Otoum, I. Al Ridhawi, Y. Jararweh, An intrusion detection system for connected vehicles in smart cities. Ad Hoc Netw. **90**, 101842 (2019)
165. S.M. Kasongo, Y. Sun, A deep learning method with filter based feature engineering for wireless intrusion detection system. IEEE Access **7**, 38597–38607 (2019)
166. A. Abeshu, N. Chilamkurti, Deep learning: the frontier for distributed attack detection in fog-to-things computing. IEEE Commun. Mag. **56**(2), 169–175 (2018)
167. T.D. Nguyen, S. Marchal, M. Miettinen, H. Fereidooni, N. Asokan, A.-R. Sadeghi, Dïot: A federated self-learning anomaly detection system for iot, in *2019 IEEE 39th International Conference on Distributed Computing Systems (ICDCS)* (IEEE, 2019), pp. 756–767
168. D. Preuveneers, V. Rimmer, I. Tsingenopoulos, J. Spooren, W. Joosen, E. Ilie-Zudor, Chained anomaly detection models for federated learning: An intrusion detection case study. Appl. Sci. **8**(12), 2663 (2018)
169. S. Samarakoon, M. Bennis, W. Saad, M. Debbah, Distributed federated learning for ultra-reliable low-latency vehicular communications. IEEE Trans. Commun. **68**(2), 1146–1159 (2019)
170. J. Schneible, A. Lu, Anomaly detection on the edge, in *MILCOM 2017-2017 IEEE Military Communications Conference (MILCOM)* (IEEE, 2017), pp. 678–682
171. Y. Lu, X. Huang, Y. Dai, S. Maharjan, Y. Zhang, Blockchain and federated learning for privacy-preserved data sharing in industrial iot. IEEE Trans. Ind. Inf. **16**(6), 4177–4186 (2019)
172. J. Ren, H. Wang, T. Hou, S. Zheng, C. Tang, Federated learning-based computation offloading optimization in edge computing-supported internet of things. IEEE Access **7**, 69194–69201 (2019)
173. S.A. Rahman, H. Tout, C. Talhi, A. Mourad, Internet of things intrusion detection: Centralized, on-device, or federated learning? IEEE Netw. **34**(6), 310–317 (2020)
174. S. Savazzi, M. Nicoli, V. Rampa, Federated learning with cooperating devices: A consensus approach for massive iot networks. IEEE Internet Things J **7**(5), 4641–4654 (2020)

Hybrid Deep Neural Architecture for Detection of DDoS Attacks in Cloud Computing

Aanshi Bhardwaj, Veenu Mangat, and Renu Vig

Abstract Detection and prevention of distributed denial of service (DDoS) attacks are considered to be a keystone of network security. Though a good number of potential solutions have been provided for the detection of attacks, but due to frequent change in attack vectors, a competent technique is essential for combating these new attacks. In this paper, we propose a hybrid method which uses deep neural network (DNN) model for distinguishing DDoS attacks in cloud environment using ant colony optimization (ACO) for learning prime or important hyperparameters for effective classification of DNN. The use of optimal parameters in DNN makes it more accurate for detection of attacks. The proposed approach is validated by comparing its performance w.r.t. parameters detection accuracy, detection rate with the results of three other recent methods based on machine learning. Experiments have been conducted on the CICIDS2017 dataset which is a new benchmark dataset in the area of network security. Proposed approach gives promising results over CICIDS2017 dataset. The detection rate and accuracy are 95.74% and 98.25%, respectively, which are better than state-of-the-art methods.

Keywords Distributed denial of service attacks · Machine learning · Deep neural networks · Ant colony optimization

1 Introduction

Cloud computing has very attractive features like elastic, on demand and fully managed computer system resources and services which makes it well suited for application in sectors such as health care, manufacturing, retail, and entertainment. Due to its distributed and dynamic nature as well as vulnerabilities in virtualization implementation, the cloud environment is prone to various cyber-attacks.[1] There

[1] https://www.imperva.com/blog/top-10-cloud-security-concerns.

A. Bhardwaj (✉) · V. Mangat · R. Vig
University Institute of Engineering and Technology, Panjab University, Chandigarh, India
e-mail: aanshibhardwaj@pu.ac.in

M. Paprzycki et al. (eds.), *Intelligent Systems, Technologies and Applications*, Advances in Intelligent Systems and Computing 1353,
https://doi.org/10.1007/978-981-16-0730-1_5

are various security issues related to cloud computing models. Some of them are inability to access data coming to and from cloud service, incapable of accessing cloud security, theft and misuse of data hosted, insufficient skilled staff, no control over sensitive data access, and advance threats like malware injection attack, wrapping attacks, distributed denial of service attack (DDoS), virtual machine escape, etc.[2]

Another major cloud security concern is the storage and computation results at the cloud end due to limitation of resources. Wei et al. [1] proposed SecCloud protocol for secure computation auditing by using novel scheme called privacy cheating discouragement using digital signature verification. A comprehensive survey is provided in [2] for detailed security challenges that occur due to valuable characteristics of cloud.

DDoS attack is very perilous attack. It is a deadly weapon which overwhelms the server or network by sending floods of packets toward it. The attack disrupts the services running on the target, thereby blocking the legitimate traffic accessing its services. The rationale for undertaking this problem is twofold. Firstly, recent statistics reveal that the potency and frequency of DDoS attacks are increasing at an alarming rate. In February 2020,[3] Amazon web services (AWS) customers suffered from a critical breakdown when DDoS attack targeted Amazon's simple service storage (S3) and other services, thereby taking them down for almost eight hours. This attack resulted in smashing DNS web service router which further effected services like elastic load balancing (ELB), elastic compute cloud (EC2), and relational database service (RDS). Secondly, attack vectors are evolving quickly. The attackers have recently introduced a new DDoS amplification protocol which has already increased the attack capacity to reach up to 350 Gbps. The attackers exploit the multicast protocol web services dynamic discovery protocol for amplifying DDoS attacks by spoofing the return IP address. The world' s largest online encyclopedia was rendered inaccessible in various countries of Europe, Africa, and the Middle East on September 6, 2019. Wikipedia is often struck by attacks, but in 2019, the attack was unusual concerning capacity and duration. According to a security week article, researchers have found out that on an average the total number of distinct DDoS attacks been suffered by the Internet is about 28,700 attacks each day.[4] Authors in [3] discussed National Cyber Security Strategies (NCSS) of various countries like India, New Zealand, Australia, etc. They also discussed challenges in standardization of cyber security. The challenges being faced are in the areas of standardization, lack of agility, economic considerations, competing set of standards, and lack of awareness.

[2]https://www.mcafee.com/enterprise/en-us/security-awareness/cloud/security-issues-in-cloud-computing.html.

[3]https://siliconangle.com/2020/06/17/aws-mitigated-record-breaking-2-3-tbps-ddos-attack-february.

[4]https://www.corero.com/blog/853-academic-research-reports-nearly-30000-dos-attacks-per-day.html.

It is predicted that DDoS attacks will increase in intensity and frequency as more and more organizations are moving toward cloud platforms, increase in adoption of IoT devices, and DDoS-for-hire services become easily available. Also, it has been studied that the number of IoT devices in use will be approximately 25 billion by 2021. The information collected by these IoT sensors are very sensitive so in order to prevent its leakage, recently, a new authentication-based method called LAM-CIoT has been proposed [4]. It offers reliable security and has low computation overhead.

There are many different solutions for defending DDoS attack because the attack categories are multitude. The DDoS attack taxonomy is given in [5]. Authors divided DDoS into two broad categories: reflection attacks and exploitation attacks. Various solutions like packet-level analyses, flow-level analyses, behavioral analyses, traffic mining, and deep packet inspection have been proposed by researchers for combating DDoS attacks [6–11]. But a detailed study of all these methods reveals that there are various issues and problems. One is the class imbalance problem for machine learning-based techniques, wherein the distribution of different classes, i.e., attacks traffic (positive class) and legitimate traffic (negative class), is extremely imbalanced. This problem will lead to overfitting of the data to the majority class and result in undesirable and misleading accuracy. There is another problem which hampers the accuracy, i.e., the selection of optimal features. The inclusion of some features is powerless in classifying the traffic when attack vectors are unknown and continuously changing. Moreover, the traffic generated by the network devices is boundless due to which a number of techniques are not capable of addressing this big data effectively. According to the report by Domo,[5] every day more than 2.5 quintillion bytes of data are generated, and this is going to increase every day. It is estimated that 1.7 MB of data every second will be generated by every person on earth in 2020. The data speed in the network, the data generation rate, variety, and complexity of data are also growing enormously. There are many challenges in learning useful information from large amount of data and also to categorize such data. The learning rate also becomes slow due to large number of features in the traffic or dataset. The other challenges being faced are handling the noisy data, high dimensionality, computational cost for such large dataset, etc.

In this work, we have proposed a scheme based on DNN and ACO for detection of attacks. ACO has been used for hyperparameter optimization of DNN. The hyperparameter optimization helps in reducing the loss function while updating the weights of neural network during backpropagation. It provides true optimum values for weight and bias and also helps in faster convergence of the network.

The major contributions of our work are:

1. We propose a novel deep learning architecture for detection of DDoS attacks from network traffic.
2. The proposed approach uses ACO optimization for hyperparameter optimization of the DNN.

[5]https://www.socialmediatoday.com/news/how-much-data-is-generated-every-minute-infographic-1/525692.

3. Validation of the proposed approach has been done using standard CICIDS2017 dataset which overcomes most disadvantages of other network intrusion datasets.
4. Performance of the proposed approach is promising when analyzed w.r.t. parameters accuracy, precision, and recall and compared with three recent deep learning techniques from literature.

To the best of our knowledge, there is no other work that applies a similar approach for detection of DDoS attacks over CICIDS2017 dataset.

2 Related Work

Fadlullah et al. [12] provided anomaly based detection technique for encrypted application level protocol attacks using monitoring stubs. Monitoring stubs analyze encrypted packets, derive important features, and prepare a normal usage behavior profile. The non-parametric CUSUM algorithm detects anomalies in the normal usage of traffic to detect attacks. The anomaly is notified by monitoring stubs. Experiments show the proposed approach is effective in lowering false alarms during flash crowd. Authors proposed an ensemble deep learning model using autoencoder (AE), deep belief network (DBN), DNN, and extreme learning methods. The approach was validated using NSL-KDD dataset which achieved detection rate as 97.95% and false alarm rate as 14.72% [13].

Distributed machine learning-based intrusion detection system (IDS) has been proposed by Idhammad et al. for cloud environment [14]. The pre-processed data is used by anomaly detection module which uses Naïve Bayes classifier for separating the network traffic data into normal or abnormal traffic. Ensemble learning classifier based on random forest performs a multi-class classification. Experimentation was performed on Google platform using CIDDS-001 dataset and showed better performance than standard random forest. Authors in [15] proposed an IDS which combines cuckoo optimization feature selection (COFS) and Naïve Bayes algorithm (NBA). COFS removes the redundant and irrelevant features, and the processed features are passed to NBA which performs the classification part. The proposed method achieved accuracy which outperformed the existing feature selection methods like information gain-based feature selection (IGFS), chi-square feature selection (CSFS), information gain ratio-based feature selection (GRFS), and one R feature selection (ORFS).

In study [16], authors gave an intelligent scheme based on DNN using hybrid optimization system. The hybrid optimization consists of improved genetic algorithm (IGA) and simulated annealing algorithm (SAA). The GA is improved by optimizing the fitness function using parallel processing and fitness value hashing strategies. Experimentation was performed using CloudSim 4.0 simulator and showed high detection rate of 99.95 and 0.05% false positive rate. Authors in [17] combined binary bat algorithm with random forest for classification of intrusions in the network.

Two new fitness functions—similarity-based fitness function (FSFF) and classifier accuracy-based fitness function are used in bat algorithm for selecting the optimal features. Experiments have been performed on UNSW-NB15 and CICIDS2017 datasets showed highest accuracy of 97.09% and false positive rate of 2.03%.

A powerful IDS which is a combination of three models multi-layer perceptron (MLP) network, artificial bee colony (ABC), and fuzzy C-means (FCM) clustering algorithm has been proposed by Hajimirzaei et al. [18]. FCM clustering prepares homogeneous subsets of training data. MLP with backpropagation classifies the intrusions and normal traffic. ABC helps MLP with optimal weights and bias values for efficient classification. Root mean square error, kappa statistic, and mean absolute error are used for comparing the proposed method with other state-of-the-art methods.

Authors proposed a hybrid technique for detection of DDoS attacks [19]. Due to limitation of resources, the computational load of detection process is divided between client and proxy side. At client side, the data is collected, features are extracted and a divergence test is performed. If the divergence results are greater than threshold it is attack and the data is sent to proxy side for further processing. At proxy side, the data is pre-processed and transformed to be fed to various classifiers. The classifiers used are random forest, MLP, Naïve Bayes, decision tree, and k-nearest neighbors (KNN) for getting best results. Experimentation showed that the proposed method gave best results among all the classifiers.

The authors in [20] proposed double particle swarm optimization (PSO)-based method for the detection of attacks. The proposed method consists of four stages, namely data pre-processing, pre-training, training, and testing. In data pre-processing, the non-numerical values are transformed to numerical values using hot-encoding method, and also the values are normalized using min–max normalization. The next stage of pre-training involves double PSO optimization algorithm which is hierarchical multi-purpose algorithm which has two levels. First level uses FPSBPSO-E (filter-based feature selection algorithm using entropy) which finds the best feature subset for optimal accuracy and the second level uses PSO for hyperparameter selection. DNN, long short-term memory-recurrent neural network (LSTM-RNN), and DBN are trained and tested on the generated results of the above two stages. NSL-KDD and CICIDS have been used for validating the results and showed high accuracy, precision, recall, and $F1$-score.

Wang et al. [21] proposed MLP-based model for feature selection and sequential backward feature selection (SBS) which is a wrapper feature selection method. The method was tested on the NSL-KDD dataset and showed that using feedback mechanism the proposed method can effectively perceive the detection errors. The accuracy achieved is 97.66% with false alarm rate of 0.62%.

3 Proposed Method

Our work has been divided into four consecutive phases: data pre-processing, pre-training, training, and testing phases. The four phases have been described in the following section.

3.1 *Data Pre-processing*

CICIDS2017 Dataset

The experiments were conducted on the CICIDS2017 dataset[6] [22]. This dataset has been selected as it is distinct and includes up-to-date attacks. Also, the features are incomparable and complete in comparison with other datasets like UNSW-NB15 [23], CAIDA,[7] AWID [24], DARPA [25], and CIDDS-001 [26]. The dataset contains benign and malicious traces of network traffic. It is a labeled dataset with total of 84 features including the last status label to identify attack or not. The features were extracted by CICFlowMeter-V3. The output of CICFlowMeter-V3 is a CSV file that includes: Flow ID (1), Source IP (2) and Destination IP (4), Source Port (3), Destination Port (4), Protocol (6), Timestamp (7), and Label (84). The above-mentioned features are the basic features and the features from number 8 to 83 are high level features because these are statistically computed as shown in Table 1.

It has 11 characteristic features which make it an authentic IDS dataset. These crucial characteristics are anonymity, attack diversity, available protocols, complete capture, complete interaction, complete network configuration, complete traffic, feature set, metadata, heterogeneity, and labeling. None of the available datasets like ISCX2012, KDD, DARPA, CAIDA, Kyoto, CDX, etc., have all of these above-mentioned features.

3.2 *Pre-processing Phase*

The pre-processing phase has two parts. CICIDS 2017 dataset has total of 84 features. The first part is to perform data normalization in which all the numeric values of all features are transformed into [0,1] range. The second part is to remove the irrelevant feature values. The rows that have the feature "Flow Packets/s" having values equal to "Infinity" or "NaN" are removed. Also, the features that have the same value for all rows, namely Bwd PSH Flags, Bwd URG Flags, Fwd Avg Bytes/Bulk, Fwd Avg Packets/Bulk, Fwd Avg Bulk/Rate, Bwd Avg Bytes/Bulk, Bwd Avg Packets/Bulk, Bwd Avg Bulk/Rate, and Fwd Avg Bytes/Bulk are removed.

[6]https://www.unb.ca/cic/datasets/ids-2017html..

[7]https://www.caida.org/data/overview.

Table 1 Listing CICIDS2017 dataset features

S. No.	Feature	S. No.	Feature	S. No.	Feature	S. No.	Feature
1	Flow ID	22	Flow packets/s	43	Fwd packets/s	64	Fwd avg bulk rate
2	Source IP	23	Flow IAT mean	44	Bwd packets/s	65	Bwd avg bytes/bulk
3	Source port	24	Flow IAT std	45	Min packet length	66	Bwd avg packets/bulk
4	Destination IP	25	Flow IAT max	46	Max packet length	67	Bwd avg bulk rate
5	Destination port	26	Flow IAT min	47	Packet length mean	68	Subflow fwd packets
6	Protocol	27	Fwd IAT total	48	Packet length std	69	Subflow fwd bytes
7	Time stamp	28	Fwd IAT mean	49	Packet len. variance	70	Subflow bwd packets
8	Flow duration	29	Fwd IAT std	50	FIN flag count	71	Subflow bwd bytes
9	Total Fwd packets	30	Fwd IAT max	51	SYN flag count	72	Init_Win_bytes_fwd
10	Total backward packets	31	Fwd IAT min	52	RST flag count	73	Act_data_pkt_fwd
11	Total length of fwd pck	32	Bwd IAT total	53	PSH flag count	74	Min_seg_size_fwd
12	Total length of bwd pck	33	Bwd IAT mean	54	ACK flag count	75	Active mean
13	Fwd packet length max	34	Bwd IAT std	55	URG flag count	76	Active std
14	Fwd packet length min	35	Bwd IAT max	56	CWE flag count	77	Active max
15	Fwd pck length mean	36	Bwd IAT min	57	ECE flag count	78	Active min
16	Fwd packet length std	37	Fwd PSH flags	58	Down/up ratio	79	Idle mean
17	Bwd packet length max	38	Bwd PSH flags	59	Avg packet size	80	Idle packet
18	Bwd packet length min	39	Fwd URG flags	60	Avg fwd segment size	81	Idle std

(continued)

Table 1 (continued)

S. No.	Feature	S. No.	Feature	S. No.	Feature	S. No.	Feature
19	Bwd packet length mean	40	Bwd URG flags	61	Avg bwd segment size	82	Idle max
20	Bwd packet length std	41	Fwd header length	62	Fwd avg bytes/bulk	83	Idle min
21	Flow bytes/s	42	Bwd header length	63	Fwd avg packets/bulk	84	Label

3.3 Pre-training Phase

Gradient descent in neural network provides the optimized weights which helps the neural network to make valid predictions. But gradient descent can make the learning process slower and miss the optimal weights. Hence, a more effective optimization technique is required which makes the learning rate faster and also helps in achieving better results. In this work, we have used ACO for optimization of the hyperparameters weights and bias of DNN, to obtain accurate results.

ACO

The ACO is a swarm intelligence technique motivated by the foraging behavior of real ant colonies [27]. ACO algorithms were initially proposed to solve hard combinatorial optimization problems in reasonable amount of computation time. In general, an ACO algorithm consists of two modes, i.e., the forward and backward modes. In the forward mode, the probabilistic solution is constructed by the population of ants based on existing pheromone trails. In the backward mode, the constructed solution is used to update pheromone value. After several iterations, the ants will converge into a near-optimum or optimum solution.

The framework of the ACO is shown below in Fig. 1.

The probability of ant k moving from neuron i to j is given by Eq. (1):

$$p_{ij}^{k} = \begin{cases} \frac{(\tau_{ij})^{\alpha}.(\eta_{ij})^{\beta}}{\sum_{l \in J_i^k} (\tau_{il})^{\alpha}.(\eta_{il})^{\beta}}, & \text{if } \ j \in J_i^k \\ 0, & \text{if } \ \in j \notin J_i^k \end{cases} \quad (1)$$

where τ_{ij} is the amount of pheromone on connection i, j, α is a parameter used to control the influence of τ_{ij}, η_{ij} is the desirability of connection i, j, and β is used to control the influence of η_{ij}.

The pheromone value is updated using Eq. (2):

Fig. 1 ACO framework

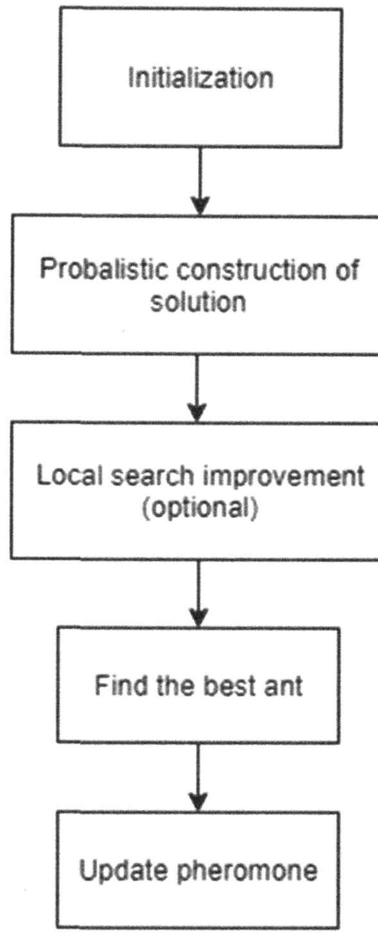

$$\Delta\tau_{ij} = (1-\rho)\tau_{ij} + \sum_{k=1}^{m} \Delta\tau_{ij}^{k} \tag{2}$$

where τ_{ij} is the amount of pheromone on connection i, j, ρ is the rate of pheromone evaporation, and $\Delta\tau_{ij}$ is the amount of pheromone deposited.

ACO is used to optimize weight and bias values of DNN for effective classification. In DNN, a potential path for ant is obtained by each neuron in every layer connected to each neuron in the successive layer. Every possible connection is provided with a base value of pheromone, and value of pheromone on each connection is stored. The initial ants select a fully connected path between the neurons to create a neural network design using the pheromone value on each connection with the specific weights and bias values. Population of the best ants and corresponding solutions with weights and bias values is maintained, and when an ant reports improved accuracy and solution, the pheromone of that connection will increase in the DNN. Periodically, it reduces pheromone levels so that solution is not stuck in local optima and there can be balance between exploration and exploitation. This strategy allows the DNNs with potentially many hidden layers and hidden nodes to evolve and choose best parameters. The objective function is to maximize the predictive accuracy and minimize loss function through hyperparameter optimization of DNN through ACO.

The pseudocode of the basic ACO procedure is as follows.

```
Random initialization of the pheromone value
Do
For each iteration
For each ant
For each parameter
           Compute the probability P as per Eq. (1)
                 Determine the Pmax
                 Deduce the value of Vi
End

Compute detection accuracy as objective function

End

Deduce the best detection accuracy
Update pheromone values according to (2)
End

Store the best solution
END
```

3.4 Training and Testing Phase

DNN model is constructed by tuning hyperparameter values using optimization technique described above. The resulting model is first trained using 75% of the CICIDS2017 dataset, and then the trained model is tested on 25% of the remaining dataset.

DNN is a new paradigm and was first executed in 2006 [28]. DL is a component of machine learning algorithms but it has more hidden layers than the latter. The algorithm learns in multiple levels and the levels correspond to multiple concepts where high level concepts are refined from lower level concepts. There are different DL methods which have been used in the field of cybersecurity. These methods are DNN, deep belief networks, deep autoencoders, LSTM, and convolutional neural network. We have chosen DNN for detection of DDoS attack from the dataset rather than any other technique of deep learning because of the favorable results as depicted in [29, 30]. Moreover, DNN is a popular method for static classification such as detection of intrusions.

Overview of DNN

DNN is simply a representation of the artificial neural network (ANN) except it has numerous deep hidden layers. Principally, DNN commonly consists of an input layer, one or more hidden layers, and an output layer. The hidden layers considered to do the most significant work in DNN. DNN is a fully connected network such that every neuron of each layer is connected to neurons of subsequent layer. Moreover, the information is processed and communicated through DNN in feed-forward manner, i.e., from the input layer to the output layer via the hidden layers.

Proposed DNN architecture

This work proposes a DNN architecture with one input layer, six hidden layers, and one output layer as shown in Fig. 2. The activation function and bias value are applied on previous layer which computes weights for the next layer. All the layers are connected layers and the activation function between the layers is ReLu whose

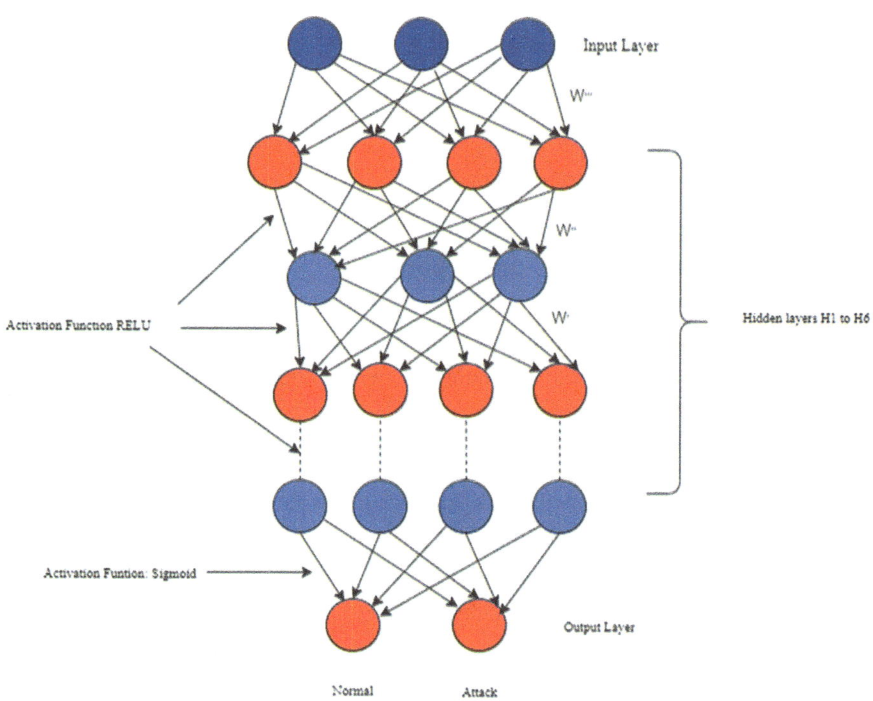

Fig. 2 Proposed DNN architecture

Table 2 Activation functions

Function	Formula	Range
Sigmoid	$S(x) = \frac{1}{1+e^{-x}}$	Between 0 and 1
ReLu	$f(x) = x^{+} = \max(0, x)$	$-\infty$ to $+\infty$
Tanh	$f(x) = \frac{e^{2x}-1}{e^{2x}+1}$	Between -1 and 1
Softmax	$S(x_i) = \frac{e^{xi}}{\sum_{j=1}^{n} e^{xj}}$	Between 0 and 1

mathematical formula is given in Table 2. The ReLu has been used because its faster than other activation functions and solves the error gradient issues. At the last layer, we have used sigmoid function because it gives best result when we have to find the probability of something which lies between the range 0 and 1. Moreover, in our case, we have to deal with the probability of whether the attack is present or not. The equations of activation function are given in the table below. The cost function to be optimized in our proposed model is generally called as loss function. The loss function used is cross-entropy or logarithmic loss whose equation is given in Eq. (3) where y is the label 1 for attack and 0 for not attack. $P(y)$ is the predicted probability of being attack for all N points. The approach for minimizing the lost function is stochastic gradient (SGD) which is computed through backpropagation using α as learning rate.

$$\text{Loss function} = \frac{-1}{N}\sum_{i=1}^{N} y_i \log(P(y_i)) + (1 - y_i) * \log(1 - P(y_i)) \tag{3}$$

4 Experimentation and Validation

The experiments have been conducted using a Windows 10–64 bits PC with 16 GB RAM and CPU Intel(R) Core-i7. For simulation of cloud, VMware workstation has been used in which a cloud server is made with different virtual machines.

The performance of the detection system is determined by how correctly the proposed system is able to classify the incoming traffic into accurate type. To evaluate our proposed method, we have used four metrics which are accuracy, precision, recall, and F-score. The description of these metrics is given below.

Detection Accuracy: It is the proportion of correctly classified instances to the total number of classified instances. It is computed using Eq. (4):

$$\text{Detection Accuracy} = \frac{\text{True Negative} + \text{True Positive}}{\text{True Negative} + \text{True Positive} + \text{False Positive} + \text{False Negative}} * 100 \tag{4}$$

Precision (*P*): It is a measure of quality of the detection method, i.e., the rate of samples that correctly labeled as an attack from all samples in the test set that were classified as an attack. It is computed using Eq. (5):

$$\text{Precision} = \frac{\text{True Positive}}{\text{True Positive} + \text{False Positive}} \tag{5}$$

Recall (*R*): It is the measure of completeness of the classifier, i.e., the rate of samples that correctly classified as an attack for all attack samples that existed in the test set. It is computed using Eq. (6):

$$\text{Recall} = \frac{\text{True Positive}}{\text{True Positive} + \text{False Negative}} \tag{6}$$

***F*-score**: It calculates the balance between precision and recall. It is considered as the harmonic mean of recall and precision and is computed using Eq. (7):

$$F\text{ - Score} = \frac{2 * P * R}{P + R} \tag{7}$$

5 Results and Discussion

Table 3 shows the results of the proposed approach applied on the CICIDS2017 dataset. The model takes fraction of seconds to compute the accuracy on pre-trained model. We can apply this methodology in real-time applications as the detection time is very low. Moreover, it shows that with increase in number of epochs, more times the number of weights is changed in neural network. When epochs are less the neural network is underfitting, as the epochs increases the neural network gives optimal results.

To evaluate our proposed model, we compared our work with recent studies in the literature. The comparison has been done on the basis of two performance metrics, i.e., detection rate/recall and detection accuracy. Table 4 shows the compar-

Table 3 Results of the proposed approach

No. of epochs	Accuracy (%)	Precision (%)	Recall (%)	*F*-score (%)
10	56.68	64.38	47.61	54.73
30	79.21	83.37	51.20	63.43
50	95.43	97.1	91.29	94.11
75	97.95	99.98	93.75	96.76
100	98.52	99.99	95.47	97.67

Table 4 Comparison between proposed DNN and recent studies

Study	Technique	Detection rate (%)	Detection accuracy
[20]	DNN + Double PSO	83	88.04% (Overall)
[30]	Decision tree and rule-based concept	99.879	96.5% (Overall)
[31]	Rough set theory and Bayes theorem	99.9	87.66% (DDoS only)
Proposed Work	DNN + ACO	95.74	98.25%

ison between our study and the previous studies. All the studies have reported their results on CICIDS2017 dataset. In [20], the authors have used two levels of PSO before applying DNN on the dataset. First level is used for optimal feature selection and second level for hyperparameter selection of DNN. Detection accuracy reported is 88.04% which is the overall accuracy for all the intrusions in CICIDS2017 dataset which are DDoS, BruteForce, Heartbleed, web attacks, etc. Authors in [14] used hierarchy of classifiers namely REP tree, JRip algorithm, and Forest PA. The output of first two is taken as input by the third classifier. The detection accuracy reported is 96.5% which is the overall accuracy for all the intrusions. The authors in [32] used rough set approximations and Bayes theorem for removing uncertain information and then classifying it. The results showed that detection accuracy for DDoS is 87.66%.

It can be seen from Table 4 that proposed DNN architecture with optimization of hyperparameters using ant colony method gives promising results in the detection of DDoS traffic in a dataset capturing cloud environment traffic. The detection accuracy of the proposed method is higher than the other recent techniques. The recall of proposed method is better that DNN + PSO but lower than decision tree-based method and rough set + Bayes theorem method. The extremely high detection rates of [31] and [32] could be because of overfitting of data. Validation over more benchmark datasets and real-cloud environments need to be carried out to ensure consistently good results of these methods.

6 Conclusion and Future Scope

Detection of DDoS attacks in cloud environment is imperative since there has been a surge in the intensity and frequency of such attacks which can potentially bring down entire computer networks such as critical networks of power grid, and health care. There is lack of classification methods that can handle imbalanced data, voluminous data, high dimensionality, and give accurate results. This work proposed a novel DNN architecture that uses ant colony method for optimization of its hyperparameters for detection of DDoS attacks over cloud dataset. The method employs a multistage process to train and test the classification model. The phases are data pre-processing involving removal of irrelevant features and computation of high level features from cloud dataset; pre-training of network parameters using ACO; training and DNN construction; and testing using performance metrics. The proposed approach has been compared with three recent approaches, viz., DNN + PSO, decision trees + rules, and rough set + Naïve Bayes, w.r.t., parameters detection rate or recall, and detection accuracy. The results of proposed approach are competitive. Our approach has achieved 98.25% accuracy which is better than the work proposed in [20, 30, 31]. However, the detection rate is less than the work done in [30] and [31], but the decision trees in [30] may not be effective on unknown data and are sensitive to small changes in data.

In future, work can be done to study the efficacy of ACO for hyperparameter optimization and compare it with other available solutions in real-world environment. Also, a study needs to be conducted into the time complexity of the various available deep learning approaches for DDoS attack detection.

Acknowledgements This research work is supported by Technical Education Quality Improvement Project III (TEQIP III) of MHRD, Government of India assisted by World Bank under Grant Number P154523 and sanctioned to UIET, Panjab University, Chandigarh (India).

References

1. L. Wei, H. Zhu, Z. Cao, X. Dong, W. Jia, Y. Chen, A.V. Vasilakos, Security and privacy for storage and computation in cloud computing. Inf. Sci. **258**, 371–386 (2014)
2. M. Ali, S.U. Khan, A.V. Vasilakos, Security in cloud computing: opportunities and challenges. Inf. Sci. **305**, 357–383 (2015)
3. J. Srinivas, A.K. Das, N. Kumar, Government regulations in cyber security: framework, standards and recommendations. Future Gener. Comput. Syst. **92**, 178–188 (2019)
4. M. Wazid, A.K. Das, V. Bhat, A.V. Vasilakos, LAM-CIoT: lightweight authentication mechanism in cloud-based IoT environment. J. Netw. Comput. Appl. **150**, 102496 (2020)
5. I. Sharafaldin, A.H. Lashkari, S. Hakak, A.A. Ghorbani, Developing realistic distributed denial of service (DDoS) attack dataset and taxonomy, in *International Carnahan Conference on Security Technology*, pp. 1–8 (2019)
6. E. Özer, M. Iskefiyeli, Detection of DDoS attack via deep packet analysis in real time systems, in *International Conference on Computer Science and Engineering*, pp. 1137–1140 (2017)
7. B. Meng, W. Andi, X. Jian, Z. Fucai, DDOS Attack detection system based on analysis of users' behaviors for application layer, in *IEEE International Conference on Computational Science and Engineering and IEEE International Conference on Embedded and Ubiquitous Computing*, Vol. 1, pp. 596–599 (2017)
8. D. Sun, K. Yang, Z. Shi, Y. Wang, A distinction method of flooding DDoS and flash crowds based on user traffic behavior, in *IEEE Trustcom/BigDataSE/ICESS*, pp. 65–72 (2017)
9. J. David, C. Thomas, Efficient DDoS flood attack detection using dynamic thresholding on flow-based network traffic. Comput. Secur. **82**, 284–295 (2019)
10. S. Saharan, V. Gupta, Prevention and Mitigation of DNS based DDoS attacks in SDN Environment, in *11th International Conference on Communication Systems & Networks*, pp. 571–573 (2019)
11. J. Hou, P. Fu, Z. Cao, A. Xu, Machine learning based DDoS detection through netflow analysis, in *IEEE Military Communications Conference*, pp. 1–6 (2018)

12. Z.M. Fadlullah, T. Taleb, A.V. Vasilakos, M. Guizani, N. Kato, DTRAB: combating against attacks on encrypted protocols through traffic-feature analysis. IEEE/ACM Trans. Networking **18**(4), 1234–1247 (2010)
13. S.A. Ludwig, Intrusion detection of multiple attack classes using a deep neural net ensemble, in *Symposium Series on Computational Intelligence*, pp. 1–7 (2017)
14. M. Idhammad, K. Afdel, M. Belouch, Distributed intrusion detection system for cloud environments based on data mining techniques. Proc. Comput. Sci. **127**, 35–41 (2018)
15. D.A.A.G. Singh, R. Priyadharshini, E.J. Leavline, Cuckoo optimization based intrusion detection system for cloud computing. Int. J. Comput. Network Inform. Secur. **10**(11), 42 (2018)
16. Z. Chiba, N. Abghour, K. Moussaid, M. Rida, Intelligent approach to build a deep neural network based IDS for cloud environment using combination of machine learning algorithms. Comput. Secur. **86**, 291–317 (2019)
17. R. Patil, H. Dudeja, C. Modi, Designing an efficient security framework for detecting intrusions in virtual network of cloud computing. Comput. Secur. **85**, 402–422 (2019)
18. B. Hajimirzaei, N.J. Navimipour, Intrusion detection for cloud computing using neural networks and artificial bee colony optimization algorithm. ICT Express **5**(1), 56–59 (2019)
19. S. Hosseini, M. Azizi, The hybrid technique for DDoS detection with supervised learning algorithms. Comput. Netw. **158**, 35–45 (2019)
20. W. Elmasry, A. Akbulut, A.H. Zaim, Evolving deep learning architectures for network intrusion detection using a double PSO metaheuristic. Comput. Netw. **168**, 107042 (2020)
21. M. Wang, Y. Lu, J. Qin, A dynamic MLP-based DDoS attack detection method using feature selection and feedback. Comput. Secur. **88**, 101645 (2020)
22. I. Sharafaldin, A.H. Lashkari, A.A. Ghorbani, A detailed analysis of the CICIDS2017 data set, in *International Conference on Information Systems Security and Privacy*, pp. 172–188 (2018)
23. N. Moustafa, J. Slay, UNSW-NB15: a comprehensive data set for network intrusion detection systems, in *Proceedings of the Military Communications and Information Systems Conference*, pp. 1–6 (2015)
24. C. Kolias, G. Kambourakis, A. Stavrou, S. Gritzalis, Intrusion detection in 802.11 networks: empirical evaluation of threats and a public dataset. IEEE Commun. Survey Tutorial **18**(1), 184–208 (2016)
25. R. Lippmann, J.W. Haines, D.J. Fried, J. Korba, K. Das, The 1999 DARPA off-line intrusion detection evaluation. Comput. Network **34**(4), 579–595 (2000)
26. M. Ring, S. Wunderlich, D. Grüdl, D. Landes, A. Hotho, Creation of flow-based data sets for intrusion detection. J. Inform. Warfare **16**, 40–53 (2017)
27. M. Dorigo, T. Stützle, Ant colony optimization: overview and recent advances, in *Handbook of metaheuristics*, pp. 311–351 (2019)
28. W. Liu, Z. Wang, X. Liu, N. Zeng, Y. Liu, F.E. Alsaadi, A survey of deep neural network architectures and their applications. Neurocomputing **234**, 11–26 (2017)
29. C.S. Wickramasinghe, D.L. Marino, K. Amarasinghe, M. Manic, Generalization of deep learning for cyber-physical system security: a survey, in *44th Annual Conference of the IEEE Industrial Electronics Society*, pp. 745–751 (2018)
30. A. Ahmim, L. Maglaras, M.A. Ferrag, M. Derdour, H. Janicke, A novel hierarchical intrusion detection system based on decision tree and rules-based models, in *2019 15th International Conference on Distributed Computing in Sensor Systems*, pp. 228–233 (2019)
31. M. Prasad, S. Tripathi, K. Dahal, An efficient feature selection based Bayesian and Rough set approach for intrusion detection. Appl. Soft Comput. **87**, 105980 (2020)

MapReduce-Driven Rough Set Fuzzy Classification Rule Generation for Big Data Processing

Hanumanthu Bhukya and M. Sadanandam

Abstract The term "big data" has become one of the essential research discussions nowadays. Due to a large amount of data availability and data processing nowadays, the topic of data science and big data becomes of prominent interest in the present research. Big data applications can be accomplished through the MapReduce programming model because these are mostly concerning scalability. The MapReduce models are intended to categorize data into various groups that are processed in parallel and whose outcome gathered to offer a single solution. There are numerous incremental models introduced by various authors to analyze and extract data from vast data sources. But, the large amount of data and diversity of the data sources there is a necessity for instant intelligent response pretense a severe problem to the current learning algorithms. Various classification models modified to this new framework, and this paper proposes a rough set fuzzy classification rule generation algorithm (RS-FCRG) to present attractive results with a MapReduce model for big data. This algorithm achieves an interpretable pattern that can handle massive data. It provides a significant accuracy with better execution time because the algorithm applies the MapReduce programming model in the Hadoop platform; it is one of the most beneficial frameworks to dispense with significant collections of data. The experiment takes on the UCI census (KDD) info dataset. The experimental results show that the proposed algorithm gets high accuracy with 95% when compared with the chi-FRBCS-Bigdata-Max algorithm.

Keywords Big data · Apache spark · Fuzzy rule-based classification system · MapReduce · Rule generation process

H. Bhukya (✉) · M. Sadanandam
Department of Computer & Science Engineering, KUCE&T, Kakatiya University, Warangal, Telangana, India

M. Sadanandam
e-mail: msadanandam@kakatiya.ac.in

M. Paprzycki et al. (eds.), *Intelligent Systems, Technologies and Applications*,
Advances in Intelligent Systems and Computing 1353,
https://doi.org/10.1007/978-981-16-0730-1_6

1 Introduction

One of the unusual big data in recent years has been known across the information technology (IT) industry as big data. The term "big data" means the research and processing of large record repositories, which traditional statistical analysis and management structures incapable of distributing [1]. This method can be recognized in different situations such as web pages, mobile devices, social networks, sensor networks, and multimedia data. With the comfort of obtaining additional information, monitoring, and information extraction procedures should be employed, and correct additional information should be collected [2]. However, traditional techniques and designs commonly used to extract information that cannot process datasets of this length [3]. Since the acquired traditional learning systems need structures that must be modified by following the advice of current guidelines that can deal with massive information, even effectively maintaining its impending capacity.

The fuzzy rule-based classification (FRBC) [4] is one of the processes to tackle big data. It is a fantastic tool available for widespread pattern recognition and mining. They can provide excellent accurate results while presenting an explainable design to the end-user by using some semantic labels. One of the challenges that make it difficult to extract valuable information from vital records is the uncertainty associated with the scope and validity inherent in big data. FRBCS genuinely deals with trade, uncertainty, or confusion, which makes it an exciting technology for handling great information, where it can reap your fundamental doubts. In the case of matters with extensive records, a variety of conditions and features are provided regularly. FRBCs reduce their performance in these places as the search area is significantly increased. This complex of growth begins the learning method from technology with problems of scalability or complexity that can end up with non-interpretable standards [5]. To this end situation, many methods have tried to increase complex systems parallel with powerful applications; however, they can focus on reducing processing time while maintaining accuracy. They cannot handle large amounts of records.

Fuzzy rule-based systems were used distinctly to control problems in large datasets [6]. One of the vital features based on the fuzzy rule-based method is their awareness because the fuzzy rule is linguistically explainable. More recently, existing systems have been applied mainly to strongly disorganize the fuzzy rules classification problems [7]. The primary way to plan FRBCS is to automate blur patterns of numerical information. Therefore, the rule-based machine to regularly arrange difficulties was a problematic part.

In many investigations, complex fuzzy sets were generated and modified using default input data for a rule-based system mainly to increase the accuracy of FRBCS classification. As explained in [8], it is possible to observe the change of association functions for predecessor fuzzy sets by weight designations, since obtaining organic properties can further degrade FRBCS interpretation potential.

The different modules of the fuzzy rule-based system (FRBS) are the rule base, fuzzification module, inference module, database, and the de-fuzzification module. Fuzzy rules are included in the rule base. The elements of the fuzzy inference system

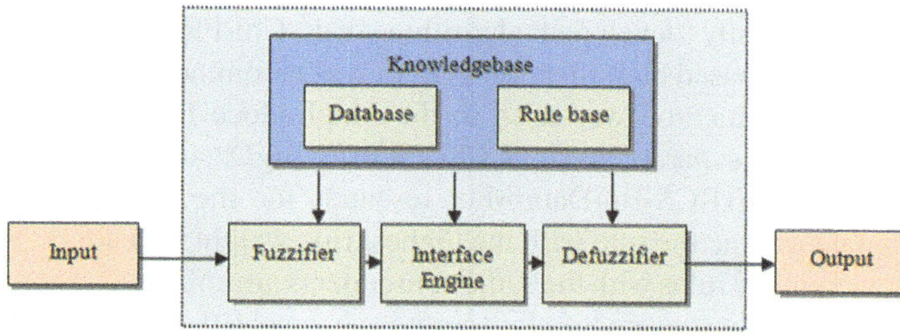

Fig. 1 Fuzzy inference system

are shown in Fig. 1 Sivanandam et al. [9]. The rule base includes fuzzy rules. The database comprises of the association functions used in the fuzzy inference system. The inference module presents the inferencing based on rules. Fuzzifier transforms crisp inputs into linguistic values. Defuzzifier converts the fuzzy score into crisp output.

In this paper, we suggested a rough set fuzzy classification rule generation algorithm (RS-FCRG) using MapReduce. It is capable of extending an interpretable version while maintaining competitive predictive accuracy within the large set of information, which has been shown as the Chi-FRBCS-BigData algorithm. This technique is wholly based on Chi et al. [10], the classical RS-FCRG method, which has been modified to deal with significant events following MapReduce technology, which varies in the "Reduce" performance and is compared to investigate how they handle extensive records.

2 Related Work

Various works discussed how a fuzzy classification rule generation could be created from very large datasets.

A set of fuzzy rules was achieved by the use of some design for the classification system. A fuzzy rule-based system and genetic algorithm were the two fundamental tools for the fuzzy rule method. Garrido et al. [11] were proposed an FRBS development model to manage difficulties in fuzzy rule generation. In that method, the number of rules of the FRBS and the description of the linguistic labels are predestined. There were two phases did in the model construction. First, the rule base was constructed, and then the linguistic labels were optimized, which preserve the interpretability of the rules [6]. Second, the heuristic systems for rule weight analysis were analyzed, which showed how each fuzzy rule's rule weight could be explicitly recognized in FRBCS. Those methods presented well in multi-class pattern classification difficulties with a lot of classes. The partition by fuzzy three-cornered sets was a homogenous fuzzy divider for an unlimited number of training patterns with the period. The association degrees for fuzzy rules are calculated, and then rule weight is determined [12].

One of the currently developed algorithms was Chi-FRBCS-BigData, and a linguistic fuzzy rule-based classification system was recommended to deal with big data. The suggested method was used in the MapReduce framework, and it has been designed in two separate versions Chi-FRBCS-BigData-Ave and chi-FRBCS-BigData-Max. Chi-FRBCS-BigData-Max research for the rules with the same precursor and determine the rule with the highest rule weight. Chi-FRBCSBigData-Ave the rule weight the rules with the same consequents are collected, and the average is calculated. The highest average rule weight is chosen [10]. Feature selection uses the ensemble to find the most accurate features [11]. There are two main classification techniques, supervised and unsupervised. The supervised classification methods are decision tree and support vector machine. The data mining method to form a classification model is CRISP-DM. A decision tree is a tool used here to generate classification rules. Some steps build the classification model [12]. First, the planning is done to get an idea, and then the data is collected and understand by some questionnaires, and the information is prepared to build a classification model using decision tree the algorithm which gives the appropriate result.

One of the most common forms of big data handling today is MapReduce, which is a new version of distributed programming that organizes calculation in two main processes: the map feature that is obtained to separate the individual dataset and address each sub-problem independently, the element boundary that collects and collects the effects of the map function.

Many researchers have started to express well-known machine learning strategies in distributed computing models that include MapReduce to overcome these challenges. This methodology quickly became very popular due to the development of open-source frameworks like Apache Hadoop2 and Apache Spark3. In recent years, several FRBCs have been introduced based on Hadoop or Spark. Although there has been a significant development, the most extreme contributions do not reach today's outcomes in terms of accuracy and interpretability. Some of them implement many near-improvements or study tactics to get an approximate global response.

2.1 Big Data and the MapReduce Programming Model

The significant period of big data is associated with exponential prosperity in generating records that the region has taken in recent years. It also generated a great distraction because of the capabilities within the development of processing data and knowledge extraction. Incredible statistics is an excellent time for combining all the large and complex data so that it will be difficult to process or check the use of traditional software tools or data processing applications. Initially, this idea was defined as a 3 V model, precisely volume, velocity, and variety.

Volume: This characteristic refers to the huge amounts of data that need to be processed to obtain helpful information.

Velocity: These items indicate that statistical treatment packages should be able to reap effects at an economic time.

Variety: This feature suggests that data can be presented in a pair of codecs: structured and unstructured, such as textual content, digital records, or multimedia, among others. Big data problems occur in various areas and sectors, such as commercial, economic and commercial sports, public administrations, national protection, or investigations.

2.2 Apache Hadoop and Apache Spark

In recent years, distributed computing has become very familiar with machine learning systems that acquire network in the open-source frameworks such as Apache Spark8 and Apache Hadoop. These structures offer a transparent exchange machine that allows the consumer to understand the most honest information processing. Hadoop Media includes a distributed reporting device that relies on the Google file system known as the Hadoop Distributed File System (HDFS) and MapReduce model implementation. The MapReduce model extension is presented as a spark.

Spark is advanced in resilient distributed dataSets (RDD), describing divided records and data transformation. The overall performance of a user-described algorithm with a level arrangement consists of several changes divided into jobs.

This information slip (Fig. 2) allows the consumer to run an unlimited number of MapReduce functions on the same important program and supports a much broader type of algorithms and strategies from Hadoop.

The implementation of user-defined algorithms involves a series of stages consisting of some adjustments that can be divided into tasks. The level consists of the most efficient variations that do not require any shuffling/splitting operations (e.g., map and filter operations).

This scrolling of records (Fig. 2) allows the user to run an unlimited range of MapReduce tasks on the same main program, helping with a much broader type of algorithms and strategies than Hadoop.

The main disadvantage of the above system is that it contains all the attributes to be completed so that the duration of the general rule is large, indicating the complexity and accuracy that are performed that it does not reach. To this end, the option is

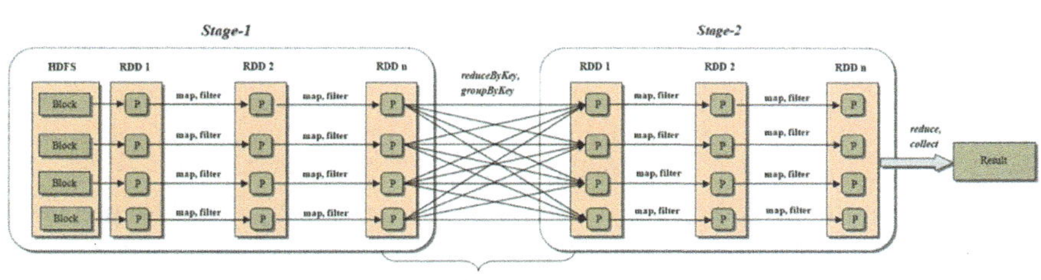

Fig. 2 Spark's data flow (P = Partition)

combined with RS-FCRG on our proposed device. The similar work with only rough sets was carried out in the following papers which could be extended with fuzzy logic for better performance [13, 14].

3 Proposed Rough Set Fuzzy Classification Rule Generation Algorithm for Big Data

Fuzzy classification rules are more expressive and allow setting the conditions of the most natural described. It is also more acceptable, as it allows symbolic information to be formulated in a natural way using linguistic terms.

Fuzzy sets represent these terms included by fuzzy rules. The members of the universe of discourse belong to a fuzzy set of certain stages of the membership, which can be described by the associated membership feature. So, fuzzy sets are usually used to address the constraints of accurate (excessive) representations by providing support for vagueness, uncertainty, ambiguity in human knowledge.

In this section, we will propose two variants of a linguistic FRBCS that control big data. First, we offer some comments associated with FRBCSs and the fuzzy learning algorithm that has been utilized in this work. Then, we will describe how this method is adapted for big data using a MapReduce system transformed to create two alternatives that will present several classification results.

3.1 Definition of Fuzzy Rule-Based Classifier

For fuzzy classification rule generation, different methods have been introduced. Let us understand that we have n training patterns $X_p = (x_{p1}, \dots, x_{pn})$, $p = 1, 2, \dots, m$ from N various classes where X_p is an n-dimensional vector of characteristics in which x_{pi} is the ith attribute value of the pth training pattern ($i = 1, 2, \dots, n$). For our N - class, n-dimensional classification difficulty, we use fuzzy "if–then" rules of the form below:

$$\text{Rule} R_q : \text{if } x_1 \text{ is } A_{q1} \text{ and } \dots \text{ and } x_n \text{ is } A_{qn} \text{ then class } C_q \text{ with } CF_q \quad (1)$$

where R_q is the label of the qth fuzzy if–then rule, $X = (x_1, \dots, x_n)$ is *an* n-dimensional vector of a pattern, A_{qi} provides a precursor fuzzy set, C_q is a class label. CF_q is the weight allocated to the qth rule.

To measure the unity degree of every training pattern X_p with the precursor element of the rule $A_q = A_{q1}, \dots, A_{qn}$ here, using the product executive as follows.

$$\mu_{Aq}(x_p) = \mu_{Aq1}(x_{p1}).\mu_{Aq2}(x_{p2}) \dots \mu_{Aqn}(x_{pn}),\ p = 1, 2, \dots, m \quad (2)$$

where $\mu_{Aq1}(x_{p1})$ is the unity degree of x_{pn} with fuzzy association function A_{qn}. To define the consistent class of the qth rule C_q, we estimate the confidence grade of the association rule.

3.2 *The Working Model of Proposed Fuzzy Rule-Based Classification Systems*

Figure 3 shows the overall block diagram of the proposed system. Training data is given as the input data to the Mapper phase. Feature selection and RS-FCRG algorithms are implemented in the Mapper phase. The output of the Mapper will be given to the reducer which fuses the fuzzy rules from the Mapper. Now the developed classifier is tested by giving test data as input. The efficiency of the classifier is computed employing accuracy.

The accuracy level is improved by the use of feature selection and bagging. Feature selection is the process of selecting a subset of related characteristics for control in model development. Here classifier divides the training dataset into equal n datasets and this individual dataset is given for feature selection. Feature selection selects the attribute which has a high relationship with the class attribute. By this method, the length of the rules is minimized and the accuracy level is improved.

Also in this paper, we suggest a learning algorithm fashioned of three steps are described below: (1) Preprocessing and Partitioning (2) Rule generation process, (3) Evolutionary rule selection.

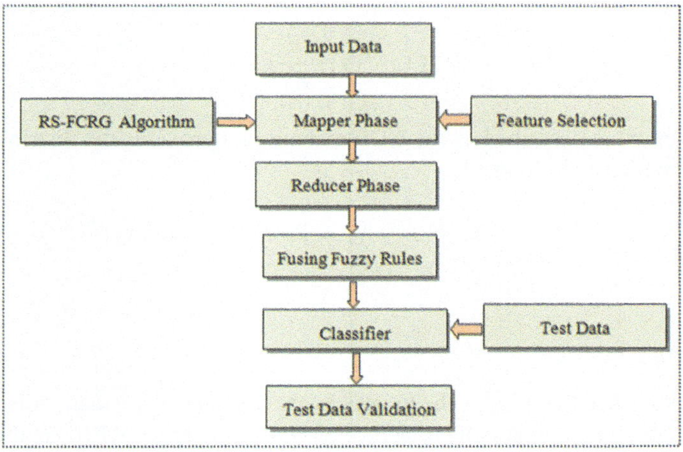

Fig. 3 Block diagram

3.2.1 Preprocessing and Partitioning

This section aims to explain fuzzy sets that meet the training data's actual appearance while keeping the number of fuzzy sets by a variable constant like low, medium, and high. This step is divided into again two parts.

Preprocessing:

The primary assignment of the training data is reorganized into a normal appearance. This transformation involves the possibility of an aggregate transform Theorem, described in Theorem 1. This theorem suggests that any dataset can be converted into a new dataset where all the variables equal a normal appearance, despite the initial distribution.

Theorem 1 *if X is a continuous random variable with cumulative distribution function (CDF).*

$F_X(x)$ *& if* $Y = F_X(x)$, *then* Y *is a uniform random variable on the interval* [0,1].

Proof Suppose $Y_g = (X)$ is a function of$X$ where$g$ is differentiable & strictly improving. Thus, its inverse g^{-1} uniquely exists. The CDF of Y can be derived using

$$F_Y(y) = \text{Prob}(Y \leq y) = \text{Prob}\big(X \leq g^{-1}(y)\big) = F_X\big(g^{-1}(y)\big) \tag{3}$$

And its density is given by

$$\begin{aligned} f_Y(y) &= \frac{d}{dy}F_Y(y) = \frac{d}{dy}F_X\big(g^{-1}(y)\big) \\ &= f_X\big(g^{-1}(y)\big).\frac{d}{dy}g^{-1}(y) \end{aligned} \tag{4}$$

This method is described as the CDF manner and provides the distribution of Y to be determined as follows

$$\begin{aligned} F_Y(y) &= \text{Prob}(Y \leq y) = \text{Prob}\big(X \leq F_X^{-1}(y)\big) \\ &= F_X\big(F_X^{-1}(y)\big) = y \end{aligned} \tag{5}$$

Partitioning

This step described fuzzy sets. The fuzzy sets are designed using three-sided association functions and automatically transformed behind the interval [0,1] in the recently modified space. It is justifying that the representation of each fuzzy set in the new space can be achieved by employing the contrary, improving distribution function or quantile function. In this condition, for every point describing the three-sided membership function, we would linearly combine the same value between the two nearest quintiles by determining the inverse of the linear function used to estimate the CDF.

3.2.2 Rule Generation Process

Later the fuzzy sets have been developed and preprocessed the training data, and the rule base is formed by employing a modern rule generation algorithm planned for big data. The rule generation algorithm process having two subsequent stages is motivated by the notions introduced in CHI-BD and a priori algorithms that stages are most promising item set search and Construction of fuzzy rules.

Most promising item set search

In this stage, we recognize each fuzzy set and nominal as items. When some of these elements are collectively developed in a distributed transaction, they create a set of elements (item set). To obtain the different potential combinations of potential elements, a three-step method is employed and explained below.

Discretization of the examples

Discretization of the examples: The complete sets of items presented in the training set are determined using the rule generation process of the CHI-BD algorithm. This method consists of deforming all samples by measuring the association measure of similarity of each value with all of the equal variables' fuzzy sets. It means every value followed by fuzzy sets reaches the most degree of membership. In the case of nominal costs, no estimate is transferred. The following example shows a detailed explanation, where the sequel determines which variable the disorganized fuzzy sets corresponds to.

Example 2 Given the following discretized example:

$$\text{Low}_1, \text{High}_2, \text{Medium}_3,$$

All the possible item sets are:

$$\{\text{Low}_1\}, \{\text{High}_2\}, \{\text{Medium}_3\},$$
$$\{\text{Low}_1, \text{High}_2\}, \{\text{Low}_1, \text{Medium}_3\},$$
$$\{\text{High}_2, \text{Medium}_3\},$$
$$\{\text{Low}_1, \text{High}_2, \text{Medium}_3\}$$

Frequent itemsets search:

The support for each set of items is counted, and only those that are between the minimum supports are kept. In the beginning Apriori algorithm, the assistance of an itemset is when the itemset rises in the training set. Here, the support of an itemset I redefined as.

$$\mathrm{supp}_{\mathrm{crisp}}(I) = \frac{\mathrm{count}(I)}{N} \tag{6}$$

where count(I) is the original support used and the number of training samples is N, it will be called as $\mathrm{supp}_{\mathrm{crisp}}(I)$ to separate the crisp support utilized in this step.

Most confident item sets selection:

Another filtering process is provided between the frequent itemsets, based on the confidence determination of the itemsets. In this step, the resolution of an itemset is described as

$$\mathrm{conf}_{\mathrm{crisp}}(I) = \frac{\max\limits_{m=1,\ldots,M}(\mathrm{countClass}(I, y_m))}{\mathrm{count}(I)} \tag{7}$$

where countClass(I, y_m) is number of sample counts belonging to the class y_m. In which item set I offered it will be called $\mathrm{conf}_{\mathrm{crisp}}(I)$ to separate this confidence from that employed in the fuzzy rules generation.

Construction of fuzzy rules

Based on the various assuring frequent itemsets obtained in the earlier step, a fuzzy rule base is designed as follows.

Conversion from item sets to candidate rules:

Each item set is transformed one or more of the candidate's rules. To this end, for a set of transmitted elements, the algorithm maintains the idea of the models in which the item set of objects appears and gets its class labels. Later, a new candidate rule was created for each of these classes.

Example 2 The following itemsets that have given the past filtering stage

$$\{\mathrm{High}_2\}, \{\mathrm{Low}_1, \mathrm{High}_2\}, \{\mathrm{Low}_1, \mathrm{High}_2, \mathrm{Medium}_3\}$$

And given the examples that have generated those itemsets:

$$\text{Low}_1, \text{High}_2, \text{Medium}_3 \rightarrow C_1$$
$$\text{Low}_1, \text{High}_2, \text{Medium}_3 \rightarrow C_2$$

we can extract the classes the item sets belongs to (C_1 and C_2) and generate the corresponding candidate rules:

IF A_2 is High THEN C_1
IF A_2 is High THEN C_2
IF A_1 is low and A_2 is High THEN C_1
IF A_1 is low and A_2 is High THEN C_2
IF A_1 is low and A_2 is High and A_3 is Medium THEN C_1
IF A_1 is low and A_2 is High and A_3 is Medium THEN C_2

Pruning and Filtering.

The rules are beginning separated based on their support and confidence. These are calculated using the following equations.

$$\text{supp}_{\text{fuzzy}}(R) = \frac{\text{matchClass} + \text{matchNotClass}}{N} \tag{8}$$

$$\text{conf}_{\text{fuzzy}}(R) = \frac{\text{matchClass}}{\text{matchClass} + \text{matchNotClass}} \tag{9}$$

Recognizing the N, matchNotClass and matchClass determined in above equations sequentially. As illustrated in the extraction of item sets, N is weighted by the cost of every class. The filtering method consists of removing those rules having support or confidence lower than a specific threshold.

3.2.3 Evolutionary Rule Selection

While the rule base has been achieved, a rule selection approach is used to make a compact and particular design. To this end, we perform the CHC evolutionary algorithm (EA) due to its strength to dispense with mixed search fields and the excellent results achieved by this EA algorithm in state-of-the-art FRBCSs like FARC-HD or IVTURS. Unlike the different approaches that offer the use of CHC

Algorithm:
Function: generate_rule_base(T, R)
IInput: A pre processed training set T R containing N labeled samples x_i
Output: *A rule base RB.*

Begin

1. **1. *Search for the most promising itemsets***
2. **1.1 *Discretization of the samples***
3. $T\,R_d \leftarrow TR.Map\ (x_i \leftarrow discretize\ (x_i))$
4. $Itemsets \leftarrow T\,R_d.Map(x_i^d \leftarrow extract\_itemsets(x_i^d))$
5. **1.2 *Search for frequent itemsets***
6. $SuppConf \leftarrow Itemsets$
7. $reduceByKey(itemset \leftarrow support\_and\_confidence(itemset)$
8. $Itemsets_{Freq} \leftarrow SuppConf.fileter(is\_frequent(itemset))$
9. **1.3 *Selection of the most confident itemsets***
10. $Itemsets_{Conf} \leftarrow Itemsets_{Freq}.filter(is\_confident(itemset)$
11. $Itemsets_{Prom} \leftarrow distributed\_pruning(Itemsets_{Conf})$
12. **2. *Construction of fuzzy rules***
13. **2.1 *Conversion from itemsets to candidate rules***
14. $Rules_{cand} \leftarrow Itemsets_{prom}.map(itemset \leftarrow rule(itemset))$
15. $Rules_{broad} \leftarrow broadcast(Rules_{cand}.collect())$
16. **2.2 *Computation of rule weights and conflict resolution***
17. $matchings \leftarrow T\,R_d.map(x_i \leftarrow matching(Rules_{broad}))$
18. $SuppConfWght \leftarrow matchings.reduceByKey(rule \leftarrow$
    $support\_confidence\_weight(rule))$
19. $Rules_{no\_conflicts} \leftarrow SuppConfw\,ght.map(rule \leftarrow$
    $resolve\_conflicts(rule))$
20. **2.3 *Filtering and pruning***
21. $Rules_{freq} \leftarrow Rules_{no\_conflicts}$
22. $filter(rule \leftarrow is\_frequent(rule))$
23. $Rules_{conf} \leftarrow Rules_{freq}.filter(rule \leftarrow is\_confident(rule))$
24. $Rules \leftarrow distributed\_pruning(Rules_{conf})$
25. ***RETURN*** $build\_rule\_base(Rules)$

The experimental study conducted shows that our new method can deal with big data problems that are acquiring the same results precisely because the original set

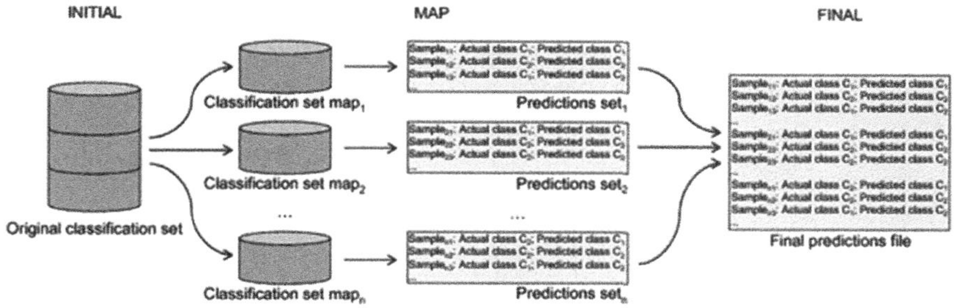

Fig. 4 Map classification sets prediction

of Chi algorithm (if they can be achieved in big data problems) regardless of the set of nodes used for execution, not like CHIBDLocal. Also, the acquired run times show that CHIBDGlobal is significantly faster than CHIBDLocal when it comes to big datasets (Fig. 4).

4 Experimental Study

This research intends to investigate the quality of the RS-FCRG Bigdata algorithm in the big data situation. For this, we will analyze the problems from the UCI dataset repository. To test the performance of the suggested approach, we have analyzed the results achieved by the RS-FCRG approach so that we can compare their behavior concerning the chosen big data difficulties.

4.1 Dataset

To evaluate the performance of the proposed method, we used the UCI Census (KDD) info dataset. The RS-FCRG-Bigdata approach is tested on UCI Census (KDD) info dataset. The data contains 41 demographic variables related to employment. The weight of the example indicates the number of people in the community represented by each record due to stratified sampling. The original table contains 199,523 rows and 42 columns. An additional column edu_year has been added to assist in the study (Table 1).

Table 1 Dataset characteristics represent

Dataset	Attribute features	No. of records	No. of attributes	No. of rows	No. of columns
Multivariate	Categorical, integer	299,285	40	199,523	42

4.2 Metrics Used

The outcome of the measures depends on the results obtained True Positive (TP), True Negative (TN), False Positive (FN), and True Negative (TN).

True Positive.

True Positive (TP): The transaction cases which are not fraud and the system model has predicted as not fraud.

True Negative (TN): The transaction cases which are fraud and the system model has predicted as a fraud.

False Positive (FN): The transaction cases which are fraud and the system model has predicted as not fraud.

True Negative (TN): The transaction cases which are not fraud and the system model has predicted as a fraud.

We have used many metrics because the set of the dataset used in this paper as markedly unbalanced. The use of the exact measurement will now be inaccurate to assess the accuracy of the method. Here we are using the below equation to calculate the accuracy of the proposed approach.

$$\text{Accuracy} = \frac{\text{TP} + \text{TN}}{\text{TP} + \text{FP} + \text{TN} + \text{FN}} \times 100 \tag{10}$$

4.3 Results and Discussions

To demonstrate how the proposed method is capable to overcome the complexity of the design by reducing the number of final rules, we display in Table 2 the number of rules generated by each mapping method and the number of final rules, when the separate RBs generated by each map are combined. To do this, we have chosen UCI Census (KDD) info dataset with 199,523 columns, 42 rows, and 40 attributes.

Table 2 Average number of rule generated for various maps

Dataset	8 Maps	16 Maps	32 Maps	64 Maps	128 Maps
Census	34,278.0	34,341.1	34,376.5	34,392.5	34,397.3

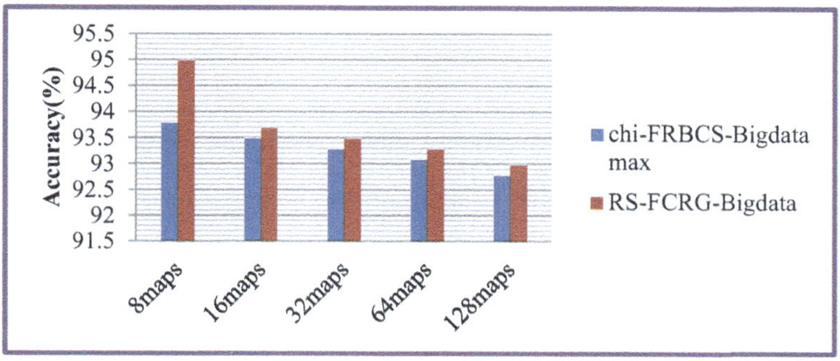

Fig. 5 Accuracy comparison between previous algorithms

In Table 2, we show the number of rules generated averagely for the proposed RS-FCRG-BigData algorithms using 8, 16, 32, 64, and 128 maps over the chosen dataset.

In Fig. 5, we represent the average results for the RSFC-RG-BigData variants, and the census datasets examined. It determines the progression of the accuracy measure when the number of maps is different. The proposed approach gets better accuracy with 95% when compared with the previous algorithm.

5 Conclusion

This paper proposed a semantic fuzzy rule-based classification algorithm for big data problems called rough set fuzzy classification rule generation algorithm (RS-FCRG). This algorithm achieved an interpretable representation that can manage big datasets, presenting good accuracy, and fast acknowledgment times. For this, the proposed approach utilized the MapReduce programming model on the Hadoop platform, and it is one of the most successful clarifications to deal with big data nowadays efficiently. This way our model distributes the calculation using the map function and then collects the results through the reduced function the experimental study took on Census-Income (KDD) dataset. The dataset is having with 199,523 rows and 42 columns, and the comparison took the chi-FRBCS-Bigdata-Max algorithm. The proposed approach got high accuracy with 95% than the previous approach.

References

1. P. Zikopoulos, C. Eaton, D. DeRoos, T. Deutsch, G. Lapis, *Understanding Big Data: Analytics for Enterprise Class Hadoop and Streaming Data* (McGraw-Hill, New York, 2011)
2. S. Madden, From databases to big data. IEEE Internet Comput. **16**(3), 4–6 (2012)

3. A. Sathi, *Big Data Analytics: Disruptive Technologies for Changing the Game* (MC Press, 2012)
4. H. Ishibuchi, T. Nakashima, M. Nii, *Classification and Modeling with Linguistic Information Granules: Advanced Approaches to Linguistic Data Mining* (Springer, Berlin, 2004)
5. T.P. Hong, Y.C. Lee, M.T. Wu, An effective parallel approach for genetic-fuzzy data mining. Expert Syst. Appl. **41**(2), 655–662 (2014)
6. O. Cordón, F. Herrera, A. Peregrín, Applicability of the fuzzy operators in the design of fuzzy logic controllers. Fuzzy Sets Syst. pp. 15–41 (1997)
7. Z. Chi, H. Yan, T. Pham, *Fuzzy Algorithms with Applications to Image Processing and Pattern Recognition* (World Scientific, New York, 1996).
8. H. Ishibuchi, T. Yamamoto, Rule weight specification in fuzzy rule-based classification systems. IEEE Trans. Fuzzy Syst. pp.428–435 (2005)
9. S.N. Sivanandam, S. Sumathi, S.N. Deepa, *Introduction to Fuzzy Logic Using Matlab* (Springer, Berlin, 2007)
10. Z. Chi, H. Yan, T. Pham, *Fuzzy Algorithms with Applications to Image Processing and Pattern Recognition* (World Scientific, Singapore, 1996)
11. J.F. Garrido, I.R. Ramos, A methodology for constructing fuzzy rule-based classification systems. Mathware Soft Comput. **7**, 432–475 (2010)
12. T. Yamamoto, H. Ishibuchi, Rule weight specification in fuzzy rule-based classification systems. IEEE Trans. Fuzzy Syst. pp. 428–435 (2005)
13. H. Bhukya, M. Sadanandam, Rough sets base associative classification rules extraction from big data. Int. J. Innov. Technol. Explor. Eng. **9**(1) (2019) ISSN: 2278-3075
14. H. Bhukya, M. Sadanandam, Rough sets base incremental associative classification rules generation on MapReduce framework. Int. J. Adv. Sci. Technol. **29**(05), 13218–13227 (2020)

Identifying Network Intrusion Using Enhanced Whale Optimization Algorithm

Anne Dickson and Ciza Thomas

Abstract Advances in networking technology and automation through numerous interfaces lead to an exponential hike in network threats and flaws in information security. Limited capabilities of intrusion detection and prevention systems enormously reduced the detection rate. Optimizing the conflicting objectives is the best way to alleviate this security challenge. Population-based search techniques are deployed for optimization by selecting the best optimal element from available set of possible solutions. Whale optimization is a neoteric methodology for solving complex problems through computational intelligence. Intelligence technologies proved that even a tiny creature like ant can be the source of inspiration. Problem statement is reducing the tradeoff between the two conflicting objectives such as true positive and false positive thereby improving the detection rate. This paper proposes a methodology to locate the attack detection point through the process of spiral updation. The motto of this work is to enhance the detection rate of IDS thereby searching the most promising regions located by search agents using an enhanced whale optimization algorithm (EWOA). NSL-KDD dataset is used for evaluating the procedure. Analysis is done using the five machine learning classifiers such as Naive Bayes, logistic, SMO, J48 and Random Forest. Empirical result shows Random Forest classifier having better performance with maximum true positive of 0.998 followed by a true positive value of 0.914 for J48 classifier through spiral updation for variant sets of random particles in the search space.

Keywords Whale optimization · Computational intelligence · Exploration · Exploitation · Spiral updation · Encircling

---

A. Dickson (✉)
Assistant Professor, Department of Computer Science, Mar Baselios College of Engineering & Technology, Trivandrum, Kerala, India
e-mail: anne.dickson@mbcet.ac.in

C. Thomas
Directorate of Technical Education, Trivandrum, Kerala, India

M. Paprzycki et al. (eds.), *Intelligent Systems, Technologies and Applications*, Advances in Intelligent Systems and Computing 1353,
https://doi.org/10.1007/978-981-16-0730-1_7

1 Introduction

Framing a flawless network design which ensures a free flow of data at affordable cost structure is the sole intention of computer network optimization. Such optimal design ensures optimal usage of network resources, thereby increasing the overall network performance. A tremendous rise in vulnerability is noticed with respect to the growth of advancement in technology. The data flow metric changed from bits to zettabytes and finally yottabytes. Intrusion detection systems and intrusion prevention systems are unable to identify the latest malware infections, hacking attempts which are known to be the root causes of current network traffic. Implementing network security is a primary challenge in front of the research community. The only thing which is forced to do is increasing the detection rate or sharpness of Intrusion Detection System (IDS) from high volume of false alerts. The dynamic nature of expanding networks through virtualization, improper network configuration management, unexpected computation overhead due to varying packet size, application level and encrypted traffic are the limiting factors of subsisting IDS. The most important limiting factor for the performance of IDS is the huge rate of false alarms. The dependency on environment, runtime limitations and the specification of detection patterns are the other causes. The level of knowledge required by intruders is getting steadily lower, yet their ability to penetrate sophisticated attacks against the survivability of systems has increased.

Many IoT applications are meant for monitoring critical cases such as intrusion detection, smoke detection, fire detection, building health monitoring and so forth. So there should not be any delay or jitter in the network. Network should be robust enough to deliver the data to the destination within a defined time and routing. Now more attention is paid to meta-heuristics for the purpose of decision making given the growing complexity of models and the needs for quick decision making in the engineering. The organization of this paper is as follows. Section 2 discusses the related works using different swarm intelligence-based optimization techniques in IDS. Insight of WOA and the proposed methodology along with data flow diagram are described in Sect. 3. Dataset used for experimental purpose along with obtained results is described in Sect. 4 followed by conclusion and future directions.

2 Related Works

Meta-heuristic algorithms fall in the category of non-conventional meta-heuristic algorithms. Diverse detection models are constructed for detecting intrusions using machine learning classifiers such as k-means [1], Naïve Bayes, support vector machine, decision tree [2], Random Forest [3] and its ensembles. Wagner et al. [4] used one-class support vector machine learning approach for IP flow-record anomaly detection and reported attacks such as DDoS, UDP and TCP flood. Adaptive intrusion detection based on AdaBoost and Naive Bayes has been implemented by Farid et al. [2]. Su [3] applied genetic algorithm and weighted k-nearest neighbour classifiers in real-time anomaly detection for both DoS and distributed DoS attack. Later genetic fuzzy systems were introduced for general attack detection in [5] by H. Sanieyei et al. Self-adaptive and dynamic clustering for online anomaly detection using self-organizing map and k-means clustering is used by Lee et al. [6].

Various computational intelligence algorithms were used for detecting network attacks. Lin et al. [7] deployed whale algorithm for feature reduction of network intrusion detection and verified using KDD CUP'99 dataset. Ant colony optimization was used by Li [8] in 2012 for designing an efficient intrusion detection based on support vector machine and feature reduction method. Baig et al. [9] and Eesa et al. [10] in 2015 proposed a novel feature selection approach based on cuttle fish optimization algorithm. Ensemble-based methods were deployed for intelligent-based classification by Zubair A. Threat analysis of IoT networks using artificial neural network was proposed by HoDo et al. [11] in 2016. Yin et al. [12] used a deep learning approach for detecting anomaly-based patterns in 2017. Hybrid-based PSO and binary PSO were also used for various attack detections. Machine learning approach for detection of nonTor traffic is done using UNB-CIC dataset by Hodo et al. [11]. But whale optimization authors [13] claim as one of the best meta-heuristic optimization method ever done in the realm of swarm intelligence with the capability of solving complex tradeoff problems.

Recent meta-analysis survey reveals the fact that WOA had better performance of exploration while tested on multimodal functions and hence improvement in convergence speed during iterations [14].

3 Experimental Setup

3.1 Background in Framing Optimization Model

An optimization algorithm is a procedure which is executed iteratively by comparing various solutions till an optimum or a satisfactory solution is found. Decision variables, constraints, and objective are the regular elements of an optimization problem. Variables which we want to optimize are called decision variables or decision vectors. Conditions subjected for optimization using inequalities are said to be constraints. Finding the values of decision variables that result in a maximum or minimum of a function called objective function. Mathematical model that attempts to maximize or minimize an objective function without violating resource constraints is also known as mathematical programming. One approach to simplify an optimization problem is to create a metamodel. It will replace the typical computational analysis by multidimensional functional approximations of objective functions and constraints. It is a mathematical approximation of computational analysis.

3.2 Whale Optimization Algorithm

A heuristic is a technique designed for solving a problem more fast when classic methods fails to find any exact solution. Mirjalili et al. [13] proposed the whale optimization algorithm inspired by the humpback whales which are known for its magical song or dialect. It lives in groups and has a peculiar hunting mechanism. WOA assumes that in '*n*' dimensional search space, out of the feasible region, the best point is the point adjacent to the optimum which is the target prey. Based on this initiative, the remaining search agents will renovate their positions based on the following relations.

$$\vec{D} = \mid \vec{C}.\vec{X}^*(n) - X(n) \mid \tag{1}$$

$$\vec{X}(n+1) = \vec{X}^*(n) - \vec{A}.\vec{D} \tag{2}$$

where '*n*' represents the number of iteration, $\vec{A}$ and $\vec{C}$ are coefficient vectors, $\vec{X}$ is the position vector and $\vec{D}$ is the distance between the whale and the prey. In our problem domain, it is the distance between the futuristic attack source and the IDS. The coactive vectors are calculated using the following conditions.

$$\vec{A} = 2\vec{a}.\vec{r} - \vec{a} \tag{3}$$

$$\vec{C} = 2.\vec{r} \tag{4}$$

$\vec{C}$ will take the values in decreasing order from 2 to 0, $\vec{r}$ is a random vector [0, 1]. Since the spiral shape converges from a big circle to small, the values have been chosen in decreasing order (Table 1; Fig. 1).

3.3 Exploration and Exploitation Phase in WOA

The two balancing factors that affect the optimization algorithms are exploration and exploitation. Exploration determines the global best by visiting new points. Improving those points so as to progress the quality of the solution is attained through

Table 1 Value of fitness function for random values in the spiral area

$\vec{a}$	$\vec{r}$	$\vec{C}$	$\vec{A}$
2	0.773	1.546	1.092
1.5	0.047	0.094	−1.359
1	0.018	0.036	−0.964
0.5	0.004	0.008	−0.496
0	0.003	0.006	0

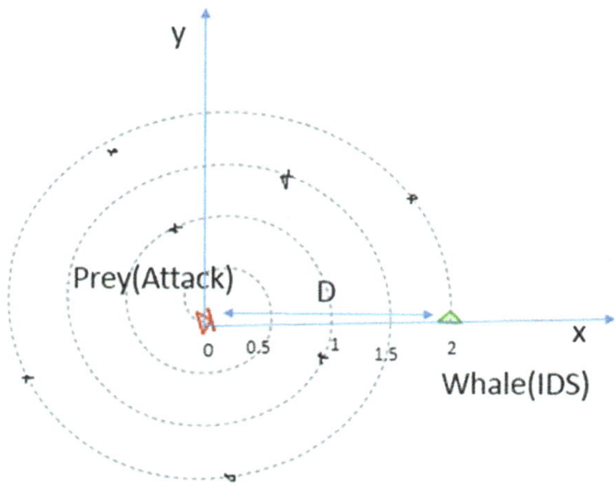

Fig. 1 Upgrading position in spiral

exploitation. Here, the exploitation phase encompasses contracting the encircling mechanism followed by spiral upgrading procedure. This encircling will gradually reduce to small circles and hence the name encircling mechanism. In order to compute this effectively, the range of parameter 'a' is reduced in a linearly decreasing fashion from 2 to 0. Hence, a spiral equation is formed with to keep down the helix-shaped motion of humpback whales as follows.

$$\vec{X}^{(n+1)} = \vec{D}^{i}.e^{bl}cos(2\pi l) + \vec{X} * n \tag{5}$$

where $\vec{D}^{i}$ is the distance, 'b' is a constant and l is a random number $-1 \leq l \leq 1$ Exploration phase [15] in WOA is used for searching the prey. Searching the optimum position in the search space for getting the attack source. Exploration is done in random fashion, and hence, it can be modeled by replacing $\vec{X}^{*}(n)$ with $\vec{X}_{rand}$ in equations (1) and (2). As far as our problem is concerned, the exploration is the ability of the IDS to find the anomaly-based intrusion detection [16] thereby increasing the detection rate. While exploration phase, the value of $\vec{A}$ can be chosen between $[-1, 1]$ so that the updation is happened based on randomly chosen search agent and its mathematical model is shown in (6) and (7).

$$\vec{D} = \left|\vec{C}.\vec{X}_{rand} - \vec{X}\right| \tag{6}$$

$$\vec{X}(n+1) = \vec{X}_{rand} - \vec{A}.\vec{D} \tag{7}$$

where $\vec{X}_{rand}$ is the random position vector chosen from current population.

3.4 Maneuver in Contracting Circle and Spiral Path

The concurrent behavior of diving through spiral path and contracting circle for catching the prey is modeled in such a way that two different updating procedures are followed. The significance of this change is reflected in the mathematical model through the constraints as inequalities. The analytical model is framed as follows:

$$\vec{X}(n+1) = \begin{cases} \vec{X}^*(n) - \vec{A}.\vec{D} & \text{if} \quad p_T < 0.5 \\ \vec{D}^i.e.^{bl}cos(2\pi l) + \vec{X}^*(n) & \text{if} \quad p_T \geq 0.5 \end{cases} \tag{8}$$

where p_T' is a user defined threshold for performing spiral updation procedure.

3.5 Proposed System Architecture

Swarm intelligence technique is a theory of simulating human behavior through computation. An ensemble of five machine learning classifiers such as Naïve Bayes, Logistic, SMO, J48 and Random Forest are taken for experimental analysis. Analyzing the best performing classifiers through ROC for the input dataset is the preliminary step. Random values of classifiers $c_1, c_2, \ldots, c_n$ are initialized as the number of particles. Dimension, number of particles, lower and upper bound values and number of iterations are given as input. Allow the 'n' number of particles to move along the 'n' dimensional search space and note its positions as $s_1, s_2, \ldots, s_n$. Check if the number of iterations are less than maximum iterations (Fig. 2).

Design the objective function based on the problem domain. WOA is not up to par in achieving global optima despite of having better convergence rate. This inadvisedly affects the convergence rate of algorithm. To reduce this, various objective functions

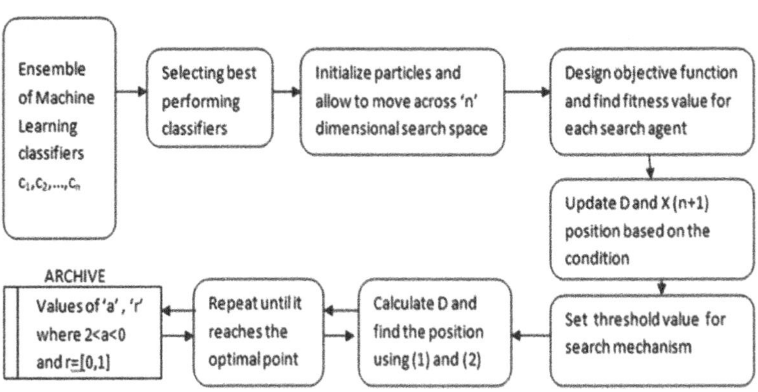

Fig. 2 Dataflow diagram showing methodology—Phase I

can be used by adjusting the parameters A and C. Here, each input data is treated as the search agent. For each $s_1, s_2, \ldots, s_n$, calculate the coefficient vectors A and C. Find a and r for different values based on the ranges $2 \leq a \leq 0$ and $r \in [0, 1]$. For each iteration, dynamic 'a' and 'r' values used can be kept in archive. Bring back the search agents if it moves beyond the search space by upgrading the values. Set the threshold value as 0.5 as $0 < (x_1, x_2) < 1$, where x_1 is false positive and x_2 is true positive. Update the values again based on whether the value is less than or greater than the p_T threshold. Calculate the value of objective function for each true positive and false positive. Continue the iterations until the optimal point is reached. Various parameters along with its value which is used in the proposed EWOA are shown in Table 2 (Fig. 3).

Enhanced whale optimization algorithm for detecting attack location

1. Generate network traffic.
2. Capture network packets using wire shark (sniffer tool).
3. Consider the fields such as Source IP, Destination IP, Source Port, Destination port, Protocol used, number of flow, duration and packet size.
4. Capture the packet flow from one instance to another instance say from 20th second to 30th second.

Table 2 Parameters of enhanced WOA

Parameter	Value and its description
$\vec{A}$	Coefficient vector
$\vec{C}$	Coefficient vector
$\vec{D}$	Distance
$\vec{X}$	Search agent
t	Number of search agents
n	Iteration
a	2 to 0
r	[0, 1]
p_T	Threshold 0.5
s	Positions $1, 2, \ldots, n$

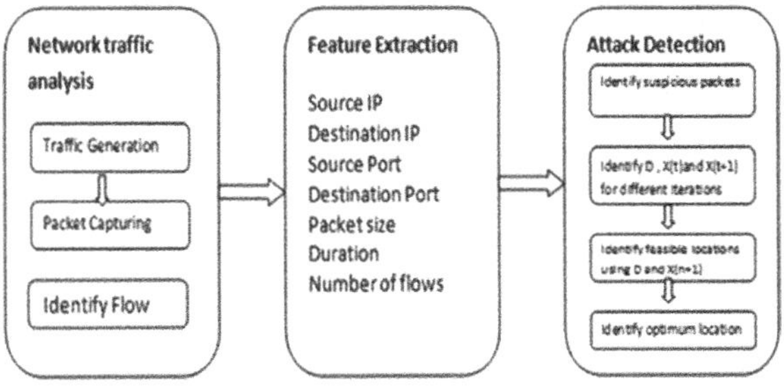

Fig. 3 Dataflow diagram showing methodology—Phase II

5. Analyze the above features and identify the suspicious packet.
6. Characteristics used to identify suspicious packets are as follows: (Since we can identify attack only from a packet source and not from inside packet.)

   a. Immediate change in the IP address range.
   b. Sudden on and off of a host.
   c. Huge byte size of the packet
   d. Check the unusual ports such as ICMP (ports other than http, ssh, https, sttp (21, 25, 80, 81, 443, 8080, 3128)
   e. Duration of active
   f. Unusual data which is coming from unknown port, we can take it as malicious. First 100 ports are the basic ports which we use, others will be using for malicious activity.
   g. Multiple packets from a same IP to multiple hosts.
   h. Sudden hike of the bandwidth.

7. If more than four conditions are satisfied, then the packet is considered as suspicious one else non-suspicious.
8. Calculate $\vec{D}$, $\vec{X}^*(n)$, $\vec{X}(n+1)$ to find the exact position of the attack in the search space based on the calculation given in (8)
9. Repeated for 50, 100, 200, 300, 400 and 500 random particles for different iterations until we find optimal point from the feasible region in the search space.

Exploration phase in WOA is used for searching the prey. Here in enhanced whale algorithm, we are searching the optimum position in the search space for getting the location of the attack. If the packet is found to be suspicious based on the given conditions in the proposed algorithm, then there is 0.75% of finding the attack [17] in that position. And that suspicious packet [18] location will be any one of the vulnerable position.

3.6 Framing an Optimization Problem with Constraints

Subsisting and recent engineering design problems in the current environment are constrained optimization problems [19]. Designing an optimization problem depends on certain conditions such as if the number of design variable is 'g' and the number of independent equality constraints is 'h', then g must be always greater than h. If $g = h$, then the equality constraints can be solved to get a solution. If $h > g$, then the number of equations will be more than unknowns. No restrictions in the number of inequality constraints. The optimization problem for optimizing true positive value with respect to our problem domain can be framed using objective function as follows.

$$\text{Maximize } L = y_2^2 - 2y_1^2 + 10y_2 - 8y_1$$

subject to the constraints

$$y_1 \leq 0.19$$
$$y_2 \leq 0.99$$
$$y_1 + y_2 < 1.2$$

where y_1 denotes the false positive and y_2 denotes true positive.

3.7 Mathematical Formulation

Standard form of a constrained optimization problem can be stated as follows: Find the design variable vector X which minimizes the objective function

$$f(x) = f(x_1, x_2, \ldots, x_n) \tag{9}$$

subject to 'm' inequality constraints $g_i(x) \leq 0$ where $i = 1, 2, 3, \ldots, m$ and 'p' equality constraint $h_j(x) = 0$ where $j = 1, 2, \ldots, p$. Here 'x' is a design vector of size $n \times 1$ with design variables. The above-mentioned equations and constraints define a vast array of practical optimization problems. If $f(x)$, $h_j(x)$ and $g(x)$ are linear in terms of x, then the problem is known as linear programming and else it will be called as nonlinear. The $g_j(x)$ can be written as $g_j(x) = -G_j(x) \leq 0$. Hence in order to convert such problem, just multiply the relation by -1. A set of all feasible solutions is called constraint set which is defined as follows.

$$S = x \begin{cases} h_j(x) = 0;\ j = 1, p \\ g_i(x) \leq= 0;\ i = 1, m \end{cases} \tag{10}$$

where 'p' stands for equality constraints and 'm' stands for inequality constraints. This constraint set defines the feasible region where the optimal point lies exactly.

4 Empirical Study with Results

Although whale optimization algorithm has the limitation of poor convergence speed, it is very competitive in performance [14] with other common optimization algorithms. On account of exploration and exploitation stages, it concurrently achieves local optima and convergence speed through iterations. The program for enhanced whale optimization algorithm is implemented in MATLAB platform with threshold value of 0.5 since algorithm says fifty percentage probability for selecting a particular prey searching method. The trustworthy dataset widely used by network security research community is NSL-KDD dataset [20], an improvement over KDD'99, in which all the redundant and duplicate records are eliminated for avoiding biased classing results. It is a rational dataset with 42 attributes where the simulated attack falls into four categories such as DoS, U2R, R2L and Probing based on basic connection features, content features, timely traffic and host features. Last attribute 42 indicates 'class' such as the particular instance connection belongs to normal or anomaly.

Random number of particles were allowed to search over n-dimensional search space with lower bound and upper bound values of zero and one with five different classifiers. Here each random value is the true positive value which we need to maximize. Each coordinate value is an ordered pair (y_1, y_2), where $y_1$is the false positive which is also called as Type I error and y_2 is the true positive, respectively. True positive is the state in which the IDS identifies an activity as an attack when the activity is really an attack. For an IDS, false positive alerts are bad, and false negatives are worse since it is an unrevealed inadmissible activity. True positives and false positives are the two conflicting objectives [21] which exists as a major challenge in IDS. When the random y_2 values are subjected to maximization, the optimum point obtained for five different classifiers [22] is shown in the tables below. Number of particles from 100 to 500 for different iterations at '*n*' dimensions were implemented. The result shows the improvement in true positive value with the number of iterations from Tables 3, 4, 5, 6 and 7.

Table 3 Optimal point for five classifiers with 100 iterations with 500 random particles using enhanced WOA

#particles	Iteration	Dim	UB	Classifier	Optimum point
500	100	30	1	Random forest	0.976
500	100	30	1	J48	0.664
500	100	30	1	Logistic	0.224
500	100	30	1	SMO	0.522
500	100	30	1	Naive Bayes	0.453

Table 4 Optimal point for five classifiers with 40 iterations with 400 random particles using enhanced WOA

#particles	Iteration	Dim	UB	Classifier	Optimum point
400	40	15	1	Random forest	0.769
400	40	15	1	J48	0.666
400	40	15	1	Logistic	0.224
400	40	15	1	SMO	0.522
400	40	15	1	Naive Bayes	0.095

Table 5 Optimal point for five classifiers with 40 iterations with 300 random particles using enhanced WOA

#particles	Iteration	Dim	UB	Classifier	Optimum point
300	40	15	1	Random forest	0.952
300	40	15	1	J48	0.906
300	40	15	1	Logistic	0.225
300	40	15	1	SMO	0.421
300	40	15	1	Naive Bayes	0.129

Table 6 Optimal point for five classifiers with 40 iterations with 200 random particles using enhanced WOA

#particles	Iteration	Dim	UB	Classifier	Optimum point
200	40	15	1	Random forest	0.976
200	40	15	1	J48	0.964
200	40	15	1	Logistic	0.571
200	40	15	1	SMO	0.521
200	40	15	1	Naive Bayes	0.134

Table 7 Optimal point for five classifiers with 40 iterations with 100 random particles using enhanced WOA

#particles	Iteration	Dim	UB	Classifier	Optimum point
100	40	15	1	Random forest	0.998
100	40	15	1	J48	0.914
100	40	15	1	Logistic	0.573
100	40	15	1	SMO	0.533
100	40	15	1	Naive Bayes	0.130

Figure 4 represents the graph of number of iterations with best score obtained so far for each machine learning classifiers like Random Forest, J48 which is the latest version of decision tree, Logistic Regression, SVM and Naïve Bayes. It shows the best performing classifier as the Random Forest with 0.998 as maximum true positive value.

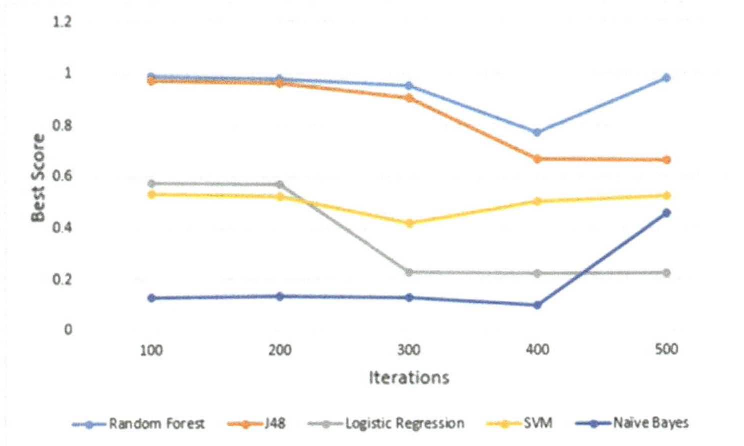

Fig. 4 Iteration versus best position with different classifiers

5 Conclusion and Future Enhancement

The quantitative output of whale optimization algorithm in various aspects shows its worthiness in solving a wide range of challenging complex problems. The coefficient vectors $\vec{A}$ and $\vec{C}$ are the two parameters that can be tuned [23]. Runtime analysis of WOA in each application makes very competitive with other optimization algorithms. Even though the value of $\vec{X}^*(n)$ is updated in each iteration, it suffers from less convergence speed which leads to an ineffective global search and hence the limitation of WOA. Depending upon the threshold 'p_T', it can turn around between spiral path and contracting circle for catching the prey. In this paper, enhanced whale optimization algorithm consider various conditions for identifying the suspicious packet. After categorizing, the exact location in spiral area is diagnosed. This helps to find the exact vulnerable location of the attack which is considered as the optimum point in the spiral area. Using random values taken from the machine learning classifier random forest, which performed best in design methodology phase I, the results obtained are fitness value of 1.992 using 0.5 threshold value with 0.998 detection rate. Proposed algorithm can be deployed successfully for improving the detection rate of any IDS in the current network scenario. In future, other control parameters along with real-time dataset which includes all types of anomaly attacks can be used for better results. Hybrid methods with evolutionary search schemes can also be utilized for future enhancement. Observed attack location can be identified as a specific attack which is coming under U2R or R2L by exploring more.

References

1. Z. Muda, M. Warusia, md N.U. Sulaiman, A Naïve Bayes and K-means learning approach for better intrusion detection. Inform. Technol. J. pp. 108–121 (2011)
2. MdD. Farid, C.M. Rahma, M.Z. Rahman, Adaptive intrusion detection based on boosting and Nave Bayesian classifier. Int. J. Comput. Appl. **24**(3), 12 (2011)
3. M.-Y. Su, Real-time anomaly detection systems for Denial-of-Service attacks by weighted k-nearest-neighbor classifiers. Expert Syst. Appl. **38**(4), 3492–3498 (2011)
4. C. Wagner, J. François, R. State, T. Engel, Machine learning approach for IP-flow record anomaly detection. LNCS **6640**, 28–39 (2011)
5. M.S. Abadeh, H.M.J. Habibi, Design and analysis of genetic fuzzy systems for intrusion detection in computer networks. Expert Syst. Appl. **38**(6), 7067–7075 (2011)
6. S. Lee, G. Kim, S. Kim, Self-adaptive and dynamic clustering for online anomaly detection. Expert Syst. Appl. **38**(12), 14891–14898 (2011)
7. H. Xu, Y. Fu, C. Fang, Q. Cao, J. Su, S. Wei, An improved binary whale optimization algorithm for feature selection of netork intrusion detection, in *4th IEEE International Symposium on Wireless Systems*, pp. 10–15 (2018)
8. Y. Yi, J. Wu, W. Xu, Incremental SVM based on reserved set for network intrusion detection. Expert Syst. Appl. **38**(6), 7698–7707 (2011)
9. Z.A. Baig, S. Sait, A.R. Shaheen, GMDH-based networks for intelligent intrusion detection. Eng. Appl. Artif. Intell. **26**(7), 1731–1740 (2013)
10. A.S. Eesa, Z. Orman, A.M. Abdulazeez, A novel feature-selection approach based on the cuttlefish optimization algorithm for intrusion detection systems. Expert Syst. Appl. **42**(5), 2670–2679 (2015)
11. E. Hodo, X. Bellekens, A. Hamilton, P.-L. Dubouilh, E. Iorkyase, C. Tachtatzis, R. Atkinson, Threat analysis of IoT networks using artificial neural network intrusion detection system, in *International Symposium on Networks, Computers and communication (ISNCC, 2016)*
12. C. Yin, Y. Zhu, X. He, *A Deep Learning Approach for Intrusion Detection Using Recruitment Neural Network* (IEEE Access Computer Science, 2017)
13. S. Mirjaliliab, A. Lewis, The Whale optimization algorithm. Adv. Eng. Software **95**, 51–67 (2016)
14. H.M. Mohammed, S.U. Umar, T.A. Rashid, A systematic and meta-analysis survey of whale optimization algorithm. J. Comput. Intell. Neural Sci. (2019)
15. S. Mirjalili, S.M. Mirjalili, A. Lewis, *Grey Wolf Optimizer* (Elsevier, 2014), pp. 46–63
16. H. Hindi, D. Brosset et al., *A Taxonomy and Survey of Intrusion Detection System Design Techniques, Network Threats and Datasets* (ACM, 2018), pp. 1–35
17. M. Shafiq, X. Yu, A.A. Laghari, L. Yao, N.K. Karn, F. Abdessamia, *Network Traffic Classification Techniques and Comparative Analysis Using Machine Learning Algorithms* (IEEE, 2016), pp. 2451–2455
18. J. Zhang, X. Chen, Y. Xiang, W. Zhou, J. Wu, Robust network traffic classification. IEEE/ACM Trans. Network. pp. 1–14 (2014)
19. S. Mirjalili, S.M. Mirjalili, A. Hatamlou, Multi-verse optimizer: a nature-inspired algorithm for global optimization. Neural Comput. Appl. pp. 1–19 (2015)
20. M. Tavallaee, E. Bagheri, W. Lu, A. Ghorbani, A detailed analysis of the KDD CUP 99 data set, in *Second IEEE Symposium on Computational Intelligence for Security and Defense Applications (CISDA)* (2009)
21. X.-S. Yang, *Nature-Inspired Optimization Algorithms* (Elsevier, Amsterdam, 2014)
22. A. Dickson, C. Thomas, Improved PSO for optimizing the performance of intrusion detection systems. J. Intell. Fuzzy Syst. pp. 1–11 (2019)
23. S. Mirjalili, P. Jangir, S. Saremi, *Multi-objective Ant Lion Optimizer: A Multi-Objective Optimization Algorithm for Solving Engineering Problems* (Springer, 2016), pp. 1–19
24. S. Mirjalili, S. Saremi, S.M. Mirjalili, L.S. Coelho, Multi-objective grey wolf optimizer: a novel algorithm for multi-criterionoptimization. Expert Syst. Appl. pp. 106–119 (2015)

25. S. Mirjalili, SCA: a sine cosine algorithm for solving optimization problems. Knowl. Based Syst. pp. 1–14 (2016)
26. F.S. Gharehchopogh, H. Gholizadeh, A comprehensive survey: whale optimization algorithm and its applications. Swarm Evol. Comput. **48**, 1–24 (2019)

Transformed WLS-Based Data Reconciliation for a Large-Scale Process Network

Boddu Chakradhar, Gautham Shanmugam, K. L. N. Sree Datta, R. Jeyanthi, Oguri Sravani, and S. M. Dhakshana

Abstract Modern process plants and industries have to modify their systems in order to constantly maintain highest performance standards. Process data are used to estimate the current state of the system, which helps in the decision-making process to ensure smooth and safe operation of the plant. These process data readings are taken from sensors but are not transmitted accurately and are inconsistent due to online errors. These erroneous measurements will affect the normal plant operation, which in turn will hamper the safety and economics of the plant. Hence, data reconciliation (DR) techniques have been implemented on the measured data to suppress the errors and obtain the actual measurements. Weighted least squares (WLS), a popular DR technique, has been used to eliminate random errors from process data. In this paper, along with conventional WLS, various modified DR techniques namely moving average WLS (MA-WLS) and exponentially transformed WLS (E-WLS) have been implemented. The reconcilability of each of these techniques has been analysed. A comparative study has been carried out to identify the best performing DR technique. The analysis is done for a large-scale process network in Python-based simulation environment.

Keywords Data Reconciliation (DR) · Weighted Least Squares (WLS) · Moving Average WLS (MA-WLS) · Exponentially transformed WLS (E-WLS)

1 Introduction

The working of modern plants and industries has a stringent need of fast and low-cost modifications to improve their performance. In the operations carried out in an industry, the process data play a very important role in process control, safety and

B. Chakradhar · G. Shanmugam · K. L. N. S. Datta · R. Jeyanthi (✉) · O. Sravani · S. M. Dhakshana
Department of Electronics and Communication Engineering, Amrita School of Engineering, Amrita Vishwa Vidyapeetham, Bengaluru, India
e-mail: r_jeyanthi@blr.amrita.edu

M. Paprzycki et al. (eds.), *Intelligent Systems, Technologies and Applications*, Advances in Intelligent Systems and Computing 1353,
https://doi.org/10.1007/978-981-16-0730-1_8

monitoring [1]. The process data are used in analyzing the current state of the system which in turn helps in the decision-making process regarding plant operations. The data for process variables like temperature, flow, pressure and level are measured in plants. The measurements obtained from sensors are not transmitted accurately and are inconsistent due to online errors [1]. These errors or deviations from the actual values occur due to surrounding process conditions, sensor inaccuracy, etc. In large-scale plants which operate continuously, it is not possible to halt the process to change sensors or instruments for long periods of time. Hence, these sensors degrade over a period of time giving inaccurate readings. Installing multiple sensors in plants for measurement of a single variable is not always feasible due to budget constraints. Errors from two sources namely gross errors and random errors lead to deviation of measured values from actual values. Data reconciliation techniques are implemented to reduce the effect of random errors on process data [1]. Reconciliation of data can be achieved by various approaches, namely multivariate statistical-based DR technique like principal component analysis (PCA), deep learning and WLS technique. Multivariate statistical methods and deep learning techniques require a very high computation time, and when a large number of data is involved, having very high computation time is not feasible. While handling large number of data, decomposition of data would be difficult in PCA [2]. On the other hand, WLS is a very popular and effective data reconciliation technique used to eliminate errors in process data, and it has a very high computation speed [3]. In a well-defined system or process network, WLS is preferred to other data reconciliation techniques. In this paper, various improvizations have been implemented on the conventional WLS method. The modified methods, namely MA-WLS and E-WLS, have been applied to the measurement data to minimize the errors even further, and then, WLS estimator is implemented. In this way, the computational speed of WLS is maintained, and the errors are also reduced. A large dimension of data can be processed at a high speed with better accuracy. These techniques have been implemented on the measurement data, and their reconcilability has been compared and analyzed.

2 Data Reconciliation (DR) Techniques

Data reconciliation is used in data correction process which involves various steps like suppression of random errors, variable diversification, parameter estimation, model identification and gross error detection. In DR techniques, principles of conservation law and data redundancy are used to correct measurements [3, 4]. One of the main requirements to apply data reconciliation is that the errors should be normally distributed [3]. The main assumption is that no systematic errors are present as it affects the robustness of the reconciliation.

2.1 Generalized Measurement Model

In a chemical process, the measurements, which are acquired, usually contain random errors and gross errors. Hence, the measured value can be written as shown in Eq. (1) [5]:

$$y(j) = x(j) + \epsilon(j) \tag{1}$$

where $x(j)$ is the true value, $\epsilon(j)$ is the random errors, and $y(j)$ is the jth measurement of N variables present in the measurements. A few assumptions are considered about random errors [3], which are

1. $\epsilon(j) \backsim N(0, \sum\epsilon)$
2. $E[\epsilon(j)\epsilon(k)^T] = 0$, where $j \neq k$
3. $E[x(j)\epsilon(j)^T] = 0$

Assumption (i) indicates that the errors are normally distributed, (ii) and (iii) show that the errors are uncorrelated, and process variables are not corelated with the errors [3]. In this paper, WLS data reconciliation techniques have been implemented on the data to eliminate the errors present in the data and to obtain accurate process values. Conventional WLS as well as novel strategies like MA-WLS and E-WLS have been implemented. The complete DR process is explained in Fig. 1. The measurement data containing the errors are pre-processed with the MA or EWMA technique to reduce the errors present in the measurement data. The WLS estimator is then applied to the pre-processed data y_i, and the reconciled data are obtained. The results obtained from each of these techniques have been compared and analysed. These techniques have been discussed in the following sections.

2.2 Pre-processing Techniques

Moving Average (MA) In the moving average technique, the mean of specified number of past data values is taken where equal weight is given to each value [6]. Mathematically, it is expressed in Eq. (2) as follows:

$$y_{ma}(k) = \frac{1}{N^*} \sum_{i=k-N^*+1}^{k} y_i \tag{2}$$

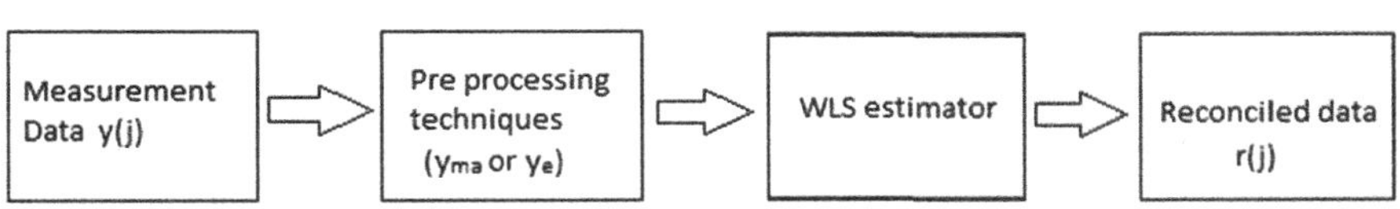

Fig. 1 Process flow diagram

y_{ma}—MA-WLS transformed value
y_i—Sample of measured variable
N^*—Number of samples that are being averaged
k—Sampling constant

Moving average method eliminates high frequency noise. It has also been implemented to eliminate the fluctuation in data [7]. This process is first applied on raw measured data. Then, WLS is implemented on the moving average data and analysed for reconcilability. MA-WLS technique is implemented on different N^* values to analyse its performance.

Exponentially weighted moving average (EWMA) In this method by improving the sample variance, more weight or less weight is given to the recent data [8]. It is used to detect changes in the process parameters.

$$y_e(k) = \alpha * y(k) + (1 - \alpha) * y_e(k - 1) \tag{3}$$

y_e—Exponentially transformed data
α—Time constant

Equation (3) indicates that it is the weighted sum of present measurement and transformed value of the previous instant. The value of α ranges from 0 to 1.

$\alpha = 1$: No transformation. ($y(k)$ is considered as transformed output).
$\alpha = 0$: The measurement is ignored.

Here, α values closer to 1 indicate that EWMA gives more weight to the recent variable, and when α is closer to 0, recent variable of data is given very less weight [8]. This method is applied to the raw data, before applying WLS technique to it.

Transformed Weighted Least Squares (WLS) Estimator It is assumed that any observation is composed of a true value along with some error which can be a random error or gross error. Random error can neither be predicted nor be explained accurately. It follows Gaussian distribution. The conventional approach to implement data reconciliation is the weighted least squares data reconciliation (WLS-DR) which is used to get accurate estimates of process variables from measurements which have noise added to it. Only random errors with known variance must be present. When gross errors are present in the measurement, WLS is not effective[9]. Gross errors spread the effect of biased measurement to the other values, and this is called smearing effect (spreading of errors throughout the data) [9]. The basic idea of this approach is to minimize the residuals in the nodes so that the estimates are much closer to true values. The objective function of conventional WLS problem can be described as the least squares between the measured data and the reconciled data [3, 5] that is

$$min \sum_{i=1}^{N} \frac{(r_i - y_i)^2}{\sigma_i^2} \tag{4}$$

In Eq. (4), r_i is the ith reconciled value, y_i is the ith pre-processed value after implementing MA or EWMA on measured data, and σ_i is the standard deviation of the measurement error. This equation can be represented in the matrix form in the following way [3, 5, 10]:

$$min[(R - Y)^T Q^{-1}(R - Y)] \tag{5}$$

$$St.AX = 0 \tag{6}$$

The objective function is solved using Lagrange multiplier, and the function for reconciled data using WLS is obtained that is shown in equation (7) [3, 5].

$$R = (I - QA^T(AQA^T)^{-1}A)Y \tag{7}$$

I is the unit matrix.
A is the constraint matrix.

3 Simulations and Results

The study of various data reconciliation techniques is implemented using a steam metering circuit. Steam networks represent a process network in which only material flows of a single component are involved. A steam metering circuit with 11 nodes and 28 streams (variables) has been considered [11, 12] in Fig. 2. In this case, the flow rate measurements have been considered. Inflow and outflow from each node have been indicated in Fig. 2.

The base values for the flow rates have been estimated from the mass balance equations and are shown in Table 1.
These base values are considered as true values (x) to which errors would be added in the form of process noise and random noise [12]. This modified data are the measured data (y). This process is implemented in Python environment. Process noise of variance 0.01 and mean 0 and the random noise with variance 1 are generated and added to the base values. This procedure is applied to generate 10,000 samples of data for each flow variable. The data generated are independently identically distributed data (i.i.d). The set of rules followed by this data is mentioned in Sect. 2.1. The material balance equations for the steam metering system can be written in matrix form as [3, 10]:

$$Ax(j) = 0 \tag{8}$$

where A is the constraint matrix which is derived from the direction of the flows to and from the nodes in the steam metering circuit, and $x(j)$ denotes true values. Constraint equations can be obtained by the mass flow balance for the inflow and outflow at each node. As there are 11 nodes, 11 constraint equations are obtained. For a node in Fig. 2, the general mass flow balance can be equated as shown in Eq. (9)

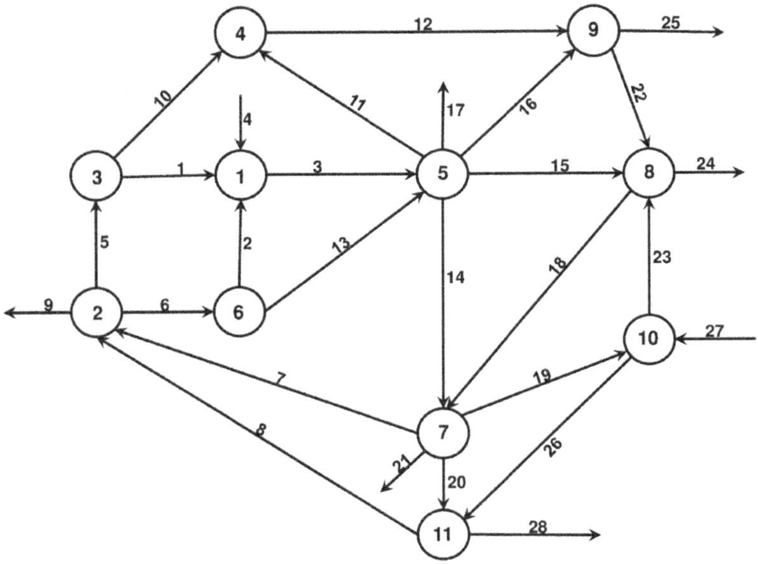

Fig. 2 Steam metering circuit

Table 1 Base values for all flow variables

Flow variables	1	2	3	4	5	6	7	8	9	10	11	12	13	14
Base values	10	10	30	10	20	20	30	20	10	10	10	20	10	10
Flow variables	15	16	17	18	19	20	21	22	23	24	25	26	27	28
Base values	10	5	5	40	5	5	10	20	20	10	5	20	45	15

$$\text{Node} = \text{Inflow} - \text{Outflow} \tag{9}$$

For node 1, the mass flow balance equation is represented in Eq. (10),

$$z_1 = F_1 + F_2 + F_4 - F_3 \tag{10}$$

where z denotes the node, and F denotes the flow variable [13]. Similarly, constraint equations can be obtained for the other ten nodes. From these constraint equations, constraint matrix A is obtained.

Then, DR techniques have been implemented on the measured data in Python environment. The efficiency of each technique has been assessed with respect to other methods. Various performance metrics are present to evaluate DR techniques, and in this paper, RMSE is considered as performance metric. It is observed that RMSE is the best metric for capturing the minute changes in noise present in the data [14]. The measurement data with the errors present are reconciled with a data reconciliation technique. Then, to check the effectiveness of the DR technique, the reconciled data are compared to the actual data containing no errors using RMSE as the performance matric. The lesser the RMSE value, more effective the DR technique [15]. When there is no RMSE after reconciliation, it indicates that all the errors have been successfully eliminated from the measurement data. This process is implemented for all the DR techniques mentioned in Sect. 2. In Table 2, the RMSE values for all the variables in the process network are tabulated for all the DR techniques. In the 'Raw data' column, the RMSE between the actual data and measurement data is present, before any DR technique is implemented. We can observe that once a DR technique is applied to the measured data, then the RMSE values reduce considerably. MA-WLS has been implemented for different sample sizes $N^* = 20$ and $N^* = 50$. The RMSE for all the techniques is analysed, and a detailed study is given in the following sections. E-WLS being the most effective DR techniques is further explored to increase its efficiency. The transformation process has been executed on the data for a range of α values from 0 to 1 with a step of 0.01. Then, the RMSE of the data has been obtained and compared. On analysis, it is confirmed that performance is best at $\alpha = 0.05$, refer Fig. 5.

In Fig. 3, the RMSE for reconciled data with respect to the actual data has been compared for all the techniques, across all the flow variables considered in the process network. It is observed that the MA-WLS and E-WLS techniques have the least RMSE values when compared with the other techniques. Hence they are most effective in reconciling the measured data. They are more effective than the conventional WLS technique.

In Fig. 4, the RMSE for all the reconciliation techniques has been plotted across all the magnitudes possible, for all the process variables present in the network. From this, it is observed that E-WLS achieves minimal RMSE consistently for all the magnitudes of flow variables. MA-WLS ($N^* = 50$) performs better than MA-WLS ($N^* = 20$). We can state that the reconcilability of E-WLS is very high when compared with the other methods.

E-WLS has further been analysed to optimize its performance. E-WLS has been implemented for various α values, and their performance is recorded in Fig. 5. From Fig. 5, we observe that E-WLS is most effective in eliminating random errors from measured data at $\alpha = 0.05$.

In Fig. 6, RMSE for flow variables with different base values has been plotted for all α values. There are 28 flow variables in our process network, denoted as F1, F2, F3, ... It can be observed for all base values that the RMSE is minimum at $\alpha = 0.05$ except for the flow variable with base value 5. We can state that E-WLS is the best out of all the data reconciliation techniques implemented previously.

Table 2 Performance of DR techniques

Flow variables/DR techniques	RMSE				
	RAW DATA	WLS	E-WLS	MA WLS (20)	MA WLS (50)
1	1.0021	0.5297	0.1373	0.0495	0.0728
2	1.0021	0.1827	0.0116	0.0508	0.2844
3	1.0021	0.5976	0.1637	0.0511	0.7993
4	1.0021	0.7007	0.1995	0.7111	0.8076
5	1.0021	0.6551	0.1275	0.0503	0.2109
6	1.0021	0.3830	0.0230	0.0996	0.3996
7	1.0021	0.2101	0.0899	0.1003	0.1897
8	1.0021	1.1161	0.2552	0.0980	0.6143
9	1.0021	0.2173	0.0543	0.0968	0.1719
10	1.0021	0.5683	0.0627	0.0995	0.1986
11	1.0021	0.4826	0.0298	0.1017	0.0945
12	1.0021	0.9145	0.2212	0.0975	0.4091
13	1.0021	0.9761	0.2530	0.0960	0.6999
14	1.0021	0.4522	0.0363	0.0967	0.8006
15	1.0021	0.2100	0.0607	0.0983	0.1150
16	1.0021	0.7331	0.127	0.1012	0.4346
17	1.0021	0.5109	0.0320	0.1476	0.0367
18	1.0021	0.9513	0.0666	0.1969	0.0238
19	1.0021	1.1241	0.1263	0.1957	0.5179
20	1.0021	1.2964	0.1941	0.1988	0.6314
21	1.0021	1.0504	0.1293	0.1994	0.1923
22	1.0021	0.9250	0.0625	0.1981	0.1980
23	1.0021	0.8827	0.0287	0.1972	0.0909
24	1.0021	1.7080	0.2526	0.2924	0.4036
25	1.0021	0.9950	0.0599	0.2896	0.5978
26	1.0021	1.1054	0.0348	0.2938	0.1102
27	1.0021	1.5258	0.0075	0.3907	0.4069
28	1.0021	1.3905	0.1286	0.4410	0.1012

4 Conclusion

In this paper, WLS, MA-WLS and E-WLS data reconciliation techniques have been implemented to reduce random errors from measured data. Conventional WLS method has been a widely accepted technique to eliminate errors, but it is effective only up to a certain extent. Hence, modifications have been devised and implemented to the WLS technique to further reduce the deviation of measured data from actual

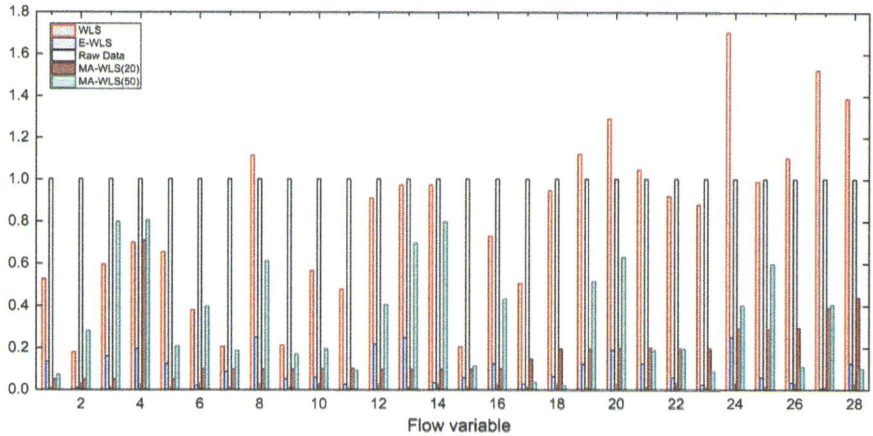

Fig. 3 Reconciliation results for all base values

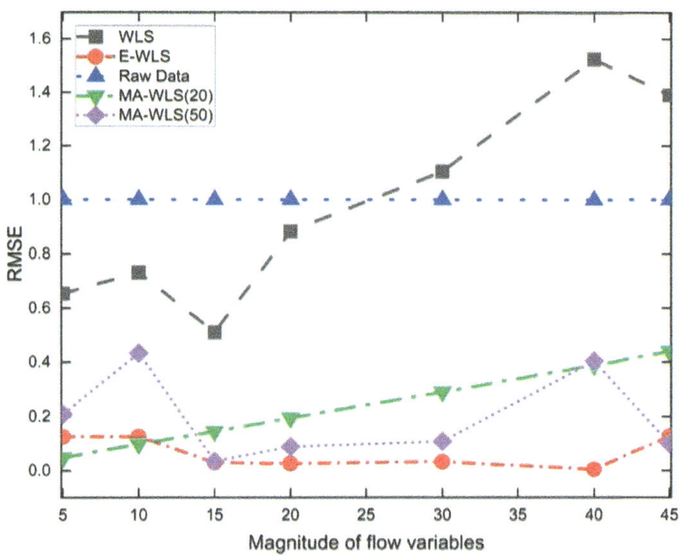

Fig. 4 Performance of DR techniques for varying magnitude

data. From analysis, it can be inferred that the modified techniques MA-WLS and E-WLS perform better than the conventional method. On further studying MA-WLS and E-WLS, it is analysed that E-WLS is more effective in eliminating errors from measured data. E-WLS is hence implemented for various α values to obtain the conditions at which optimum performance is established. This study is based on the assumptions that the errors are independently and identically distributed and have the same variance. In real-time systems, data might not necessarily have constant

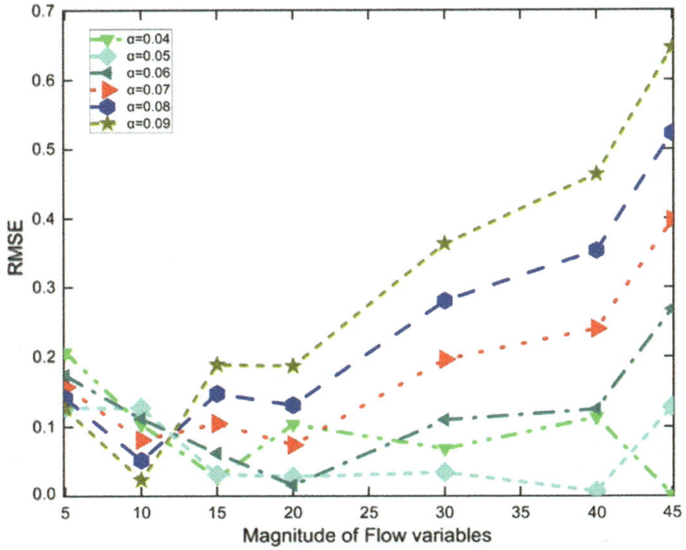

Fig. 5 Performance of E-WLS DR for different α values

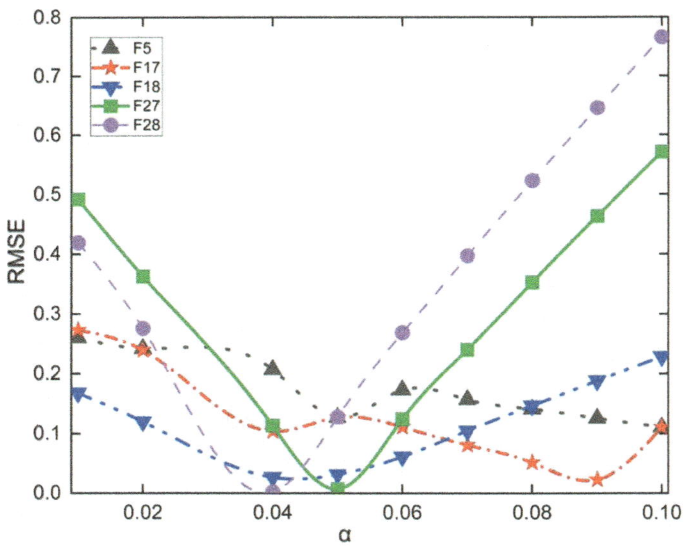

Fig. 6 Performance of E-WLS DR for flow variables across all α values

variance and will have gross errors as well. Gross error deduction and analysis of DR techniques considering data with non-uniform variance can be taken as future scope of the study.

References

1. S. Xie, C. Yang, X. Yuan, X. Wang, Y. Xie, A novel robust data reconciliation method for industrial processes. Control Eng. Pract. **83**, 203–212 (2019)
2. S. Lane, E.B. Martin, A.J. Morris, P. Gower, Application of exponentially weighted principal component analysis for the monitoring of a polymer film manufacturing process. Trans. Inst. Meas. Control **25**(1), 17–15 (2003)
3. W. Shengxi, Q. Ye, C. Chen, G. Xingsheng, Research on data reconciliation based on generalized T distribution with historical data. Neuro Comput. **175**, 808–815 (2016)
4. J. Chen, Y. Peng, J.C. Munoz, Correntropy estimator for data reconciliation. Chem. Eng. Sci. **104**, 1019–1027 (2013)
5. V.S. Yellapu, A.P. Tiwari, S.B. Degweker, Application of data reconciliation for fault detection and isolation of in-core self-powered neutron detectors using iterative principal component test. Progress Nucl. Eng. **100**, 326–343 (2017)
6. M. Hubert, E. Schmitt, A review of PCA-based statistical process monitoring methods for time-dependent, high-dimensional data, Department of Mathematics, December 20, 2013
7. R. Ouysse, Comparison of Bayesian moving average and principal component forecasts for large dimensional factor models, Australian School of Business Research Paper No. 2012 ECON
8. D.E. Seborg, T.F. Edgar, D.A. Mellichamp, F.J. Doyle III, *Process Dynamics and Control*, 3rd edn. (Wiley, Hoboken, 2010), pp. 320–321
9. X. Jiang, P. Liu, Z. Li, Data reconciliation and gross error detection for operational data in power plants. Energy pp. 1–10 (2014)
10. F.Z. Lingke, Y. Feng, Data reconciliation based on robust estimator and MT-NT method, in *Proceedings of the 35th Chinese Control Conference*, July 27–29 (Chengdu, China, 2016)
11. P. Zhang, G.R.Y. Wang, A new method of redundancy analysis in data reconciliation and its application. Comput. Chem. Eng. **25**, 941–949 (2001)
12. C. Varshith Reddy, J. Reddy Rishika, S. Ganesh, R. Jeyanthi, Principal component analysis-based data reconciliation for a steam metering circuit. Soft Comput. Sign. Process. pp. 619-626 (2019)
13. A. Rafiee, F. Behrouzshad, Data reconciliation with application to a natural gas processing plant. J. Nat. Gas Sci. Eng. **31**, 538–545 (2016)
14. T. Chai, R.R. Draxler, Root mean square error (RMSE) or mean absolute error (MAE)? Arguments against avoiding RMSE in the literature. Geosci. Model Dev. **7**, 1247–1250 (2014)
15. D.M. Prata, J.C. Pinto, E.L. Lima, Comparative analysis of robust estimators on nonlinear dynamic data reconciliation. Comput. Aided Chem. Eng. **25**, 501 (2008)

Data-Driven-Based Disruption Prediction in GOLEM Tokamak with Missing Values

Jayakumar Chandrasekaran, Surendar Madhawa, and J. Sangeetha

Abstract The sensor readings from the vessel of the Tokamak are inconsistent where some sensors always produce output; others provide output occasionally resulting in missing values on entire data. Typically, imputation methods are applied to both train and test data while training the model offline. But in the real-time application where the decision must be taken based on the instream sensor data, imputation techniques are not practical. Hence, in this paper, a data-driven approach on algorithms that inherently handle missing values and algorithms that have the provision to deal with missing values through a replacement or technique is employed. Individual, bagging and boosting algorithms are utilized to classify normal and disruption charges on the GOLEM Tokamak dataset, which consists of 117 normal and 70 disruptive shots. Boosting algorithms having an inbuilt feature to handle missing values provided better results amongst other algorithms and Categorical Boosting (CatBoost) with its ordered boosting feature gave the best metrics. Optimal thresholds for receiver operating characteristics (ROC) and precision–recall (PR) curves on the models are determined. The optimal PR values are utilized to get improved results. A comparison with the widely employed stand-alone machine learning algorithms and ensemble algorithms is illustrated. The results show the excellent performance of the CatBoost model with an $F1$ score of 0.943 with optimal PR values. The developed predictive model would be capable of warning the human operator with feedback about the feature(s) causing the disruption.

Keywords GOLEM Tokamak · Boosting · CatBoost algorithm · Optimal thresholds · Disruption prediction

J. Chandrasekaran · S. Madhawa · J. Sangeetha (✉)
School of Computing, SASTRA Deemed University, Thirumalaisamudram, Thanjavur, Tamilnadu 613401, India
e-mail: sangeetha@it.sastra.edu

M. Paprzycki et al. (eds.), *Intelligent Systems, Technologies and Applications*, Advances in Intelligent Systems and Computing 1353,
https://doi.org/10.1007/978-981-16-0730-1_9

1 Introduction

In a Tokamak, disruption can be defined as the uncontrolled and sudden loss of plasma current. When a disruption occurs, it leads to loss of thermal energy in a few milliseconds, which exposes the components facing the plasma to suffer from extreme thermal loads. Also, huge electromagnetic forces are experienced by the conductive structures. Before every disruption instability, there is a precursor phase that is characterized by the change in the state of plasma current. The change occurs due to discharge that evolves along with the instabilities of MHD. The dominant modes among the instabilities are modes (2, 1) which are often referred to as the tearing modes or kink modes in specific high-β plasmas.

Many researchers are working on developing disruption prediction algorithms for the available fusion reactors like are ADITYA [1, 2], SST-1 [3], ASDEX-U [4–6], Alcator C-Mod [7], EAST [8], DIII-D [9], JET [10–14], JT60-U [15, 16], J-TEXT [17] and NSTX [18]. The empirical causes leading to disruptions were recently examined in the Joint European Torus (JET). In case of unintentional disruptions [19, 20], the significant factors are mode-locking amplitude (20.1%), instabilities at edge radiations (51.8%), instability at vertical positions (3.9%), internal kink mode growth (4.4%) and very low safety factor Q (2.5%). The primary carrier or influencer for disruptions is mode-locking, which can be caused due to one of the three classes: error fields with low density (10.3%), locked neoclassical tearing mode (6.7%) and current ramp-up formed during instabilities at edge kink modes (3.1%). Our work addresses disruptions that occur due to factors such as mode-locking and instabilities at edge radiation.

Signals are not always received from all the sensors in the Tokamak reactor. In the GOLEM Tokamak, only nine sensors provide the output at all times, and five sensors provide output sporadically. The missing or absence of data in Tokamak is by design as some signals are not monitored always. Two approaches can be followed to build a model based on the sensor readings. The first is to eliminate the readings from the five sensors which provide occasional output and build a model on the nine sensors. The drawback of this method is that the missing values from the five sensors could provide useful information in building the model. Excluding the sensor readings would result in an inferior model that does not generalize with unknown data. The second method is to consider the output from all the sensors and perform imputation methods on the data to get a model. The imputations are applied on the train and test data separately to build a model where the entire test data is accessible. But in real time, the readings from the sensor in a GOLEM Tokamak arrive at around 50 microseconds interval, and all the test data are not available. Hence, the imputation methods are not practical for the GOLEM environment and hence are not taken into consideration.

In this work, algorithms, such as light gradient boosting machine (LGBM) [21], Xtreme gradient boosting (XGBoost) [22] and categorical boosting (CatBoost) [23], which inherently handle missing values are employed. The algorithms that do not handle missing values by itself, such as CART, random forest (RF) and logistics regression, Not a Number (NaN), are used to replace the missing values. The algorithms are trained after getting the best hyperparameters through grid search. The trained model is validated with test data, and optimal ROC and PR threshold values are obtained. Optimal ROC is used as a diagnostic tool to find the trade-off between true positive rate (TPR) and flse positive rate (FPR). Optimal PR values are used to enhance the performance of the models. The data is trained again with the optimal PR values and validated with the test data, and the results indicate improvement in the results of most of the algorithms. CatBoost performed the best amongst all the algorithms. The unique features of CatBoost are handling categorical values, generating symmetric trees and following ordered boosting to mitigate prediction shift (target leakage). The dataset derived from GOLEM Tokamak [24] is applied in this work. The derived 187 shots, 117 non-disruptive and 70 disruptive shots are utilized.

The remainder of the paper is organized as follows. Section 2 analyses the existing techniques used for fusion reactors. The basic setup of GOLEM Tokamak with the description of features is presented in Sect. 3. The methodology followed, and the details of Boosting algorithms are given in Sect. 4. In Sect. 5, the results obtained and the performance metrics are listed. Finally, Sect. 6 provides the concluding remarks and probable ideas for enhancing the work.

2 Related Work/Literature Overview

In DIII-D Tokamak [25], a machine learning (ML) algorithm that employs feedback in real time is implemented and tested to avoid tearing mode disruptions. The system is designed to prevent high-current disruptions by employing ramp down during tearing modes dynamically. The authors also compared other predicting algorithms, and it was found that extremely randomized tree, bagging and RF algorithms produced better results and provided excellent performance for the dataset used. The major drawback over this approach is it concentrates only on tearing mode disruptions. In J-Text, a disruption predictor based on artificial neural network (ANN) and anomaly detection is analysed. For a large training dataset with predominant disruption samples, excessive damage is faced by current Tokamak's. The authors claim that their method avoids damage and act as a good predictor. The authors utilized only non-disruptive shots to extract the hidden features using convolutional neural network and LSTM-based recurrent neural network. The model is designed as one class classification using only non-disruptive pulses for train and utilizing non-disrupted and disrupted pulses for testing. The authors claim that their system can predict 83% of disruptions with 18% false alarms, which is a limitation.

In [26], a single diagnostic system is known as electron cyclotron emission imaging (ECEi) from the DIII-D which is used for disruption prediction and avoidance. The authors proposed a deep convolutional neural network using long-range sequence learning idea for more computational efficiency. The authors claim that their technique has achieved an $F1$ score of 0.91 on individual time slices. The drawback of the proposed method is they have utilized a single diagnostic system instead of all available diagnostic techniques which are widely preferred by researchers around the globe, which makes the system vulnerable. A generic disruption predictor is proposed in which signal increments were modelled using a linear equation to which the input was plasma current lock signal in normalized mode. The authors claim that for disruption prediction in JET, the centroid approach is most reliable. In this proposed system, the valid alarms are at 90% and a false alarm rate of 8% with a mean warning duration of 443 ms.

In [27], the authors performed a survival analysis of a predictor using a statistical framework to time an event. A predictor was designed using RF classifier for prediction along with Kaplan–Meier estimator for survival probability analysis. The technique was applied to three plasma discharges of Alcator C-MOD and tested through triggering alarms of the threshold-based scheme. Utilization of a single statistical method for survival probability makes it application narrower. In [28], high-β plasma data is used for predicting disruptions in JT-60U that are based on sparse modelling using an exhaustive approach for SVM classifier. The SVM is used as a linear discriminator for two-class disruptive and non-disruptive pulses. The author used two different sets of features like instantaneous and time derivative features for modelling wherein, K-sparse modelling is incorporated to extract the best features for modelling a disruption predictor.

A real-time ML-based disruption predictor [29] was designed for DIII-D reactor using RF (DPRF) that is placed in a plasma control system in DIII-D. The DPRF system is made to run in real time for around 900 discharges. Based on optimized performance measures, the detected disruption ranges from 70 to 96% with an increased false alarm of 13–28%. The system produces a permissible result of 95% accuracy, but as mentioned by the author, they may not be suitable for ITER. A disruption predictor in JET is designed using a multidimensional linear model for disruption prediction. A data-driven model is derived from the interpretable linear equation for offline analysis and real-time monitoring system. The best subset of features to be applied for ML is selected using the genetic algorithm. A total of 1310 non-disruptive shot and 131 disrupted shots were used. SVM classifier is used for classification purposes producing a 94.66% of success rate with false alarm of 3.51%.

Disruption prediction in DIII-D and ALCATOR C-MOD [30–33] using ML techniques are performed using RF. The authors chose to use the same feature from two devices to predict the disruption by training on one device's data and testing on the other using the same classifier. The poor performance of the classifier is visualized in C-MOD while using molybdenum penetration. However, the limitation of the system proves that the universal classifier cannot be used for multiple devices trained for a specific system.

In [34], a ML-based technique for predicting Ayurveda-based constituents in the human body is performed. The authors created a dataset comprising of 807 healthy persons in the age group of 20–60 consisting of 62.1% men and 37.9% women. The random generator is used to randomly select the 405 persons where 324 samples are for train data and 81 for test samples. Oversampling is used to resolve the class imbalance issue. Various ML techniques like Naïve Bayes (NB), support vector machine (SVM), K-nearest neighbour (KNN), ANN and decision tree (DT) were used to train the model. On performing various techniques, CatBoost is found to produce a highly permissible result with the best score with precision, recall and F-score of around 0.95 and accuracy of 95%. In [35], a ML approach is applied with a delimited anti-jamming protocol for tracking vehicular traffic environments. CatBoost algorithm is utilized for predicting the location of jamming vehicles using VANETs. The proposed technique is found to produce an accuracy of around 97% when compared with XGBoost with K-Means, SOM, ANN and SVM techniques. The proposed ML technique achieved a 99.91% accuracy when connected with VANET for avoiding jamming effect in the Internet of Vehicle (IoV) system.

The authors [36] proposed a heart disease diagnosis technique using multiple classifiers. The XGBoost algorithm is used as a sub-classifier for processing massive ECG data that focus on heart diseases. The authors utilized the UCI dataset for their research and have used multiple classifiers for predicting the arrhythmia. Classifiers such as XGBoost, SVM, C4.5, naïve Bayes, logistic regression and RF were experimented. Even though C4.5 produced slightly better results, XGBoost is preferred due to its capability to handle imbalance data. An auto-insurance premium pricing is generated using the Markov-Modulated tree-based gradient boosting (MMGB) technique [37]. The auto-insurance policies in African countries of the Sahara region were provided on the basis of regulation and legislative dynamics. The proposed MMGB model is a tweedie general linear model (GLM)-based Poisson Gamma distribution where the rate varies with accident risk using Markovian process. The authors [38] proposed a hybrid regression algorithm with reinforced extreme gradient boosting (EGB) for examination, prediction and enhancement of crop production. The EGB is capable of creating new models by learning from the mistakes of their predecessor trees and by optimizing the loss function. The proposed technique is found to produce an accuracy of 94.15% compared to traditional models like ANN, deep Q-network, gradient boosting, RF and decision tree.

An ensemble-based technique for approval of life insurance based on risk assessment is performed using the ML technique [39]. The authors utilized a real-world dataset comprising of 128 attributes to study the risk criteria associated with an applicant for the policy. ANN and XGBoost algorithms were employed with optimized hyperparameters. The results were compared using quadratic weighted Kappa score, which shows XGBoost produced the best results. A ML-based technique using a recommender system for passengers looking for a bicycle in their nearest available dock stations is proposed [40]. The authors utilized a SMVP-based real-world traffic flow dataset that includes the temporal, spatial, historical and meteorological factors. The K-means algorithm is used along with traditional factors to form a new cluster station. The distance between the two stations was utilized to avoid imbalance. XGBoost is used to train the models to provide public bicycle traffic flow to users. The authors claim that their technique implemented in SMVP has improved by 25.58% in Hangzhou when compared with the traditional stacking and single model ML approaches. A study model for prediction of credit card users to predict whether they pay their debts or not is performed using various ML techniques [41]. Widely proven algorithms such as ANN, classification and regression tree (CART), support vector machine and C4.5 algorithms are used. Real-world dataset archived from a bank in Taiwan consisting of 10,713 customer records of relevant information were utilized as major attributes. Cross-validation and hold-out techniques for dividing the dataset into train and test are demonstrated. The CART algorithm provided an accuracy of 86% in predicting the customer's payment towards next month.

The objective of this work is to implement classification of normal and disruption readings from all the sensors, which includes occasional missing data (by the design of the GOLEM Tokomak reactor). Machine learning algorithms with missing values are employed in both train and test datasets. Imputation techniques are not used as it would not be applicable in a practical environment where the instream sensor readings are uncertain. Further, a minimum number of data (shots) are required for imputation, which is not feasible, and hence, imputation methods are not considered in this work. Instead, in our implementation, for most of the algorithms, the missing values are deemed as NaN and ensemble learners like LGBM, XGB and CatBoost which has an inbuilt mechanism to deal with missing values are employed. Among all algorithms, CatBoost performed the best. Further, the optimal threshold for ROC and PR curves are found, and the models are trained again with the new optimal threshold value as the class discriminator. Models trained with optimal PR values showed improved results in comparison with the threshold fixed by the algorithms.

3 GOLEM Tokamak

GOLEM Tokamak [24, 42] is commissioned at the Czech Technical University in Prague with its exclusive benefit of remote handling system through an Internet connection. The GOLEM Tokamak is operational at the Faculty of Nuclear Physics and Physical Engineering (FNPPE), Czech Technical University in Prague. The major radius of the Tokamak vessel is $R = 0.4$ m and minor radius $b = 0.1$ m. GOLEM Tokamak is a stainless-steel vessel with a well-equipped poloidal limiter (Molybdenum) of radius $a = 0.085$ m. The Tokamak is provided with an established set of the diagnostic system, to measure the plasma current, loop voltage, visible emission and toroidal magnetic field. Figure 1 depicts the schematic representation of GOLEM Tokamak, which is well-equipped with an array of Mirnov coils, bolometers, a visible spectrometer, a fast camera, and 12 Langmuir probes in a radial array. GOLEM Tokamak functions at a determined toroidal magnetic field of a range up to 0.8 T. The central electron temperature is less than 100 eV, with maximum plasma density average ~1019 m^{-3}, the maximum pulse length is around $t < 30$ ms.

3.1 *Signals in GOLEM Tokamak*

GOLEM Tokamak [44] is functional at the Faculty of Nuclear Physics and Physical Engineering (FNPPE), Czech Technical University in Prague with its exclusive

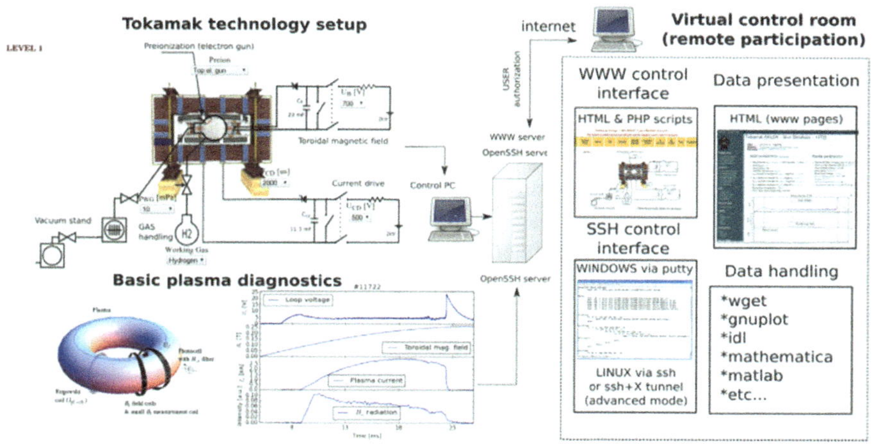

Fig. 1 Schematic representation [43] of GOLEM Tokamak experimenter

benefit of handling remote system through the Internet. The Tokamak vessel has a major radius of $R = 0.4$ m and minor radius $b = 0.1$ m. GOLEM Tokamak is a stainless-steel vessel on a well-equipped poloidal limiter (Molybdenum) of radius $a = 0.085$ m. The Tokamak is provided with an established set of the diagnostic system to measure signals such as plasma current, loop voltage, visible emission and toroidal magnetic field. Figure 1 depicts the schematic representation of the GOLEM Tokamak. The reactor is equipped with an array of Mirnov coils, bolometers, a visible spectrometer, a fast camera and twelve Langmuir probes in a radial array. GOLEM Tokamak functions at a determined toroidal magnetic field of a range up to 0.8 T. The temperature of the central electron is below 100 eV, with maximum plasma density average ~1019 m^{-3}, the length of the maximum pulse is around $t < 30$ ms. Table 1 gives the list of signals that are considered for classification on GOLEM Tokamak.

Table 1 Signals in GOLEM Tokomak

S. No.	Signal description	Acronym	Units
1	Total input power	P_{total}	W
2	Plasma current	I_{plasma}	A
3	Plasma density	Dens	m^{-3}
4	Line integrated density	L_{Dens}	m^{-2}
5	Electron temperature	Te_Probe	T [ev]
6	Safety factor q_{95}	q_{95}	s^{-1}
7	Greenwald density	$ne_{\text{Greenwald}}$	D
8	Toroidal magnetic field	B_t	T
9	Vertical plasma position	PVP	Z_m
10	Horizontal plasma position	PHP	Z_m
11–14	Mirnov coil (4-coils)	Rad	W

4 Methodology

Figure 2 shows the functional processes of this work. The missing values constitute around 15% and 13% from the train and test datasets, respectively. Among the 187 shots, 132 shots consisting of 82 normal and 50 disruptive shots are used for training. Individual algorithms such as CART, C4.5 and logistics regression along with bagging algorithms such as RF and boosting algorithms such as LightGBM (LGBM), Xtreme gradient boosting (XGB) and CatBoost are employed. Grid search is utilized for all the algorithm to get a close approximation of the best hyperparameters. From the results obtained, the optimal threshold for PR values is determined. The algorithms are then trained with optimal PR values to get better results. The validation of the resulting model is performed on the unseen data. The test data or unseen data contains 55 shots that include 35 normal and 20 disruptive shots. Since imputation methods are not used, results are obtained even with one shot of test data. To get consistency with the results, both default and optimal models utilize the same dataset for training and testing, and the comparison is listed.

4.1 *Gradient Boosting*

Robert Schapire [45] introduced the concept of combining several weak learners to form a single strong learner that is referenced as ensemble learning. Techniques such as voting, average, weighted average and stacking utilize simple algorithms as weak learners. Bagging and boosting are special techniques where bagging generates trees

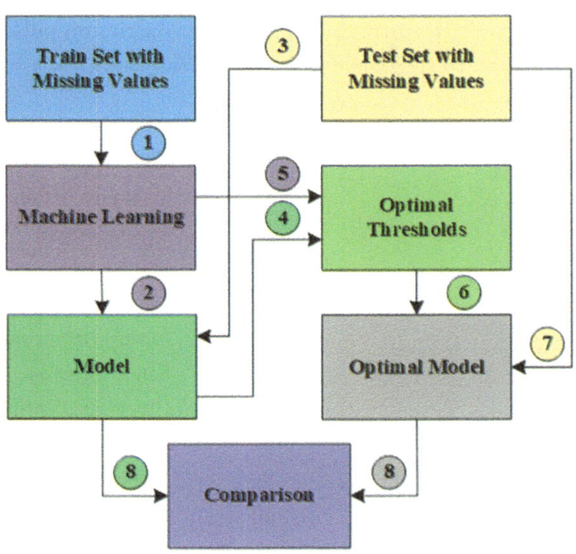

Fig. 2 Workflow to handle missing values

in parallel while boosting forms trees in an additive manner. Gradient boosting is an ensemble algorithm in which several weak learners combine to provide results that are effective in several applications.

For the GOLEM dataset, $D = \{(x_i, y_i)\} \quad i = 1, \ldots, m$, where $x_i = (x_i^1, \ldots, x_m^n)$ is a vector of m instances and $y_i \in R$ is the target categorical response and label encoded as binary. The instances (x_i, y_i) are independent, and the objective of the learning algorithm is to determine the train function $F(x_i)$ or $\hat{y}_i$, the predicted probability.

Each tree t or weak learner attempts to minimize the residual error resulting from the previous tree to create a perfect fit for the training data. In gradient boosting, a leaf is first constructed after which subsequent trees $(t = 1, 2, \ldots, T)$ are generated. The predicted probability of the leaf is given by

$$F_0(x) = \arg\min \sum_{i=1}^{m} \mathcal{L}(y_i, \hat{y}_i) \tag{1}$$

where y_i, $\hat{y}_i$ are observed and predicted probability outputs, respectively.

For the binary classification problem in this work, the initial residual error is the difference between the predicted probability and observed probabilities for each sample of the input. For $i = i, \ldots, m$

$$r_{it} = -\left[\frac{\partial \mathcal{L}(y_i, \hat{y}_i)}{\partial \hat{y}_i}\right]_{F(x)=F_{t-1}(x)} \tag{2}$$

In the GOLEM dataset, the observed probabilities are 1 for normal and 0 for disruption data. The predicted probabilities that minimize the loss function for all the input samples are calculated using the logistic function.

For the first tree $t = 1$, the tree $h_t(x)$ is formed with a training set $D = \{(x_i, r_{it})\}; \quad i = 1, \ldots, m$.

$$h_t(x) = \sum_{j=1}^{J_t} b_{jt} 1R_{jt}(x) \tag{3}$$

$1R_{jt}(x)$ is known as the indicator function and equals to 1 if $x \in R_{jt}$ and b_{jt} are the leaf values.

$$1R_{jt}(x) = \begin{cases} 1, & \text{if } x \in 1R_{jt}, \\ 0, & \text{if } x \notin 1R_{jt}. \end{cases}$$

$$\hat{y}_t = \arg\min_{\hat{y}_i} \sum_{i=1}^{m} \mathcal{L}\big(y_i, F_{t-1}(x_i) + \hat{y}_i h_t(x_i)\big) \tag{4}$$

Solving Eq. 3 for Taylor's second-order polynomial and simplifying we get

$$\hat{y}_t = \frac{\sum \text{Residual}_i}{\sum(\text{Pre.Prob.}_i \times (1 - \text{Pre.Prob.}_i))} \quad (5)$$

Prev. Prob. denotes prior probability.

The scaled output from the first tree is added to the prediction probability the initial leaf to update or boost the prediction values.

$$F_t(x) = F_{t-1}(x) + \alpha \sum_{j=1}^{t} \hat{y}_{jt} h_t(x) \quad (6)$$

The scaling factor α is known as the learning rate, which achieves a small step improvement in the prediction. Subsequent trees follow the same procedure as the first tree to update the prediction values, and the process repeats until the desired fit is obtained or a fixed number of trees or iterations are processed. The small steps in the learning rate result in better Prediction with a test dataset and hence avoid overfitting or high variance.

4.2 CatBoost

CatBoost, a variant of boosting algorithm functions similar to gradient boosting except for (a) uniquely encoding the categorical features and (b) mitigating the prediction shift issue through ordered boosting. Since the GOLEM dataset does not contain any categorical features, dealing with it is not discussed in this work. Ordered boosting, which is based on random permutation, minimizes the effect of prediction shift. The limitations of CatBoost are (a) it cannot handle sparse data, and with predominant numerical features, (b) the training is higher than other boosting algorithms. The GOLEM dataset employed in this work does not have sparse data as hence the limitation does not apply. The dataset contains only numerical features, but the training time is around 40 s which is excellent considering the results and the fact that training is done offline.

4.2.1 Handling Missing Values

Except for LGBM, XGBoost and CatBoost, the missing values in all the algorithms were considered as NaN. LGBM and XGBoost inherently handle missing values as NaN with the option of replacing it with a float value (XGBoost) or zero (LightGBM). In CatBoost, missing values can be replaced by either the minimum or maximum values in comparison with all the other values and a split that distinguish the missing values is taken into account when choosing trees.

4.2.2 Symmetric Trees

CatBoost, unlike any other ensemble algorithms, generates oblivious (forget prior decisions) or symmetric decision trees [46]. Through empirical results, it has been verified that the symmetric tree helps in getting the optimized hyperparameters. The variation in the results with and without grid search or random search is minimal. The weak learners in the CatBoost ensemble are made even weaker through symmetric trees and hence generalization to new data, and lower variance is achieved. Typically, symmetric trees ensure that the algorithm converges default hyperparameters to provide stable output and the final model is achieved in the first run itself.

In symmetric trees, the index of each leaf index is encoded as a binary vector with its length equaling the depth of the tree. All the float and categorical features are binarized to determine the model predictions. All the binary features of all instances are saved in a continuous vector V and the leaf values in float vectors of size 2^d, d is the depth of the tree. A binary vector $\sum_{e=0}^{d-1} 2^e.V(x, f(t, e))$ is built to determine the leaf index, for instance, x of tth tree, where $f(t, e)$ are the total binary features from tth tree on depth e. $B(x, f)$ is the binary feature f on instance x that is read from the vector V, and faster output can be obtained by parallelizing the vectors. The advantages of using symmetric trees are: (a) Tree construction is restricted to one feature split per level which reduces the complexity of the algorithm and (b) Only a single feature split is carried out instead of the typical k splits on numerical features which makes the training process faster.

4.2.3 Ordered Boosting

Classical boosting is prone to overfitting, which affects the performance of the model. The leaf value, which is the estimate of the gradient of all data in classical boosting, is biased as the estimate is made on the same data on which the model was built. The bias leads to a shift in the distribution of the gradients estimate in any domain when compared to the true distribution of the gradients [47]. The working of gradient boosting algorithms such as XGBoost and LightGBM (compared with CatBoost in this work) is based on the following stages:

i. Selecting a tree structure and determining the leaves once the tree construction is finalized.
ii. The algorithm iterates through various splits to select the best tree construction and builds the tree with the selected splits.
iii. The values and scores are set in the leaf's and trees, respectively, and the best split is chosen.

The leaf values in all the stages are calculated as gradients [22] or Newton approximations. In CatBoost, the first stage is operated in a different way to utilize unbiased estimates of the Gradient, and the next two stages comply with the functioning of XGBoost and LightGBM. Algorithm 1 describes the ordered boosting procedure, and Algorithm 2 details the steps followed in building a tree.

Let F^t be the model produced after constructing the first t trees and $h^t(x_k, y_k)$ be the gradient on kth training instance after constructing t trees. The model F^t has to be trained without the observation x_k to make the gradient $h^t(x_k, y_k)$ corresponding to F^t unbiased. Since unbiased gradients are required, no instances might be used for training F^t, which is impossible to process. To mitigate the issue, a separate model M_k is trained for each example x_k which is never updated using a gradient estimate. With the model M_k, , the gradient on x_k is estimated that is applied to score the resulting tree.

Algorithm* 1: *Ordered Boosting

input: $\{(x_k, y_k)\}_{k=1}^{m}, I$

$\sigma \leftarrow$ *random permutation of* $[1, m]$

$M_i \leftarrow 0$ *for* $i = 1, \ldots, m$

for $t \leftarrow 1$ *to* I:

for $i \leftarrow 1$ *to* m:

$r_i \leftarrow y_i - M_{\sigma(i)-1}(x_i)$

for $i \leftarrow 1$ *to* m:

$\Delta M \leftarrow LearnModel\big((x_j, r_j): \sigma(j) \le i\big)$

$M_i \leftarrow M_i + \Delta M$

return M_m

Figure 3 depicts a part of the symmetrical CatBoost trees generated with a tree depth of six. In CatBoost, all models of tree M_t have the same construction and s random permutations of the training dataset are generated to achieve stable and effective output. A random permutation is sampled to obtain gradients, and the same permutation method is employed for the computation of statistics on categorical features. The different models M_t obtained through training on each permutation σ result in low variance.

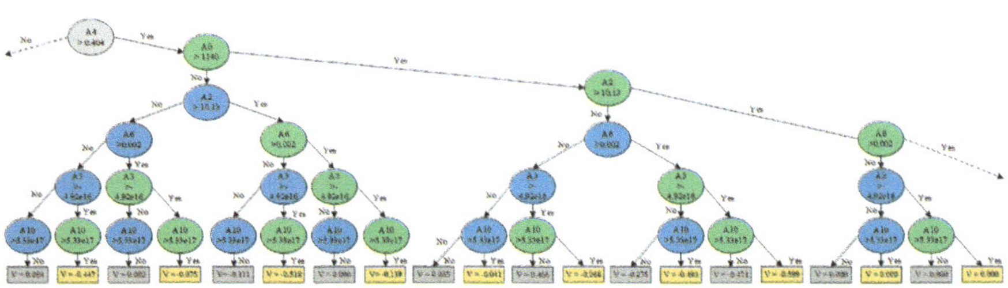

Fig. 3 Part of the generated CatBoost tree

Algorithm 2: CatBoost – Building a Tree

input: $M, \{(x_k, y_k)\}_{k=1}^{m}, \alpha, L, \{\sigma_i\}_{i=1}^{s}, Mode\ grad \leftarrow Calc.Gradient(L, M, y)$

$r \leftarrow random\ (1, s)$

  if $Mode = Plain$

    $G \leftarrow (grad_r(i)\ for\ i = 1,..,m)$

  if $Mode = Ordered$:

    $G \leftarrow (grad_{r,\sigma_r(i)-1}(i)\ for\ i = 1,..,m)$

$T \leftarrow emptytree$

  foreach *step of top – down procedure*

    foreach *candidate split c*:

      $T_c \leftarrow add\ split\ c\ to\ T$

      if $Mode = Plain$:

        $\Delta(i) \leftarrow avg\ (grad_r(p)\ for\ p: leaf_r(p) = leaf_r(i))\ for\ i = 1,..,m$

  if $Mode = Ordered$:

        $\Delta(i) \leftarrow avg\ (grad_{r,\sigma_r(i)-1}(p)$ for $p: leaf_r(p) = leaf_r(i), \sigma_r(p) < \sigma_r(i))\ for\ i = 1,..,m$

      $loss(T_c) \leftarrow cos(\Delta, G)$

    $T \leftarrow argmin_{T_c}(loss(T_c))$

if $Mode = Plain$:

    $M_{r'}(i) \leftarrow M_{r'}(i) - \alpha\ avg\ (grad_{r'}(p)\ for\ p: leaf_{r'}(p) = leaf_{r'}(i))\ for\ r' = 1,..s, i = 1,..,m$

if $Mode = Ordered$:

    $M_{r',j}(i) \leftarrow M_{r',j}(i) - \alpha\ avg\ (grad_{r'}(p)\ for\ p: leaf_{r'}(p) = leaf_{r'}(i), \sigma_{r'}(p) \leq j)/$
    $for\ r' = 1,..s,\ i = 1,..,m, j \geq \sigma_{r'}(i) - 1$

return T, M

4.2.4 Prediction Shift and Analysis

Gradients at each step are estimated utilizing the target values of the same data on which the current model F^{t-1} was built. For a training instance x_k, the conditional probability $F^{t-1}((x_k) \mid x_k)$ is shifted from the test data distribution $F^{t-1}((x) \mid x)$. The shift in prediction or target leakage in the conditional distribution of the gradient affects the generalization of the trained model to unknown or new data. The prediction shift is predominant in small datasets. Since prediction shift is present in both regression and classification tasks, a simple regression task is taken for analysis for which the loss function is $\mathcal{L}(y, \hat{y}) = (y - \hat{y})^2$. The residual function is given by $r^{t-1}(x_k, y_k) := y_k - F^{t-1}(x_k)$. If only two features ($m = 2$) are considered, then x^1, x^2 are the independent and identically distributed Bernoulli random variables that can take only two values ($0\ and\ 1$) with $p = \frac{1}{2}$, $y = f^*(x) = c_1 x^1 + c_2 x^2$. If $N = 2$ (steps of gradient boosting) with tree $depth = 1$ (stumps) and $\alpha = 1$ (step size) then the model obtained is $F = F^2 = h^1 + h^2$. Without loss of generality, it is assumed that h^1, h^2 are based on x^1, x^2, respectively, when $|c_1| \rangle |c_2|$.

4.2.5 Computational Complexity

For each permutation σ, different T models M_t are trained and to build a tree with m instances $O(m^2)$ approximations are required for storage and recalculations. For each model M_t, $M_t(x_1), \ldots, M_t(x_t)$ have to be updated which results in $O(s.m^2)$ complexity. To reduce the complexity, instead of storing and updating $O(m^2)$ values for $M_t(x_j)$, $M_t^{'}(x_j)$ values are stored where $t = 1, \ldots, [\log_2(m)]$, $j < 2^{t+1}$ and $M_t^{'}(x_j)$ is the approximation for instance j for the first 2^t trees. The complexity for calculating gradient and updating models for t iterations is reduced to $O(s.m)$. The complexity for building all the T trees is $O(|C|.m)$ while calculating all leaf values is $O(m)$. The number of predictions is given by

$$M_t^{'}(x_j) < \left(\sum_{0 \leq \log_2(n)} 2^{t+1} \right) < 4n \quad (7)$$

Thus, the gradient on the instance x_k for selecting a tree is based on the approximation $M_t^{'}(x_j)$ where $t = [log_2(k)]$.

5 Results and Discussion

The investigation was conducted on the Tokamak dataset with training—testing split of 70–30%, thereby utilizing 132 shots for training and 55 shots for testing. In general, ensemble classifiers performed better in comparison with individual algorithms. Missing values were considered as NaN in all the algorithms except CatBoost, where the minimum of all the values replaces the missing values.

5.1 CatBoost—Tuning and Training

Figure 4 shows grid search results that were used to find the optimized set of hyperparameters as per the recommendation from the developers of CatBoost. The number of estimators (trees), the tree depth, the learning rate, L2 regularization and bagging temperature. Except for L2 regularization, other hyperparameters showed significant changes in terms of log loss. The values with minimum log loss are employed for training and further tuned to get maximum results. However, it should be noted the sum of false positives and false negatives with optimized hyperparameters are less only by about 30. The default hyperparameters gave better results in comparison with other algorithms which prove that CatBoost does not require much tuning. Figure 5 shows the train and validation plot for the CatBoost algorithm with the tuned hyperparameters.

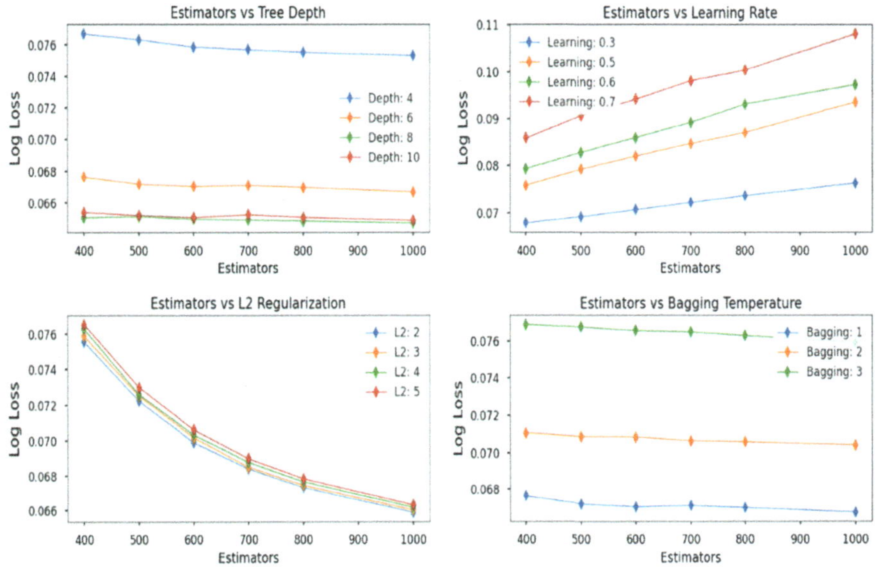

Fig. 4 Grid search results of CatBoost

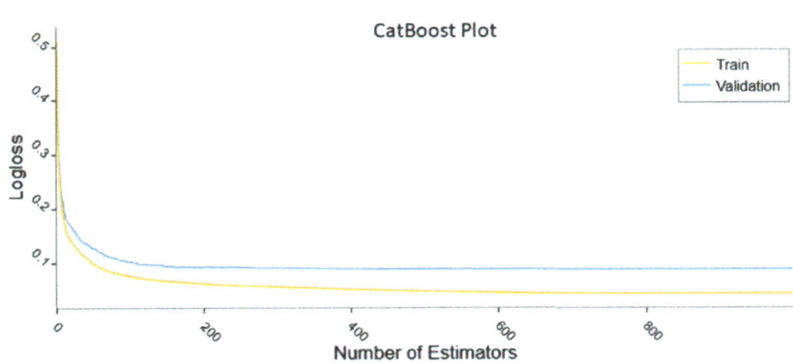

Fig. 5 Train and validation plot of CatBoost

5.2 Optimal Threshold

Many algorithms can interpret definite class prediction from the class probabilities known as the decision threshold or threshold. Default threshold may not be optimum due to factors such as skewness in class distribution, the priority of classification among the classes and insufficient calibration by the algorithms. Tuning the thresholds to find optimal values can mitigate the issues and improve the performance of the model. The typical threshold value for the normalized predicted probabilities is

0.5 or values between 0 and 1. For binary classification, values of predicted probabilities greater than 1 are assigned to *class 1* and values lesser than 1 are assigned to *class 0*.

5.2.1 ROC Threshold

For a range of increasing thresholds, ROC curve gives a predicted probabilities plot to find the best balance between the FPR (X-axis) and TPR (Y-axis).

$$\text{FPR} = \frac{\text{False Positives}}{(\text{False Positives} + \text{True Positives})} \tag{8}$$

$$\text{TPR/Sensitivity/Recall} = \frac{\text{True Positives}}{(\text{False Positives} + \text{False Negatives})} \tag{9}$$

Area under curve (AUC) values of the ROC curve represent the diagnostic capacity of the model in a single value. Typically, AUC value greater than 0.9 is considered good for a classifier. The optimal ROC threshold can be found with TPR (sensitivity/recall) and specificity (selectivity/true negative rate) metrics.

$$\text{Specificity} = \frac{\text{True Negatives}}{(\text{False Positives} + \text{True Negatives})} = 1 - \text{FPR} \tag{10}$$

Geometric mean (G-mean) gives optimal ROC threshold value is calculated as

$$\text{G - mean} = \sqrt{\text{Sensitivity} \times \text{Specificity}} \tag{11}$$

5.2.2 Precision–Recall (PR) Threshold

A classifiers performance in terms of the minority class is defined by the PR curve where precision is given by

$$\text{TPR/Sensitivity/Recall} = \frac{\text{True Positives}}{(\text{True Positves} + \text{False Positives})} \tag{12}$$

The definite PR values for increasing thresholds are calculated and plotted with recall (x-axis) and precision (y-axis) values for the corresponding thresholds. Figures 6 and 7 depict the ROC and PR curves, along with the optimal threshold and AUC values for all the algorithms employed in this work. The optimal PR threshold value obtained can be used to determine the class labels and improve the performance of the classifier, which would improve the $F1$-measure.

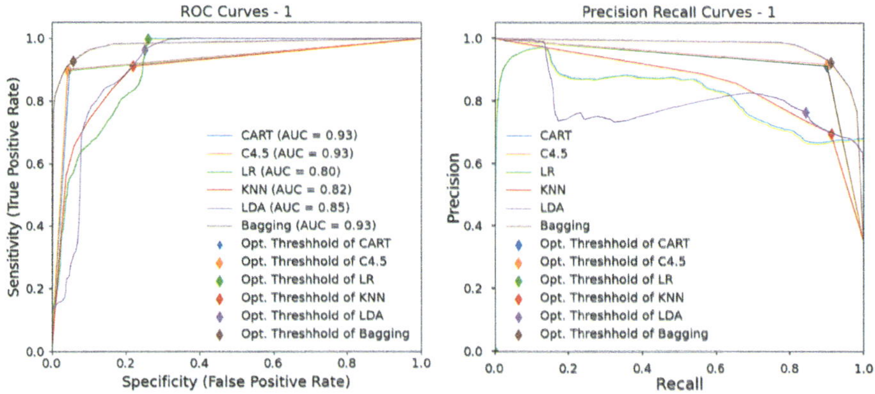

Fig. 6 ROC and PR curves of Algorithms-1

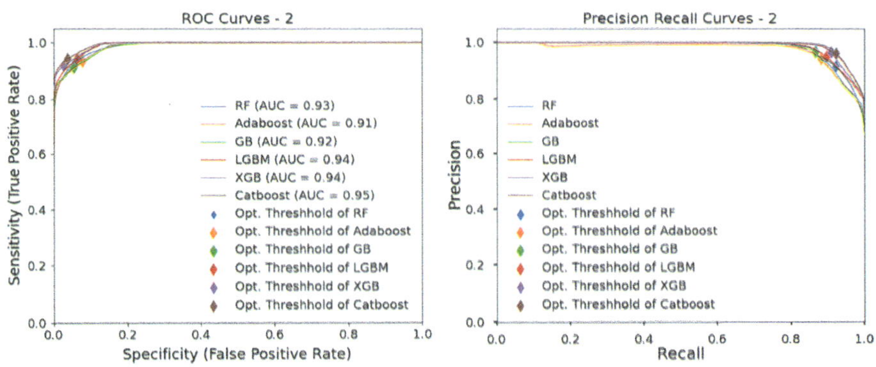

Fig. 7 ROC and PR curves of Algorithms-2

$$F1\text{ - measure} = 2 \times \left(\frac{\text{Precision} \times \text{Recall}}{\text{Precision} + \text{Recall}} \right) \tag{13}$$

Table 2 presents a results comparison of the machine learning algorithms. It is shown that employing the optimal PR threshold value to distinguish the class labels gave improved results for most of the algorithms. Only CART and logistics regression have been already tuned to the optimal value, and hence, the values are the same. It is seen from the results that CatBoost outperformed all the algorithms. The first nine algorithms in Table 4 were simulated with MATLAB with all the missing values considered as NaN. Scikit-learn machine language library was used for LGBM, XGBoost and CatBoost simulation as the algorithms inherently handle missing values.

Table 2 Comparison of results with optimal thresholds

Algorithm	Precision	Recall	$F1$	Optimal PR threshold	With optimal PR threshold values		
					Precision	Recall	$F1$
CART	0.913	0.899	0.906	0.880	0.913	0.899	0.906
C4.5	0.920	0.901	0.910	0.890	0.920	0.906	0.913
Logistics regression	0.717	0.766	0.740	0.717	0.717	0.766	0.740
KNN	0.804	0.740	0.771	0.200	0.742	0.835	0.786
LDA	0.706	0.897	0.791	0.520	0.765	0.841	0.801
Bagging	0.946	0.881	0.912	0.600	0.964	0.869	0.914
RF	0.938	0.896	0.916	0.471	0.946	0.897	0.921
AdaBoost	0.750	0.997	0.856	0.503	0.946	0.883	0.913
Gradient boosting	0.952	0.872	0.910	0.526	0.970	0.865	0.914
LGBM	0.943	0.904	0.923	0.597	0.937	0.909	0.924
XGBoost	0.971	0.904	0.936	0.482	0.967	0.908	0.937
CatBoost	0.957	0.926	0.941	0.533	0.963	0.923	0.943

6 Conclusion and Future Works

Disruptions in a Tokamak system must be detected at the earliest to avoid significant damage to the components and the vessel. Due to the nature of the Tokamak, the absence of data is a process by design and neglecting the signals is not desirable. The imputation methods are not appropriate, and approaches dealing with such a scenario are presented in this work. The missing values are substituted by NaN for widely used algorithms which provided good results. Algorithms that by nature deal with the absence of data such as LGBM, XGB and CatBoost fared better than NaN substitution. Though categorical features are not used, CatBoost demonstrated superior performance over other algorithms because of its unique feature in constructing the trees. It can be ascertained that ordered boosting is the reason for the efficiency of CatBoost. The stability of our model is substantiated with its minor accuracy variation over thirty iterations of train and test. A comparative study comprising decisive metrics executed over the standard train–test split of the dataset is demonstrated with often used algorithms. However, it must be noted that the stacking is implemented on the GOLEM Tokamak dataset. Performance of CatBoost might vary with other datasets and may not generalize to other applications. Probable future works are (a) experimenting CatBoost classifier in a real-time environment and datasets from other domain (b) to implement neural networks with more data (c) operate on raw sensor signals and provide an end-to-end solution through deep learning techniques (d) exploring the transformation of the dataset to a time-series forecasting problem (e) determining effective ways to give feedback to the plasma control system.

References

1. A. Sengupta, P. Ranjan, Prediction of density limit disruption boundaries from diagnostic signals using neural networks. Nucl. Fusion **41**, 487–501 (2001). https://doi.org/10.1088/0029-5515/41/5/302
2. A. Sengupta, P. Ranjan, Forecasting disruptions in the ADITYA Tokamak using neural networks. Nucl. Fusion **40**, 1993–2008 (2000). https://doi.org/10.1088/0029-5515/40/12/304
3. A. Sengupta, P. Ranjan, Identification of plasma parameters and optimization of magnetic sensors in the superconducting steady-state Tokamak-1 using neural networks. Fusion Technol. **39**, 1–17 (2001). https://doi.org/10.13182/FST01-A146
4. R. Aledda, B. Cannas, A. Fanni, A. Pau, G. Sias, and the A.U. Team, Improvements in disruption prediction at ASDEX Upgrade. Fusion Eng. Des. **96–97**, 698–702 (2015). https://doi.org/10.1016/j.fusengdes.2015.03.045
5. B. Cannas, A. Fanni, G. Pautasso, G. Sias, P. Sonato, An adaptive real-time disruption predictor for ASDEX upgrade. Nucl. Fusion. **50** (2010). https://doi.org/10.1088/0029-5515/50/7/075004
6. B. Cannas, A. Fanni, G. Pautasso, G. Sias, Disruption prediction with adaptive neural networks for ASDEX Upgrade. Fusion Eng. Des. **86**, 1039–1044 (2011). https://doi.org/10.1016/j.fusengdes.2011.01.069
7. C. Rea, R.S. Granetz, K. Montes, R.A. Tinguely, N. Eidietis, J.M. Hanson, B. Sammuli, Disruption prediction investigations using machine learning tools on DIII-D and Alcator C-Mod. Plasma Phys. Controlled Fusion **60**, 1–13 (2018). https://doi.org/10.1088/1361-6587/aac7fe
8. R.S. Granetz, C. Rea, K.J. Montes, R.A. Tinguely, N. Eidietis, O.M. Meneghini, D.L. Chen, B. Shen, B.J. Xiao, Machine Learning for Disruption Warning on Alcator C-Mod DIII-D, and East Tokamas. Nucl. Fusion **59**, 1–8 (2019)
9. C. Rea, K.J. Montes, K.G. Erickson, R.S. Granetz, R.A. Tinguely, A real-time machine learning-based disruption predictor in DIII-D. Nucl. Fusion **59**, 1–10 (2019). https://doi.org/10.1088/1741-4326/ab28bf
10. A. Murari, M. Lungaroni, E. Peluso, P. Gaudio, J. Vega, S. Dormido-Canto, M.M. Baruzzo, M. Gelfusa, J. Contributors, Adaptive predictors based on probabilistic SVM for real time disruption mitigation on JET. Nucl. Fusion **58**, 1–31 (2018)
11. B. Cannas, A. Fanni, E. Marongiu, P. Sonato, Disruption forecasting at JET using neural networks. Nucl. Fusion **44**, 68–76 (2004). https://doi.org/10.1088/0029-5515/44/1/008
12. B. Cannas, A. Fanni, P. Sonato, M.K. Zedda, A prediction tool for real-time application in the disruption protection system at JET. Nucl. Fusion **47**, 1559–1569 (2007). https://doi.org/10.1088/0029-5515/47/11/018
13. G.A. Ratt, J. Vega, A. Murari, G. Vagliasindi, M.F. Johnson, P.C. De Vries, An advanced disruption predictor for JET tested in a simulated real-time environment. Nucl. Fusion **50**, 1–17 (2010). https://doi.org/10.1088/0029-5515/50/2/025005
14. A. Murari, J. Vega, G.A. Ratt, G. Vagliasindi, M.F. Johnson, S.H. Hong, Unbiased and non-supervised learning methods for disruption prediction at JET. Nucl. Fusion **49** (2009). https://doi.org/10.1088/0029-5515/49/5/055028
15. R. Yoshino, Neural-net disruption predictor in JT-60U. Nucl. Fusion **43**, 1771–1786 (2003). https://doi.org/10.1088/0029-5515/43/12/021
16. R. Yoshino, Neural-net predictor for beta limit disruptions in JT-60U. Nucl. Fusion **45**, 1232–1246 (2005). https://doi.org/10.1088/0029-5515/45/11/003
17. S.Y. Wang, Z.Y. Chen, D.W. Huang, R. Tong, W. Yan, Y.N. Wei, T.K. Ma, Prediction of density limit disruptions on the J-TEXT Tokamak. Plasma Phys. Controlled Fusion **58**, 1–10 (2016). https://doi.org/10.1088/0741-3335/58/5/055014
18. S.P. Gerhardt, D.S. Darrow, R.E. Bell, B.P. Leblanc, J.E. Menard, D. Mueller, A.L. Roquemore, S.A. Sabbagh, H. Yuh, Detection of disruptions in the high-β spherical torus NSTX. Nucl. Fusion **53**, 1–19 (2013). https://doi.org/10.1088/0029-5515/53/6/063021
19. A.H. Boozer, Theory of tokamak disruptions. Phys. Plasmas **19** (2012). https://doi.org/10.1063/1.3703327

20. L.E. Zakharov, S.A. Galkin, S.N. Gerasimov, Understanding disruptions in tokamaks. Phys. Plasmas **19** (2012). https://doi.org/10.1063/1.4705694
21. G. Ke, Q. Meng, T. Finley, T. Wang, W. Chen, W. Ma, Q. Ye, T.Y. Liu, LightGBM: a highly efficient gradient boosting decision tree. Adv. Neural Inform. Process. Syst. **2017-December** (2017), 3147–3155
22. T. Chen, C. Guestrin, XGBoost: a scalable tree boosting system, in *Proceedings of the ACM SIGKDD International Conference on Knowledge Discovery and Data Mining*. **13–17-August-2016** (2016), pp. 785–794. https://doi.org/10.1145/2939672.2939785
23. L. Prokhorenkova, G. Gusev, A. Vorobev, A.V. Dorogush, A. Gulin, Catboost: unbiased boosting with categorical features. Adv. Neural Inform. Process. Syst. **2018-December,** 6638–6648 (2018).
24. GOLAM Tokamak, (n.d.). https://golem.fjfi.cvut.cz/
25. Y. Fu, D. Eldon, K. Erickson, K. Kleijwegt, L. Lupin-Jimenez, M.D. Boyer, N. Eidietis, N. Barbour, O. Izacard, E. Kolemen, Machine learning control for disruption and tearing mode avoidance. Phys. Plasmas **27** (2020). https://doi.org/10.1063/1.5125581
26. R.M. Churchill, B. Tobias, Y. Zhu, Deep convolutional neural networks for multi-scale time-series classification and application to tokamak disruption prediction using raw, high temporal resolution diagnostic data. Phys. Plasmas. **27** (2020). https://doi.org/10.1063/1.5144458
27. J. Vega, A. Murari, S. Dormido-Canto, F. Hernández, T. Cruz, D. Gadariya, G.A. Rattá, A linear equation based on signal increments to predict disruptive behaviours and the time to disruption on jet. Nucl. Fusion. **60** (2020). https://doi.org/10.1088/1741-4326/ab5880
28. W. Zheng, Q.Q. Wu, M. Zhang, Z.Y. Chen, Y.X. Shang, J.N. Fan, Y. Pan, and J-TEXT Team, Disruption predictor based on neural network and anomaly detection on J-TEXT. Plasma Phys. Controlled Fusion (In Press) (2020)
29. G.A. Rattá, J. Vega, A. Murari, A multidimensional linear model for disruption prediction in JET. Fusion Eng. Des. **146**, 2393–2396 (2019). https://doi.org/10.1016/j.fusengdes.2019.03.198
30. T. Yokoyama, Y. Miyoshi, R. Hiwatari, A. Isayama, G. Matsunaga, N. Oyama, Y. Igarashi, M. Okada, Y. Ogawa, Prediction of high-beta disruptions in JT-60U based on sparse modeling using exhaustive search. Fusion Eng. Des. **140**, 67–80 (2019). https://doi.org/10.1016/j.fusengdes.2019.01.128
31. C. Rea, R.S. Granetz, K. Montes, R.A. Tinguely, N. Eidietis, J.M. Hanson, B. Sammuli, Disruption prediction investigations using machine learning tools on DIII-D and Alcator C-Mod. Plasma Phys. Controlled Fusion **60** (2018). https://doi.org/10.1088/1361-6587/aac7fe
32. A.R.A. Tinguely, K.J. Montes, C. Rea, R. Sweeney, R.S. Granetz, An application of survival analysis to disruption prediction via Random Forests. Plasma Phys. Controlled Fusion **61**, 1–24 (2019)
33. C. Rea, K.J. Montes, K.G. Erickson, R.S. Granetz, A real-time machine learning-based disruption predictor in DIII-D (2019)
34. V. Madaan, A. Goyal, Predicting ayurveda-based constituent balancing in human body using machine learning methods. IEEE Access **8**, 65060–65070 (2020). https://doi.org/10.1109/ACCESS.2020.2985717
35. S. Kumar, K. Singh, S. Kumar, O. Kaiwartya, Y. Cao, H. Zhou, Delimitated anti jammer scheme for internet of vehicle: machine learning based security approach. IEEE Access **7**, 113311–113323 (2019). https://doi.org/10.1109/access.2019.2934632
36. H. Li, B. Pu, Y. Kang, C.Y. Lu, Research on massive ECG data in XGBoost. J. Intell. Fuzzy Syst. **36**, 1161–1169 (2019). https://doi.org/10.3233/JIFS-169890
37. D. Arku, K. Doku-Amponsah, N.K. Howard, A Markov-modulated tree-based gradient boosting model for auto-insurance risk premium pricing. Risk Decis. Anal. **8**, 1–13 (2020). https://doi.org/10.3233/RDA-180050
38. D. Elavarasan, D.R. Vincent, Reinforced XGBoost machine learning model for sustainable intelligent agrarian applications. J. Intell. Fuzzy Syst. **Preprint** (2020) 1–16. https://doi.org/10.3233/JIFS-200862

39. R. Jain, J.A. Alzubi, N. Jain, P. Joshi, Assessing risk in life insurance using ensemble learning. J. Intell. Fuzzy Syst. **37**, 2969–2980 (2019). https://doi.org/10.3233/JIFS-190078
40. F. Lin, J. Jiang, J. Fan, S. Wang, A stacking model for variation prediction of public bicycle traffic flow. Intell. Data Anal. **22**, 911–933 (2018). https://doi.org/10.3233/IDA-173443
41. M. Yontar, Ö.H. Namli, S. Yanik, Using machine learning techniques to develop prediction models for detecting unpaid credit card customers. J. Intell. Fuzzy Syst. **Preprint** (2020) 1–15. https://doi.org/10.3233/JIFS-189080
42. O. Grover, J. Kocman, M. Odstrcil, T. Odstrcil, M. Matusu, J. Stöckel, V. Svoboda, G. Vondrasek, J. Zara, Remote operation of the GOLEM tokamak for fusion education. Fusion Eng. Des. **112**, 1038–1044 (2016). https://doi.org/10.1016/j.fusengdes.2016.05.009
43. V. Svoboda, B. Huang, J. Mlynář, G.I. Pokol, J. Stöckel, G. Vondrášek, Multi-mode remote participation on the GOLEM tokamak. Fusion Eng. Des. **86**, 1310–1314 (2011). https://doi.org/10.1016/j.fusengdes.2011.02.069
44. V. Svoboda, A. Dvornova, R. Dejarnac, M. Prochazka, S. Zaprianov, R. Akhmethanov, M. Bogdanova, Remote operation of the GOLEM tokamak with hydrogen and helium plasmas. J. Phys.: Conf. Ser. **768** (2016). https://doi.org/10.1088/1742-6596/768/1/012002
45. F. Yoav, S. Robert, A decision-theoretic generalization of on-line learning and an application to boosting. J. Comput. Syst. Sci. (1997) 119–139. https://doi.org/10.1145/2818346.2823306
46. M. Ferov, M. Modrý, Enhancing LambdaMART using oblivious trees (2016). https://arxiv.org/abs/1609.05610
47. B. Leon, L.C. Yann, Large scale online learning, in *Advances in Neural Information Processing Systems*, pp. 217–224 (2004). https://users.cecs.anu.edu.au/~jpetterson/

IoT-Based Home Vertical Farming

V. S. Abhay, V. H. Fahida, T. R. Reshma, C. K. Sajan, and Siddharth Shelly

Abstract Worldwide, around three million hectares of agricultural land are lost each year as a result of soil degradation and conversion for various development purposes, which in turn reduced the crop yield. Vertical farming is a type of farming method which cultivates crops using vertical structures and controlled environment which maximize the production and efficiency from minimum area. The proposed system is cultivating in a carefully controlled environment where saplings will be planted on nutrient media on chambers illuminated with LED lights. This study will also integrate the vertical farming structure with the Internet of Things. Hence, automate the farm activities with less human intervention and eradicate all the issues that may arise in between and control the farming activities from remote places using the mobile application. Crop yield forecasting features use convolutional neural networks (CNNs) and recurrent neural networks (RNNs) to improve the yield.

Keywords Vertical farming · Internet of things · Grow tower · Neural networks

1 Introduction

According to the United Nations World Food Programme, approximately 1 billion people among the world population are undernourished. About 42% of these people live in two of the world's most populous nations, India and China. By 2050, the number of people living in urban areas is expected to rise to more than 6 billion, 90% of them in developing countries. This tremendous increase in population urges for an exponential increase in food production [1]. But, agricultural land is increasingly becoming a scarce resource due to soil degradation and extensive urbanisation. Each year, around three million hectares of land are losing, due to soil degradation which leads to erosion of topsoil [2]. In addition, four million hectares are also lost each year

V. S. Abhay · V. H. Fahida · T. R. Reshma · C. K. Sajan · S. Shelly (✉)
Department of Electronics & Communication Engineering, Mar Athanasius College of Engineering, Kothamangalam, Ernakulam, Kerala, India
e-mail: sidhushelly@mace.ac.in

M. Paprzycki et al. (eds.), *Intelligent Systems, Technologies and Applications*, Advances in Intelligent Systems and Computing 1353,
https://doi.org/10.1007/978-981-16-0730-1_10

due to unscientific conversion of arable land into highways, factories, and various other constructions. As more land is lost the need gap between the food production and the need for a growing population becomes wider [3]. Therefore, we require various new approaches integrated with new age technologies toward food production for sustainable development [4].

Vertical farming is the modern farming technique in which the saplings are planted on small pots or similar objects ordered on a soil-free nutrient media vertical structures in such a way that maximum yield will be obtained from minimum area. Plants are grown on an automatically controlled environment, where less human intervention is needed in the whole farming process. Vertical farming system is a viable alternative for conventional methods of farming by optimizing the land use efficiency. The air, light, humidity, nutrient supply, watering, etc. are artificially maintained using different equipment. The system can eliminate the effects of seasonal changes on crops as plants are grown on using the methods installed to regulate the ambient temperature and humidity. Since the plants are already isolated from external climate, vertical farming system can also eliminate the potential attacks by harmful pests and other weeds without using or by minimal use of pesticides or herbicides [5].

2 Literature Review

The Internet of Things is a technology which creates a network where almost all things in this world will be connected to the Internet for the purpose of the exchange of the information. IoT has been used extensively for identifying, tracking, locating, and managing various systems. Using a Wi-Fi module, the vertical farming system is integrated into IoT. Saraswathi et al. in [6] proposes the use of IoT in hydroponics(farming using water) using temperature, humidity, pressure, and PH sensor. This work focuses on two tasks—the first one is to automate the green house environment monitoring and the subsequent is to automate the PH level and maintenance of electrical conductivity. Bin Ismail et al. in [7] propose the use of IoT for the watering system in an indoor vertical farm and introduced various programming platforms to integrate IoT devices to cloud computing storage. Authors built a system to monitor the soil moisture and to control water content through the web browser on the laptop, mobile phone, and other handheld and compact devices. Vidhya and Valarmathi in [8] depict the studies conducted to know how much, due to an increase in industrialization, there is a notable decrease in agricultural farmland and traditional farming where planting with soil consumes longer time. Growing a plant without soil helps to increase the yield of crops in short duration for any climatic condition. From this idea, pest and weed can be controlled and labour cost is reduced. The survey spreads idea on technology and techniques to collect a review on automatic monitoring of hydroponics farms using IoT. Various parameters like temperature, humidity, air, etc. inside the system will undergo real-time monitoring using various sensors installed in the system, and the condition can be accessed remotely using a mobile application. Using this, optimal conditions can be maintained for maximum growth using minimal human intervention [2]. In this study, a mobile application is used to manage the system from remote places using smartphones. Using the android application, it is easy to remotely access the data from sensors, control the optimal conditions, and know the crop yield forecast more accurately and easily. The Wi-Fi module is used as a web server from where data will be fetched for mobile application. And we are also deploying the crop yield forecast which can be also deployed in our mobile application using convolutional neural networks and recurrent neural networks. Using previous datasets, prediction models can be trained.

3 Vertical Farming System

Vertical farming is the proposed indoor farming technique where vertical racks are used for farming in a carefully controlled atmosphere. A well-planned vertical farming system is designed for the study using polyvinyl chloride (PVC) pipes and plastic containers. Hydroponic grow tower is the most important part of the system, and the plant saplings will be planted in it. Hydroponics is the process of growing plants by the use of water as the medium where the nutrients needed for our plant will be provided through the aqua medium. A hydroponic grow tower is made as shown in Fig. 1. The container on the bottom of the grow tower accommodates the nutrient solution for the plant growth. We can use a plastic container with a round bottom with 25 cm diameter as the container. The supply of water and nutrients into the grow tower will be done by a water-lifting submersible pump. In this study, we will use a 14 W pump. The grow tower is made up of PVC pipes (20 cm diameter) which have small cup size cavities as shown in Fig. 2. There are incisions on the top part of the cup which goes inside the tower to receive the water falling from above. The grow tower consists of two tubes, there is a tube of 0.6 cm diameter inside the main tube which is connected to the outlet of the submersible motor placed in the container. The motor pump will output water through this small tube to the top, and the holes on top of the covering plate of the grow tower will make the waterfall on the top incisions made on the grow cups. These incisions go up to the center of the bottom of the cup, so that water is held only up to the bottom center.

The vertical grow tower will be placed inside the chamber where the controlled environment is maintained. The chamber consists of a square prism-like structure with a dimension of 50 * 50 * 120 cm. Since we are adopting controlled environment agriculture, the effects of seasonality can be minimalized to a great extent. 24-h LED illumination will give the scope for the same level of photosynthesis all the day over the year. Ambient temperature and humidity can be maintained inside the system using an exhaust fan. Plants are grown on a soil-free nutrient solution, where the roots are directly bathed in the solution. In this study, we are focusing on the hydroponics system, it is a cultivation system which involves growing plants in a nutrient solution. Cooper's solution is the most preferred one for nutrients [9] which was found to result in early harvest, a greater number of leaves and larger average leaf length for spinach in a hydroponic system. The whole farming process will

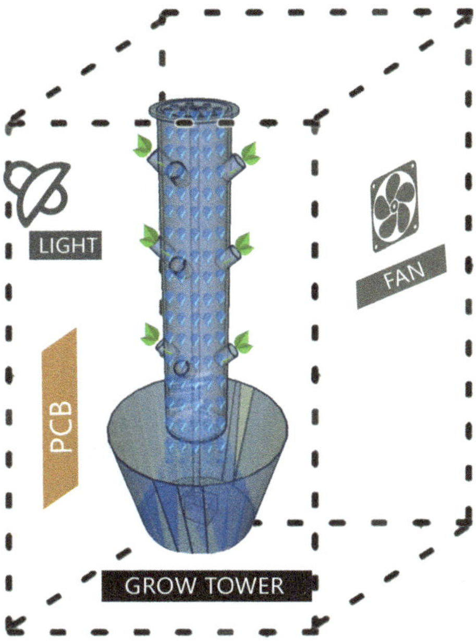

Fig. 1 Proposed vertical farming system

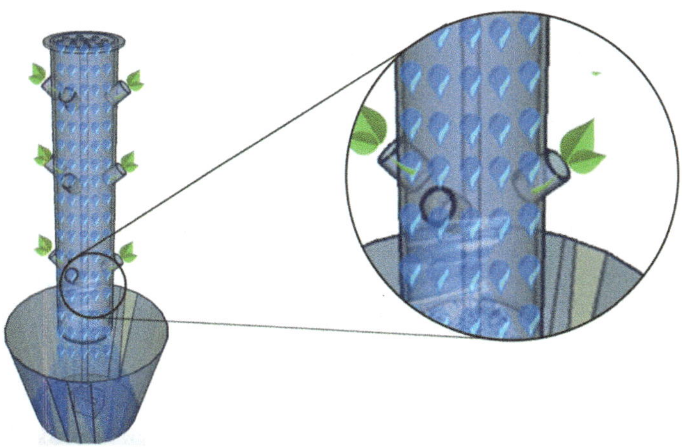

Fig. 2 Grow tower

be independent of the changing seasons, unpredictable weather and all the exterior conditions, which lead to the production of the maximum yield. Because of the extensive work, during all the time careful management is also needed. Since it is covered from external conditions, the potential attack by pests and weeds can be eliminated to a great extent, thereby the use of harmful pesticides can also be avoided [10].

4 IoT Integration and Mobile Application for System

IoT is the new-age technology for the interconnection of systems into the Internet, and it can be accessed through mobile applications [7]. In this study, ATmega 32 is used as the microcontroller, which is responsible for accessing data from the sensors and controlling the automatic system. ATmega 32 is a high-performance, low-power Microchip 8-bit AVR RISC-based microcontroller combines 32 KB ISP flash memory with read-while-write capabilities, 1 KB EEPROM, 54/69 general purpose I/O lines, 32 general purpose working registers. The temperature and humidity sensor and the Wi-Fi module will be connected directly, fan through the SPST relay and submersible motor followed by an SPST to the microcontroller and ATmega 32 will monitor and control the system [11].

The circuit diagram and the block diagram are shown in Figs. 3 and 4, respectively. The automation of the system will work in a way that motor will automatically switch on at particular intervals of time and provide nutrient solution to the roots of the plant. LED light will illuminate all day around in order to facilitate the photosynthesis, and the user can switch on the fan according to the temperature and humidity levels from the sensor.

DHT11 is the temperature and humidity sensor used here, which is a single-wire sensor that can provide humidity and temperature values serially. DHT11 will measure the relative humidity in percentage (20–90% RH) and temperature in degree Celsius in the range of 0–50°C (degree Celsius). The sensor has 4 pins of which 2 pins are used for supply, 1 is not used and the last one is used for the data. This is

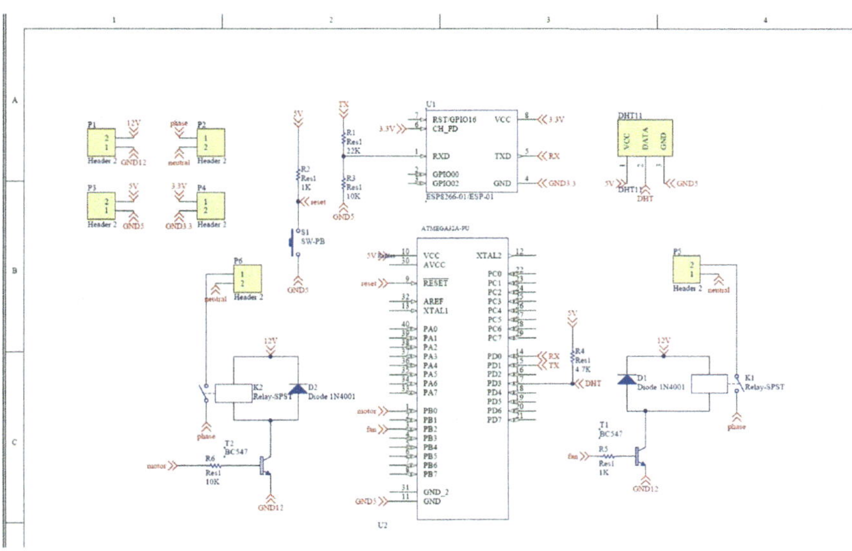

Fig. 3 Circuit diagram

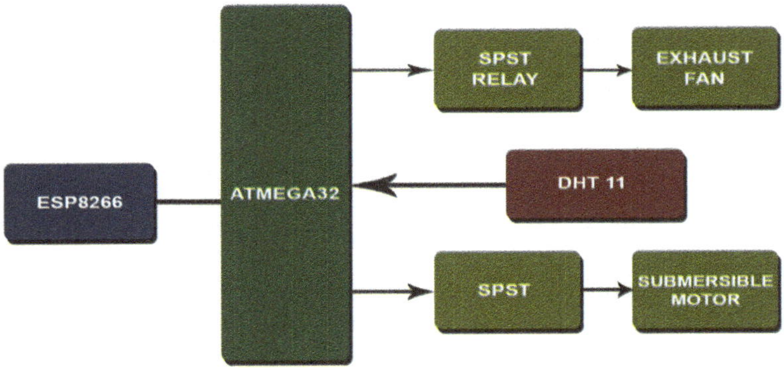

Fig. 4 Block diagram of the automation system

the only pin used for communication, and the data will be received as 40-bit data serially. The data format consists of 8-bit integral RH data + 8-bit decimal RH data + 8-bit integral Temperature data + 8-bit decimal Temperature data + 8-bit checksum [12]. We will interface DHT 11 with ATmega 32, MCU will send a start pulse to the sensor and DHT11 changes from the low power consumption mode to the running mode. Then, DHT11 will send a response signal of 40-bit serial data. Without the start signal from the microcontroller Unit, DHT11 will not collect temperature and humidity information spontaneously.

As mentioned earlier, the vertical farming system will be connected to the internet by using a Wi-Fi module, ESP8266, so that we will be able to monitor and control the vertical farming system from remote devices. The Wi-Fi module will be set up as a web server by connecting the module into the router. In this study, Arduino IDE with ESP8266 add-on for the IDE and we will get the ESP IP address [13]. Using this IP, data can be accessed from the MCU. Here, ESP is receiving an HTTP request from a new client, in this case from our mobile application.

An android application is used in the setup in order to communicate with the ESP8266 module and thereby with the entire system. The application uses Wi-Fi Manager to connect the web server we had set up using the Wi-Fi module. Using this application, data can be accessed from our system using our android system. There is provision to monitor the current status of the system including the data from sensors, working of fan, light, etc. from remote places, and the user can also view the atmosphere from his place. In this application, accessing previous data about the specific crop is also possible. And this enables the user for the crop yield prediction using the previous data. Figure 6 gives the prototype of our mobile application users can access the farming system from remote places through the application.

5 Developed PCB

Given in Fig. 5 is the developed PCB for the project. The board comprises of:

- ATmega32 Microcontroller
- 2 SPDT relay
- 3 AC connector
- And and ESP8266 Wi-Fi module

From the app, the motor is controlled via the ESP8266 module. The command sent from the app is sent to the microcontroller via the Wi-Fi module, and depending on the state of the command (whether it is on/off), the motor is switched on//off via the SPDT relay. The power voltage for each component is as follows:

- ATmega32 Microcontroller—5 V
- ESP8266—5 V
- SPDT relay—12 V.

6 Power Supply

The power supply we are using is manufactured by www.Robu.in, and it is actually a DC-DC buck converter which down converts 12V DC output to 5V and also one of the output port is directly connected to the input, i.e. the 12 V DC. Figure 5 represents the developed PCB and power supply.

Fig. 5 Developed PCB and power supply board

7 Crop Yield Forecasting Using Machine Learning

Crop yield forecasting using machine learning is one of the attractive features of our study. It is extremely hard to predict the yield due to the dependence of many factors including environment factors, management factors. The model in this study is designed to analyze the trend of crop yield using the dataset available. The prediction model will be using convolutional neural networks (CNNs) and recurrent neural networks (RNNs) for crop yield prediction based on previous agricultural data [14]. The proposed CNN-RNN model has proved to achieve a root-mean-square-error (RMSE) of 8–9%, which is the best performance among the other models. Weighted convolutional neural networks (WCNNs)-sparse convolutional neural networks (SCNNs) models were designed to capture the linear and nonlinear effects of the weather and soil data, respectively [15]. The WCNN model used one-dimensional convolution to analyze the temporal dependencies of weather data. Fully connected layer is used to combine the high-level features weather components and soil conditions extracted by WCNN-SCNN where the RNN is designed to capture the dependencies over a number of years. Since the system in this study is a vertical farming structure, it is independent of soil properties, rather the nutrient solution will affect instead of soil and weather dependency also give way to the atmosphere inside the closed chamber. RNN models predict the crop yield of country for years, using information from a number of years. In this study, the prediction model is using the data set consists of season, crop, area of cultivation and production yield from different states and union territories from 1997 to 2014.

The CNN-RNN has mainly three features which enable us to use them efficiently for crop prediction. (1) The CNN-RNN model was designed to capture the time dependencies of environmental factors and the genetic improvement of seeds over time without having their genotype information. (2) The model can generalize the yield prediction to untested environments without significant drop in the prediction accuracy. (3) The model could reveal the extent to which weather conditions, accuracy of weather predictions, soil conditions, and management practices were able to explain the variation in the crop yields.

8 Future Improvements

The system in the study consists of only one LED lamp. Instead of that, lights with different wavelengths can be used for different crops and a provision to vary the brightness of the light should be enabled so that, the yield can be increased and. pH is one of the main things that affect the crops, and it is better if a pH meter can be integrated into the system to analyze the pH of the nutrient solution more effectively. A camera attached to the system will enable the remote surveillance of the system from our application. In addition, intruder detection sensors prevent the disturbances across the system, and water level sensor to automatically detect the level of nutrient solution can also be implemented (Fig. 6).

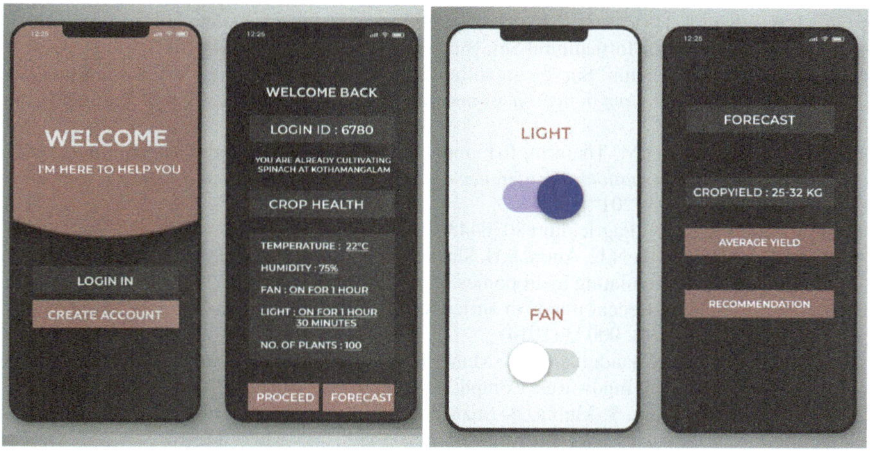

Fig. 6 Prototype of mobile application

9 Conclusion

In this paper, vertical farming was presented as an alternative for conventional farming system. Vertical farming system is connected to the Internet using IoT. The system can be controlled and monitored using a mobile application by setting up the Wi-Fi module as the web server. User will be also able to remotely observe the farming system. The system will also provide the crop yield prediction using machine learning techniques.

Acknowledgements The authors would like to acknowledge the financial assistance for the project by Centre for Engineering Research & Development (CERD), of A. P. J. Abdul Kalam Technological University Thiruvananthapuram, Kerala, India, with File ref. No: KTU/RESEARCH 2/4664/2019 and the CERD order No: U.O.No. 218/2020/KTU.

References

1. http://www.fao.org/3/a-i4646e.pdf
2. C. Peuchpanngarm, P. Srinitiworawong, W. Samerjai, T. Sunetnanta, DIY sensor-based automatic control mobile application for hydroponics, in *Fifth ICT International Student Project Conference (ICT-ISPC)*, Nakhon Pathom, pp. 57–60 (2016). https://doi.org/10.1109/ICT-ISPC.2016.7519235
3. D. Gwynn-Jones, H. Dunne, I. Donnison, P. Robson, G.M. Sanfratello, B. Schlarb-Ridley, K. Hughes, P. Convey, Can the optimisation of pop-up agriculture in remote communities help feed the world? Global Food Security **18**, 35 (2018)
4. I.A. Lakhiar, J. Gao, T.N. Syed, F.A. Chandio, N.A. Buttar, Modern plant cultivation technologies in agriculture under controlled environment: a review on aeroponics. J. Plant Interact. **13**(1), 338–352 (2018). https://doi.org/10.1080/17429145.2018.1472308

5. A.M. Beacham, L.H. Vickers, J.M. Monaghan, Vertical farming: a summary of approaches to growing skywards. J. Horticultural Sci. Biotechnol. **94**(3), 277–283 (2019)
6. R. Vidhya, K. Valarmathi, Survey on automatic monitoring of hydroponics farms using IoT, in *3rd International Conference on Communication and Electronics Systems (ICCES)*, pp. 125–128 (2018)
7. M.I.H. Bin Ismail, N.M. Thamrin, IoT implementation for indoor vertical farming watering system, in *IEEE International Conference on Electrical, Electronics and System Engineering (ICEESE)*, pp. 89–94 (2017)
8. https://doi.org/10.3390/agriculture3030443
9. A.H. Shah, S.U. Munir, N.U. Amin, S.H. Shah, Evaluation of two nutrient solutions for growing tomatoes in a non-circulating hydroponics system. Sarhad J. Agric. **27**(4), 557–567 (2011)
10. A. Garg, R. Balodi, Recent trends in agriculture: vertical farming and organic farming. Adv. Plants Agric. Res. **1**(4), 00023 (2014)
11. A.M. Abhishek, R. Chandrakanth, C. Mahesh, M.M. Marigiri, B. Ramya, K5, smart vertical farm using IOT. Int. J. Innov. Res. Comput. Commun. Eng. **6**(4), 3593–3597 (2018)
12. J. Nuvvula, S. Adiraju, S. Mubin, B. Shahana, V. Valisetty, Environmental smart agriculture monitoring system using internet of things. Int. J. Pure Appl. Math. **115**(6), 313–320 (2017)
13. G.N. Balaji, V. Nandhini, S. Mithra, N. Priya, R. Naveena, IOT based smart crop monitoring in farm land. Imperial J. Interdisciplinary Res. (IJIR) **4**, 88–92 (2018)
14. M. Mehra, S. Saxena, S. Sankaranarayanan, R.J. Tom, M. Veeramanikandan, IoT based hydroponics system using deep neural networks. Comput. Electron. Agric. **155**, 473 (2018)
15. S. Khaki, L. Wang, S.V. Archontoulis, A CNN-RNN framework for crop yield prediction. Front Plant Sci. **10**, 1750 (2020)

A Scalable Multi-disease Modeled CDSS Based on Bayesian Network Approach for Commonly Occurring Diseases with a NLP-Based GUI

P. Laxmi, Deepa Gupta, Radhakrishnan Gopalapillai, J. Amudha, and Kshitij Sharma

Abstract A clinical decision support system (CDSS) is a useful support system for healthcare professionals in initial diagnosis of diseases. Symptoms described by patients are key inputs in this decision-making process. The proposed work creates a model of commonly occurring Indian diseases along with laboratory tests identified and medications for the diseases diagnosed. A Bayesian network is automatically generated from the inputs provided by the experts in medical domain which is used to predict the probabilities of possible diseases. The model also has the ability to learn from past experiences. The proposed work also provides a graphical user interface (GUI) framework in designing CDSS having an end-to-end design of CDSS. The GUI for this model uses natural language processing techniques for text processing.

Keywords Clinical decision support system · Natural language processing · Artificial intelligence · Machine learning

1 Introduction

A strong healthcare industry is vital for the health and happiness of people. One of the key aspects of providing right healthcare is correct and preferably less expensive diagnosis of the disease. A clinical decision support system (CDSS) is a useful aid

P. Laxmi · D. Gupta · R. Gopalapillai (✉) · J. Amudha
Department of Computer Science and Engineering, Amrita School of Engineering, Bengaluru
Amrita Vishwa Vidyapeetham, Bengaluru, India
e-mail: g_radhakrishnan@blr.amrita.edu

D. Gupta
e-mail: g_deepa@blr.amrita.edu

J. Amudha
e-mail: j_amudha@blr.amrita.edu

K. Sharma
Paralaxiom Technologies Private Limited, Bengaluru, India
e-mail: kshitij.sharma@paralaxiom.com

M. Paprzycki et al. (eds.), *Intelligent Systems, Technologies and Applications*,
Advances in Intelligent Systems and Computing 1353,
https://doi.org/10.1007/978-981-16-0730-1_11

in this process. Decision support systems using mathematical models are useful in providing assistance to make right decision. CDSS can be broadly categorized into knowledge-based and non-knowledge-based systems. Traditional knowledge-based decision systems typically have a knowledge base and an inference engine. The knowledge base is created with inputs from experts in medical domain. These systems can have either a text-based user interface or a graphical user interface (GUI). Recent developments in artificial intelligence and machine learning have given a thrust to decision support systems.

This paper focuses on the design of a clinical decision support system for certain diseases commonly seen in India based on common symptoms. The inference engine is based on Bayesian networks. Interviews with healthcare experts are used to prepare the knowledge base. The proposed work also provides a GUI framework for end-to-end design of a CDSS.

The remainder of this paper is organized as follows: Sect. 2 discusses the literature survey done in the field of CDSS. Section 3 discusses the architecture and implementation details of the proposed work. Section 4 brings forth the experimental results, and Sect. 5 has the conclusion, limitations, and future work of the proposed work.

2 Literature Survey

Clinical decision support systems vary based on the type of knowledge acquired, the inference mechanism used, and the type of diseases diagnosed. There are rule-based CDSS which works on a set of If-Else rules. These rule-based systems have been commonly used in detecting general diseases, developing a scalable multi-disease diagnosis system, and diagnosing hematological disorder [1–3].

Patient discharge summaries are a good source of information for diagnosis and treatment. Natural language processing can be used to extract such information automatically [4, 5]. The symptoms described by a patient may not be very crisp and accurate. The fuzziness in the description can be taken care by fuzzy logic-based CDSS [6, 7]. Fuzzy systems have been used in diagnosing people with obstructive pulmonary diseases [8]. Fuzzy-based systems are also used for multi-disease diagnosis [9]. CDSS systems based on ontologies can be implemented without explicit inputs from healthcare professionals [10] CDSS systems can also incorporate clinical process guidelines (CPG) [11].

Machine learning techniques are used in many areas of healthcare systems such as assessing emotional support [12]. Machine learning (ML) methods using decision trees, support vector machines, and self-organizing neuro-fuzzy logic have all been used in CDSS [13–15]. Many CDSS implementations have used probabilistic approaches. Jiang et al. [16] have used a multi-layer naïve Bayes methodology. Bayesian network-based approach has been used in detection of Alzheimer's disease, dementia, and mild cognitive impairment [17]. Their architecture components at three layers consist of demographic data, clinical inputs, and symptoms. The network was created statically with inputs from healthcare professionals. A similar architecture with parameters inferred from clinical dataset has been used in the detection of dengue fever [18]. Bayesian networks are used for detection of breast cancer [19] and other diseases [20]. Bayesian networks capable of predicting multiple diseases also have been reported [21].

The work described in this paper also uses a Bayesian network methodology. The work proposes a GUI framework which incorporates Bayesian methodology for disease prediction. Scalability in terms of design is the major focus of this work. The work also includes laboratory test checking and medicine recommendation. Most of the work in literature mentions a single disease modeling using Bayesian approach. Very few works have been done in multi-disease modeling this methodology. The proposed work attempts multi-disease modeling for commonly occurring Indian diseases. A two-layer approach having symptoms and diseases was selected for the implementation of CDSS.

3 The Proposed System for CDSS

The proposed work is an attempt to model commonly occurring Indian diseases along with the laboratory test identification for the diagnosed diseases and medications for the diseases diagnosed. Learning module is also incorporated in the model which records all the data entered into the system. The model is adapted to learn from past experiences. The high-level architecture of this system consists of four major components which are shown in Fig. 1. The four components are:

- Training module: The training module refers to the automatic creation of the Bayesian network. This module is divided into training from Microsoft excel sheets and training from database.
- Create Patient module: This module refers to the creation of new patients to the system.
- Symptom Checker module: This module takes care of the symptoms entered by the user, processing of the symptoms, and the probable diseases prediction.
- Laboratory Test Checker module: This module handles the confirmation of disease based on the laboratory test values entered.

Each of the components is explained in detail in the following subsections.

3.1 *Training Module*

The training module refers to the Bayesian network creation and Bayesian parameter estimation from the clinical data. The Bayesian network is automatically created without the help of experts. The training of the system can be done by either including external data into the CDSS data or taking only CDSS data from the application. The

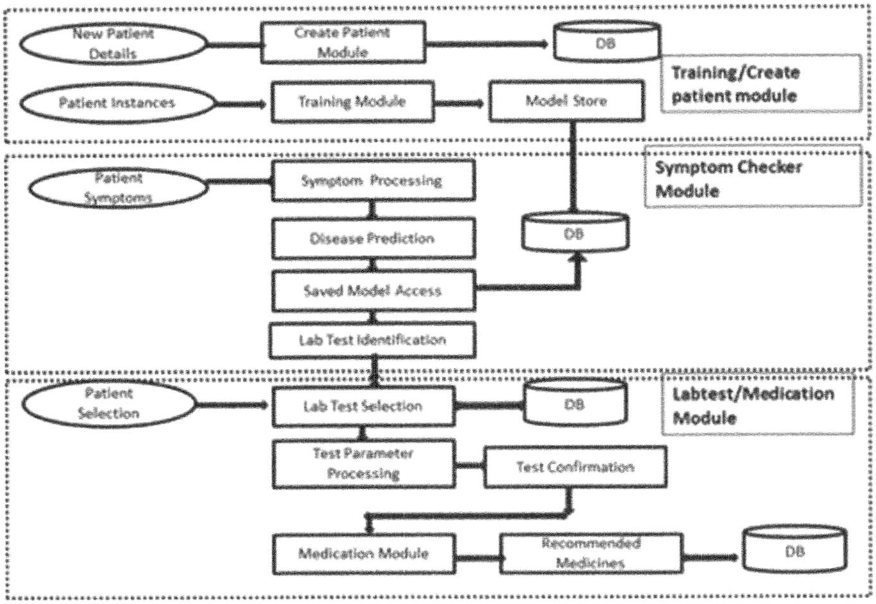

Fig. 1 High-level architecture of CDSS

Microsoft excel data/database data act as training data for the CDSS. The pseudocode for automatic network creation is mentioned in Fig. 2.

After the creation of the Bayesian network, the maximum likelihood estimator (MLE) algorithm is used for parameter creation from the training data.

PseudoCode 1 : ***Automatic Network Creation***

```
1:Procedure AddNewColumn(DataFrame):
        1.  Search for new columns in Database
        2.  If present add the new columns in Database
2:Procedure BayesianNetworkCreation(DataFrame):
        1.  For all columns marked <i> as disease:
                1.  For all columns marked <j> as symptom:
                2.  Set numYes=if i==Y then count(j==Y) else 0
                3.  Set numNo=if i==Y then count(j==N) else 0
                4.  If (numYes>=numNo):
                        1.  Create an edge between disease and symptom
3:Call AddNewColumn(excel data /database data)
4: Call BayesianNetworkCreation (excel data /database data)
```

Fig. 2 Pseudocode for automatic network creation

```
PseudoCode 2 : Create patient module
1: Procedure createPatient(name,sex,weight,gender):
        1. Query the database table for max patient ID
        2. Create a new record in the patient table with max patientID +1
        3. Update the record with name, sex,weight,gender
2: Call createPatient (name.sex,weight,gender)
```

Fig. 3 Pseudocode for create patient module

3.2 Create Patient Module

Create patient module undertakes addition of new patients in the CDSS application. The system registers a patient after entering basic information about the patient. The patient is supplied with a patient id which is further used for all process. The pseudocode for this module is mentioned in Fig. 3.

3.3 Symptom Checker Module

The symptom checker module is an important module as it deals with the symptoms experienced by a patient, and the disease is diagnosed based on the symptoms. Natural language processing is used for processing the symptoms entered. SNOMED-CT which is a nomenclature of medical terms is also used in validation of symptoms entered. The architecture of the symptom checker module is shown in Fig. 4. The details of the components are discussed in Fig. 4.

- **Dictionary Information**: The user of the CDSS is provided with a dictionary to select the words which he needs to enter. The dictionary is a combination of the WordNet dictionary and in-built symptom corpus. This is used to avoid the spelling errors of the user. The symptoms which are already a part of the model are also included in the dictionary.
- **Preprocessing symptoms**: NLP techniques are used for preprocessing the symptoms entered by the user. At first, the symptom entered is converted into tokens using tokenization functions which are part of NLTK. The tokens generated are processed again to remove the stop words like 'the', 'a', 'an', etc.
- **Symptom selection and validation**: All the tokens generated as a part of previous phase is analyzed to check whether these tokens are any symptom nodes in the Bayesian model of CDSS. Synonyms of the tokens are also taken into consideration while checking the model nodes. If the tokens are a part of Bayesian nodes, it is taken for probable disease inference. For all other tokens which are not a part of the Bayesian model, the tokens are validated with SNOMED-CT to ensure that they are a valid medical terms. Those tokens are stored in the CDSS for future training of the model.

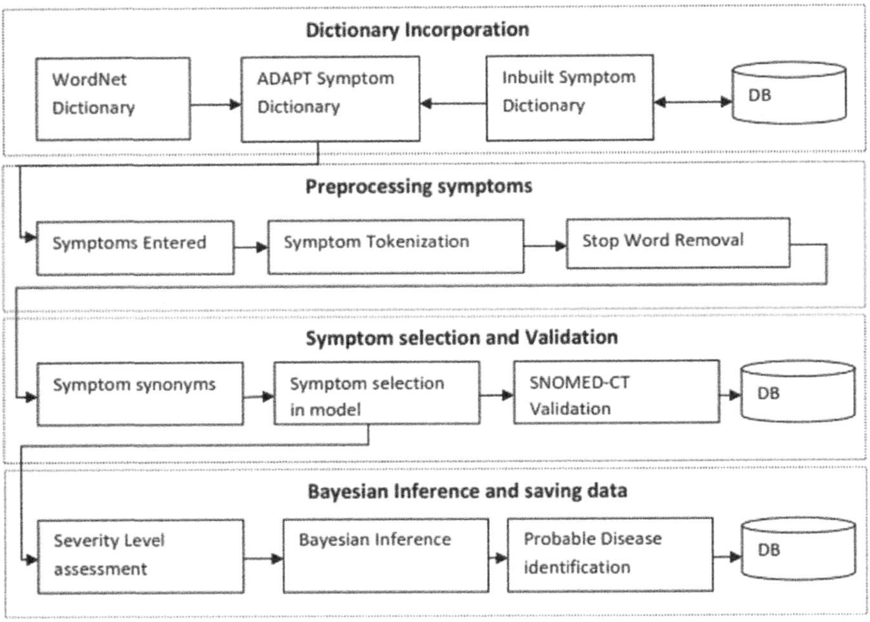

Fig. 4 Symptom checker architecture

- **Bayesian Inference and saving data**: As most of the nodes of Bayesian network have values yes (Y) or no (N), the presence of a symptom token will take the value 'Y'. As of now the Bayesian nodes are limited for Y/N and H/M/L as values. After severity assessment (High/Medium/Low), the symptom tokens which are a part of the Bayesian network are passed to inference engine for giving a probable set of diseases. The inference engine which is used in the proposed work uses variable elimination algorithm. Variable elimination tries eliminating variables to reduce the computation involved in finding posterior probabilities based on evidence.

3.4 Laboratory Test Checker Module

This module deals with the disease confirmation of a patient by entering the laboratory test values. The symptom checker module will return probable diseases diagnosed for a patient. The laboratory test checker module will display the laboratory tests recommended for the diseases diagnosed for a patient. The laboratory test data is stored in database tables. The database is looked up for the tests which has to be displayed to the user. The appropriate test is selected by the doctor, and the input values are entered. The test results are based on the input test values. If a patient is diagnosed positively for the test, the data is updated in the table. The flowchart for this module is mentioned in Fig. 5.

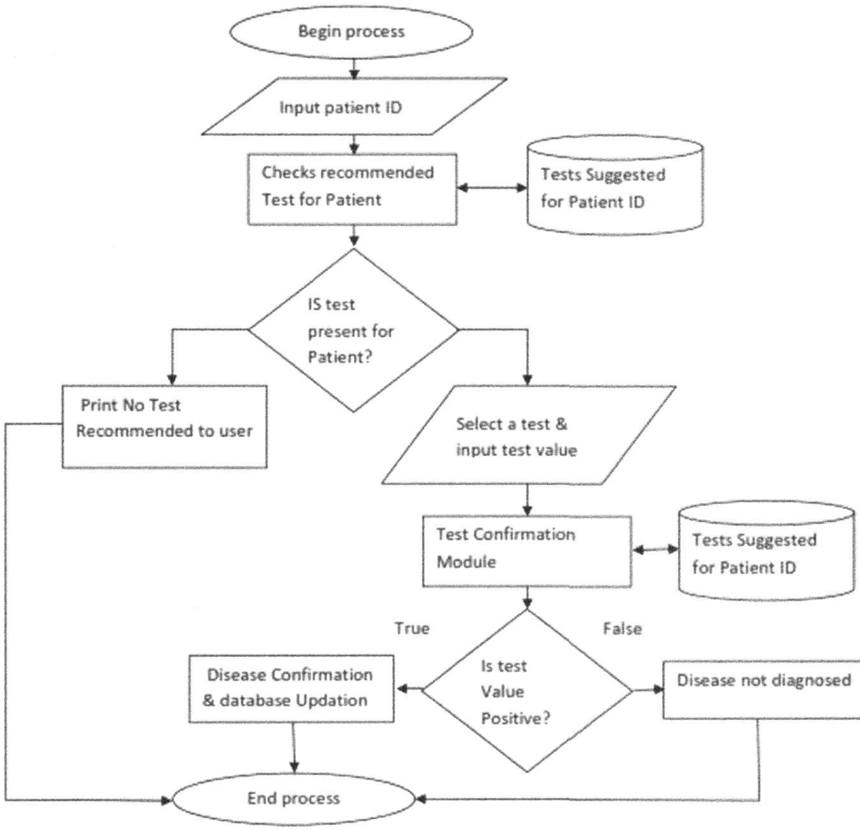

Fig. 5 Laboratory test checker flowchart

4 Experimental Results and Discussion

The proposed work is an end-to-end clinical decision support system with an NLP-based GUI. The main page of the work has all the options as mentioned in Fig. 6.

The symptom checker page has the NLP-based autosuggestion feature as mentioned in Sect. 3.1. Image of symptom checker page is shown in Fig. 7.

The probable diseases diagnosed by the CDSS on the click of the button will provide the form as mentioned in Fig. 8.

The home screen also provides other options for training, laboratory test checking, and adding patients with the system. The training of the Bayesian network has been done with a dataset of 185 records collected through consultation with experts working in the healthcare industry. A screen shot of the two-layer Bayesian network structure obtained is shown in Fig. 9. The precision and recall recorded for the experiments conducted are shown in Table 1.

Fig. 6 Home page

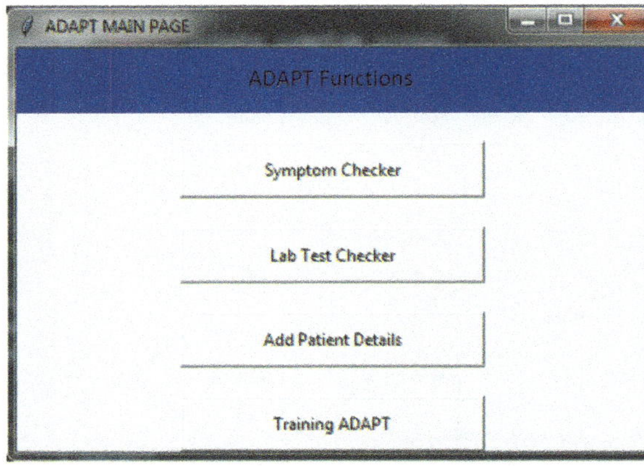

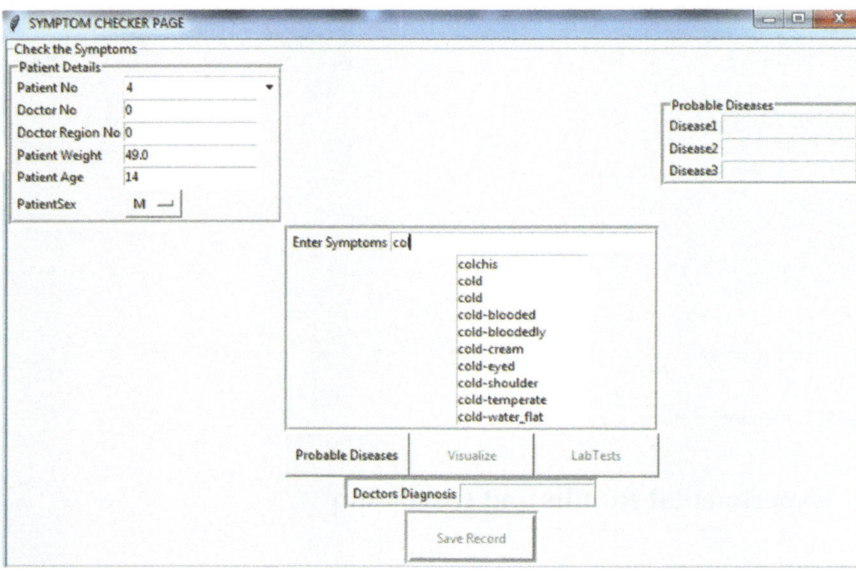

Fig. 7 Symptom checker page

The scalability of the system is also tested for adding new symptom and new disease to the system. The new symptoms which are not a part of the CDSS were added and tested. The model could very well handle new additions.

The implementation of the framework is done in Python language. The results indicate that this framework can be deployed in clinics for symptom assessment.

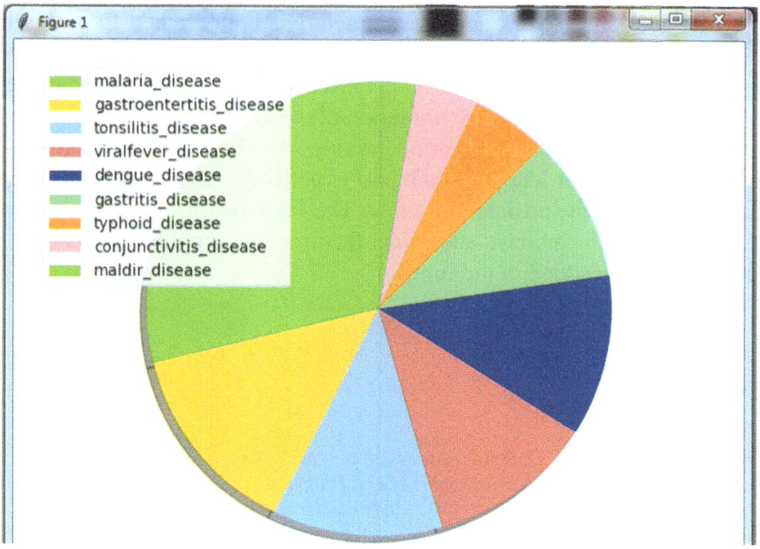

Fig. 8 Visualization of probable diseases

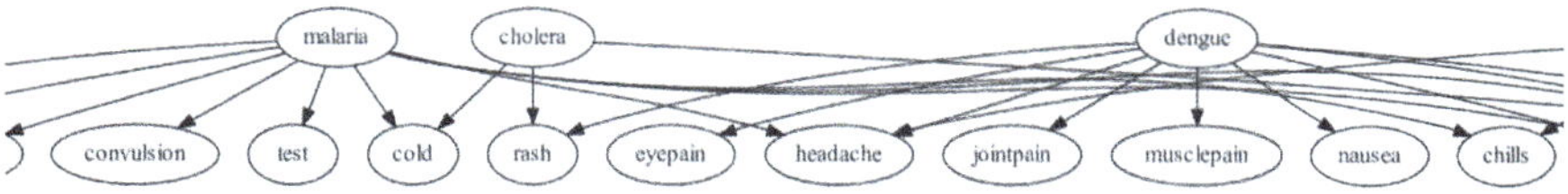

Fig. 9 Bayesian network generated by training

Table 1 Precision and recall of diseases identified

Disease	Precision	Recall
Malaria	0.6	0.6
Typhoid	1	1
Dengue	0.5	0.5
Conjunctivitis	1	1
Gastritis	1	1
Viral fever	1	1
Tonsillitis	1	1
Gastroentertitis	1	1

5 Conclusion and Future Work

A Bayesian network-based clinical decision support system for commonly occurring diseases in India has been presented in this work. The model is a scalable one as any number of diseases and symptoms can be added to the model. The network structure of Bayesian network and its parameters is both created from the dataset. A rule-based methodology is used for modeling laboratory tests and prescription generation module. The GUI for this model has also been designed, and it uses NLP techniques for text processing. The limitation of the system is the non-availability of an extensive pre-defined dataset for undertaking modeling of multiple diseases. As the model is designed for learning from past experiences, the work can be deployed in clinics for assessment and data collection. The future work can be done in the NLP part of the GUI which can have additional features for symptom processing. Many more diseases can be added in the future.

References

1. Y. Zhang, H. Li, H. Duan, Q. Shang, An integration profile of rule engines for clinical decision support systems. in *International Conference on Progress in Informatics and Computing (PIC)* (2016), pp. 762–766. https://doi.org/10.1109/PIC.2016.7949601
2. H. Chu, Y. Yang, Q. Li, Y. Xu, H. Wei, A scalable clinical intelligent decision support system. in *Proceedings of the 14th International Conference on Inclusive Smart Cities and Digital Health*, vol 9677, (2016), pp. 159–165. https://doi.org/10.1007/978-3-319-39601-9_14
3. Y.Y. Chen, K.N. Goh, K. Chong, Rule based clinical decision support system for hematological disorder. in *IEEE 4th International Conference on Software Engineering and Service Science* (2013), pp. 43–48. https://doi.org/10.1109/ICSESS.2013.6615252
4. P.C. Nair, D. Gupta, B.I. Devi A survey of text mining approaches, techniques, and tools on discharge summaries. in *Advances in Computational Intelligence and Communication Technology*, ed. by X.Z. Gao, S. Tiwari, M. Trivedi, K. Mishra. Advances in Intelligent Systems and Computing, vol 1086, (Springer, Singapore, 2021)
5. P.C. Nair, D. Gupta, B.I. Devi, N.R. Bhat, Automated clinical concept-value pair extraction from discharge summary of pituitary adenoma patients. in *2019 9th International Conference on Advances in Computing and Communication (ICACC)* (Kochi, India, 2019), pp. 258–264. https://doi.org/10.1109/ICACC48162.2019.8986171
6. B. Malmir, M. Amini, S.I. Chang, A medical decision support system for disease diagnosis under uncertainty. Expert Syst. Appl. **88**, 95–108 (2017). https://doi.org/10.1016/j.eswa.2017.06.031
7. T. Samad-Soltani, M. Ghanei, M. Langarizadeh, Development of a fuzzy decision support system to determine the severity of obstructive pulmonary in chemical injured victims. Acta Inform. Med. **23**(3), 138–141 (2015). https://doi.org/10.5455/aim.2015.23.138-141
8. S. El-Sappagh, F. Ali, A. Ali, A. Hendawi, F.A. Badria, D.Y. Su, Clinical decision support system for liver fibrosis prediction in hepatitis patients: a case comparison of two soft computing techniques. IEEE Access **6**, 52911–52929 (2018). https://doi.org/10.1109/ACCESS.2018.2868802
9. S. Tandra, D. Gupta, J. Amudha, K. Sharma, A fuzzy-neuro-based clinical decision support system for disease diagnosis using symptom severity. in *Soft Computing and Signal Processing. ICSCSP 2019*, ed. by V. Reddy, V. Prasad, J. Wang, K. Reddy. Advances in Intelligent Systems and Computing, vol. 1118 (Springer, Singapore, 2020)
10. Y. Shen, K. Yuan, D. Chen, J. Colloc, M. Yang, Y. Li, K. Lei, An ontology-driven clinical decision support system (IDDAP) for infectious disease diagnosis and antibiotic prescription. Artif. Intell. Med. **86**, 20–32 (2018). https://doi.org/10.1016/j.artmed.2018.01.003
11. A. Galopin, J. Bouaud, S. Pereira, B. Seroussi, An ontology-based clinical decision support system for the management of patients with multiple chronic disorders. Stud Health Technol Inform. **216**, 275–279 (2015). https://doi.org/10.3233/978-1-61499-564-7-275

12. P. Naveen, P.C. Nair, D. Gupta, Predicting the degree of emotional support in an online health forum for HIV using data mining techniques. in Emerging Trends in Electrical, Communications, and Information Technologies, ed. by T. Hitendra Sarma, V. Sankar, R. Shaik. Lecture Notes in Electrical Engineering, vol. 569 (Springer, Singapore, 2020)
13. Subiyanto, A. Mulwinda, D. Andriani, Intelligent diagnosis system for acute respiratory infection in infants. in *3rd International Conference on Science in Information Technology (ICSITech)* (2017), pp 558–562. https://doi.org/10.1109/ICSITech.2017.8257175
14. M. Ahmad, V. Tundjungsari, D. Widianti, P. Amalia, U.A. Rachmawati, Diagnostic decision support system of chronic kidney disease using support vector machine. in *Second International Conference on Informatics and Computing (ICIC)* (IEEE, 2017), pp. 1–4. https://doi.org/10.1109/IAC.2017.8280576
15. M.M. Baig, H.G. Hosseini, M. Lindén, Machine learning-based clinical decision support system for early diagnosis from real-time physiological data. in *2016 IEEE Region 10 Conference (TENCON)* (2016), pp. 2943–2946. https://doi.org/10.1109/TENCON.2016.7848584
16. Y. Jiang, B. Qiu, C. Xu, C. Li, The research of clinical decision support system based on three-layer knowledge base model. J. Healthc. Eng. **2017**, 6535286 (2017). https://doi.org/10.1155/2017/6535286
17. F.L. Seixas, B. Zadrozny, J. Laks, A. Conci, C.M. Saade, A Bayesian network decision model for supporting the diagnosis of dementia, Alzheimer's disease and mild cognitive impairment. Comput. Biol. Med. **51**, 140–158 (2014). https://doi.org/10.1016/j.compbiomed.2014.04.010
18. C. Sa-Ngamuang, P. Haddawy, V. Luvira, W. Piyaphanee, S. Iamsirithaworn, S. Lawpoolsri, Accuracy of dengue clinical diagnosis with and without NS1 antigen rapid test: comparison between human and Bayesian network model decision. PLoS Negl Trop Dis. **12**(6), e0006573 (2018). https://doi.org/10.1371/journal.pntd.0006573
19. S. Liu, J. Zeng, H. Gong, H. Yang, J. Zhai, Y. Cao, Quantitative analysis of breast cancer diagnosis using a probabilistic modelling approach. Comput. Biol. Med. **92**, 168–175 (2016). https://doi.org/10.1016/j.compbiomed.2017.11.014
20. A.C. Constantinou, N. Fenton, W. Marsh, L. Radlinski, From complex questionnaire and interviewing data to intelligent Bayesian network models for medical decision support. Artif. Intell. Med. **67**, 75–93 (2016). https://doi.org/10.1016/j.artmed.2016.01.002
21. P. Laxmi, D. Gupta, G. Radhakrishnan, J. Amudha, K. Sharma, Automatic multi-disease diagnosis and prescription system using Bayesian network approach for clinical decision making. in *Advances in Artificial Intelligence and Data Engineering*, ed. by N. Chiplunkar, T. Fukao. Advances in Intelligent Systems and Computing, vol. 1133 (Springer, Singapore, 2021)

Analyze and Visualize the Correlation Between Heart and Cancer Diseases Using Data Mining Techniques

Nuha S. Varier, Sriraj Vuppala, Rohan Boggarapu, Ankita Mohapatra, and Sangita Khare

Abstract Cure for diseases involves analyzing the right cause, so that a treatment can be done by observing the symptoms. Accurate symptoms can be obtained by conducting appropriate medical checkup. Due to the quality of present livelihood, it is essential to diagnose diseases in regular intervals through routine checkups and to have knowledge of how one disease can lead to another. There are different types of data mining techniques, which can be efficiently utilized to recognize heart and cancer diseases. The result can be used to detect the presence or recurrence of a disease. This report brings out the correlation between heart and cancer diseases by identifying the common community health status indicators (CHSI). Productive conclusion on heart and cancer diseases is given by testing the various data mining algorithms. The major aim of the work is to implement data mining techniques to cluster states of the USA on the basis of deaths due to heart and cancer disease and to find out the main cause of the death due to these diseases.

Keywords Clustering · Classification · CHSI dataset · Tableau tool · Visualization · Correlation · Data mining · Cancer · Heart disease

1 Introduction

In the past few years, there has been a huge change in the healthcare sector. The reason being new trends in technology, technology is giving a new dimension to the healthcare domain. Information technology is making people's life easier and also helping in proper decision making. Using the current technology trend, we can take better health precautions and can make the right decision. Considering technology as an aspect, there is broad improvement in healthcare sectors. Technology in health care is helping the patients and doctors and is also of great help in medical laboratories. Various chronic diseases like cancer, heart, diabetes, arthritis, etc., require various

N. S. Varier · S. Vuppala · R. Boggarapu · A. Mohapatra · S. Khare (✉)
Department of Computer Science and Engineering, Amrita School of Engineering, Amrita Vishwa Vidyapeetham, Bengaluru, India
e-mail: k_sangita@blr.amrita.edu

M. Paprzycki et al. (eds.), *Intelligent Systems, Technologies and Applications*, Advances in Intelligent Systems and Computing 1353,
https://doi.org/10.1007/978-981-16-0730-1_12

tests, lab reports, and proper diagnosis, which involves high cost and time for the patients as well as doctors. Therefore, it is important to do proper study about these, so as to minimize the overall time, cost, and effort involved to diagnose chronic diseases [1]. One of the possible solutions is to use current technology such as machine learning techniques in the healthcare domain.

Machine learning has become a key focus and has a lot of scope in the current technological world. There are different machine learning algorithms, which can be applied, based on the requirements. It is believed that machine learning techniques are worth implementing in the field of research, which gives appropriate results in decision making and analysis. There are various machine learning techniques available such as classification, clustering, regression, recommendation techniques, etc.

Early diagnosis is best for better treatment [2]. An absence of early diagnosis may lead to the death of a person. If diseases like cancer can be predicted in the early stages, it can be cured. If the disease cannot be diagnosed at initial stages, then the disease may grow day by day, and it can also cause weakening of the immune system in the body. In developing countries like India, Bangladesh, and Indonesia, people living in metropolitan cities have easier access to good medical facilities like doctors, diagnosis machines, etc., which is not the case with people living in rural areas. Diseases like cancer, obesity, heart attack, diabetes are easier to diagnose if known beforehand; but due to lack of knowledge about proper diagnosis, many people have lost their life.

While diagnosing diseases in pathology, it is very difficult to identify patients who are infected with cancer because the symptoms that can be seen in a person affected with cancer will be identical to other diseases such as flu and heart disease. Present medical system includes many processes for diagnosing a disease. Some of the processes are clinical report, microscopy test, radiography, and biopsy, but there are few limitations in some of these like consumption of too much time, cost, etc. If the diagnosis is wrong due to information collected from the patient, there is a chance of adverse health effects on the patient. Data mining algorithms can play a major role in such situations.

Data mining (DM) algorithms can be used for classification, clustering, finding association among features, etc., in medical domain, apart from various other applications in other domains [3]. Some of the data mining algorithms that are generally used for classification and clustering are K-means, decision tree, Bayesian networks, etc. [4, 5]. During the time of diagnosis, there might be a problem of some missing values, but techniques from DM can be used to predict the missing values. Data mining also helps in ranking the attributes.

Few existing studies that are similar to the work done have been explored and explained in Sect. 2. Dataset used for this work is explained in Sect. 3, followed by the proposed design in Sect. 4 where each section implemented in this work is explained in detail. Methodology, where stepwise procedure to get to the result obtained, is explained in Sect. 4. Result obtained is mentioned in Sect. 5 which is followed by the conclusion in Sect. 6. The future scope of this work is explained in Sect. 7.

2 Literature Survey

The primary objective of this section is to understand the type of work that is being carried out on the medical data and how analysis and visualization is done on the data. Many machine learning and data mining algorithms have been explored in healthcare for analyzing and predicting the possibility of occurrence of a disease. Classification and clustering algorithms are most popularly used for this. Few algorithms such as decision tree, K-nearest neighbor (KNN), neural networks, and SVM are used for analyzing similar patterns and to effectively predict a disease [6–9].

Data mining techniques will help healthcare industries to reduce the number of tests. Classification is one methodology in DM which is used for healthcare data observation. The main objective for classification is to precisely predict the target class [10]. Various DM techniques have been implemented on heart datasets [11]. Techniques such as decision tree, Adaboost, KNN, and K-means are used as classification algorithms to check their accuracy. It has been concluded that accuracy of a technique depends on the dataset used and the extent to which it has been modified. DM techniques can be used to detect cancer in early stages which helps in prevention of the disease which in turn helps in reducing the number of deaths caused by cancer [12]. A multi-layered method combination using clustering and decision tree is developed to predict cancer in this paper.

Visualization reduces time of action, as one can spend less time to comprehend the data. Visualization tools such as tableau, ArcGIS, CiteSpace, and WEKA [13, 14] have been used previously to explore healthcare data. In one of the works [15], the major aim is to find similar kinds of patterns in data using clustering and visualize various states of India based on the factors for finding the cause of malnutrition in kids. To cluster the Indian states, K-means clustering is executed, and value of k is found through the elbow method. After clustering, WEKA is used for visualization. Tableau tool provides interactive visualizations which helps in better understanding of the data [16]. This aspect of tableau helps in revealing deeper meaning in healthcare data.

Few works are done to predict heart- and cancer-related problems in diabetic patients using data mining techniques [17, 18]. Presence of those chronic diseases is predicted with the help of factors like obesity, age, and some other habits. Particle swarm optimization is used for identification of attributes and extracting them. An improved K-nearest neighbor is used for classification purpose which helps in getting good accuracy. A similar paper [19] concentrates on prediction of diabetes, heart disease, and breast cancer. Techniques such as association rule mining, clustering, and classification have been used.

After going through all the various models used in various datasets, this work can add on by using some of these algorithms on a dataset that has not been explored. The dataset undergoes preprocessing which is further exposed to algorithms like elbow method, K-means clustering, and decision tree. After applying preprocessing techniques and implementation of clustering and classification algorithms, visualization is done using tableau. Rather than focusing on a particular type of heart or cancer disease, this work takes into consideration a general case of both the diseases.

3 Data Description

The source of the dataset is CHSI—community health status indicators [20] which contains the information about heart, cancer, and obesity diseases, out of which this work is using heart- and cancer-related data. The data used in this work is publicly available. The dataset has different files. The files included are as follows: Data element description, demographics, defined data value, measures of birth and death, leading causes of death, summary measures of health, risk factors and access to care, healthy people 2010, relative health importance, preventive services use, and vulnerable pops and env health. From these files, only two files are used for this project- 'Leading causes of death' and 'Risk factors and access to care.' All the files present in the CHSI dataset contains county-wise data. But, as per the requirement and necessity, the data has been converted to state-wise data.

LEADING_CAUSES_OF_DEATH.csv—This file has 738,369 instances and 235 attributes. It contains the percentage of patients, based on age in each county who died due to diseases like heart and cancer. The data gives us the number of patients in each state and each county. RISK_FACTORS_AND_ACCESS_TO_CARE.csv—This file contains 97,402 instances and 31 attributes. This file contains data about the various factors that lead to the death of patients suffering from heart and cancer disease and percentage of people affected by those risk factors.
DATA_ELEMENT_DESCRIPTION.csv—This folder contains the description of each and every element of every file in the CHSI dataset.
DEFINED_DATA_VALUE.csv—This folder contains the description of every missing value and its meaning.

Out of 235 attributes in Leading_causes_of_death.csv file, 33 attributes have been taken into consideration which are shown in Table 1. Out of 31 attributes in Risk_factors_and_access_to_care file, 8 attributes have been taken into consideration which are mentioned in Table 2.

Attribute of Leading_causes_of_death.csv file contains three parts. First part describes the age group of a patient.

- B stands for people from 1 to 14 years
- C stands for people from 15 to 24 years
- D stands for people from 25 to 44 years
- E stands for people from 45 to 64 years.
- F stands for people of 65 + years.

Second part describes the type of person.

Table 1 Attributes in Leading_causes_of_death.csv file

Attributes	Attributes	Attributes
CHSI_State_Name	F_Hi_HeartDis	D_Bl_Cancer
D_Wh_HeartDis	F_Ot_HeartDis	D_Hi_Cancer
D_Bl_HeartDis	B_Wh_Cancer	D_Ot_Cancer
D_Hi_HeartDis	B_Bl_Cancer	E_Wh_Cancer
D_Ot_HeartDis	B_Hi_Cancer	E_Bl_Cancer
E_Wh_HeartDis	B_Ot_Cancer	E_Hi_Cancer
E_Bl_HeartDis	C_Wh_Cancer	E_Ot_Cancer
E_Hi_HeartDis	C_Bl_Cancer	F_Wh_Cancer
E_Ot_HeartDis	C_Hi_Cancer	F_Bl_Cancer
F_Wh_HeartDis	C_Ot_Cancer	F_Hi_Cancer
F_Bl_HeartDis	D_Wh_Cancer	F_Ot_Cancer

Table 2 Attributes in Risk_factors_and_access_to_care

Attributes	Attributes
CHSI_State_Name	High_Blood_Pres
CHSI_County_Name	Smoker
Few_fruit_veg	Obesity
No_Exercise	Diabetics

- 'Wh' represents white person
- 'Bl' represents black person
- 'Hi' represents Hispanic person
- 'Ot' represents person belonging to the rest of the categories.

Third part describes the type of disease.

- HeartDis represents heart disease
- Cancer represents cancer disease.

The attribute CHSI_State_Name and CHSI_County_Name contain the state of the USA and counties in those states, respectively. Few_Fruit_Veg signifies the availability of fresh fruits and vegetables in that region. The percentage of people who do exercise regularly is mentioned in the attribute No_Exercise. High blood pressure patients, count of smokers, obese people, and percentage of diabetic patients are specified in the attribute High_Blood_Pres, smoker, obesity, and diabetics, respectively.

4 Proposed Methodology

The architecture follows a top–down method in executing the algorithms. Each different branch originating from a same given node represents a different algorithm used on the particular dataset. The whole original dataset is split into heart and cancer.

Machine learning and data mining techniques [21] play a vital role in describing several aspects of cancer and heart diseases. In Fig. 1, the architecture diagram shows different phases in the overall work done. There are total four phases, which are explained here.

Phase 1: The first phase in this work is preprocessing of the dataset.
Phase 2: In the second phase, the elbow method is implemented followed by the K-means clustering algorithm.
Phase 3: In phase three, the decision tree algorithm has been applied and attributes that are a risk factor for heart and cancer disease are found separately.
Phase 4: In the fourth phase, which is the final phase, visualization of the cluster of states based on various attributes is done.

4.1 Data Preprocessing

The key step to any data mining and machine learning experiment is data preprocessing. As the CHSI dataset is a huge volume dataset, choosing the right set of files and attributes for analyzing and visualizing is essential. All the data including attributes, files, and tuples that are required to achieve the problem statement is identified. In this step, division of the identified dataset is done into two parts—heart disease dataset and cancer disease dataset. Other datasets like age-wise datasets are also created for both heart and cancer disease using R programming language. Also, a dataset regarding the risk factors is created using R for both the diseases. The data has many missing or non-recorded values which have been dealt by complete case analysis or list-wise deletion. In this method, the tuples, which have missing values, are eliminated, and the rest of the data is used. This whole process is known as data cleaning.

4.2 K-Means Clustering

Before implementing the K-means clustering algorithm, it is very crucial to figure out the value of k which is number of clusters to be found in the dataset. Elbow method is implemented to find the most suitable value of k [22]. The elbow of the plot is considered the most appropriate k value for a given data. The plot has k values on the x-axis and sum of squared errors on the y-axis. In elbow method, series of

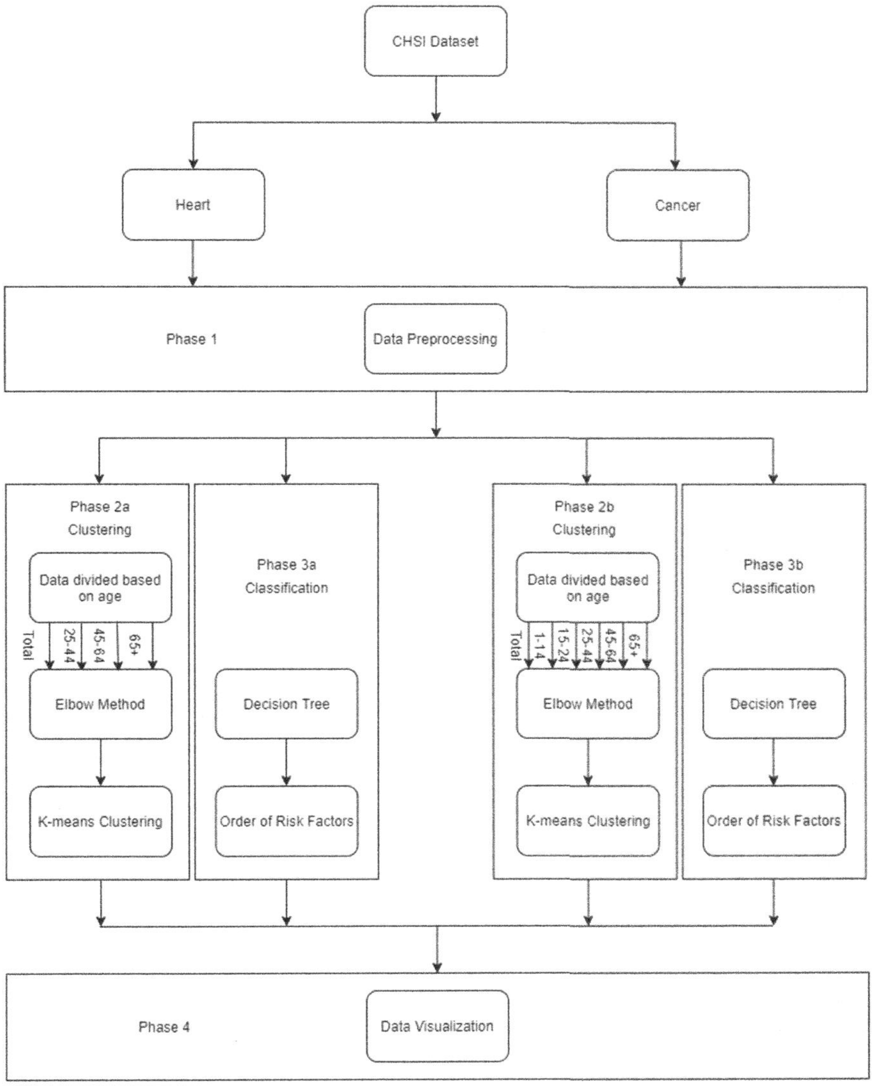

Fig. 1 System architecture

k-means clustering is done on a range of k values, and the best k value among them is chosen.

The elbow method is separately implemented for age-wise data and the total number of deaths for both the diseases. A total of ten elbow plots have been implemented. The elbow showed k as four for the heart dataset which contains the total number of deaths, for three different age groups 25–44, 45–64, and 65+. Similarly, the elbow showed k as four for the cancer data which contains the total number of deaths, for three age groups 15–24, 45–64, and 65+ and k as 3 for two age groups 1–14 and 25–44.

K-means clustering is an algorithm implemented with a dataset where the class label is not specified. The main focus is to find out collections or categories in the dataset [23–25]. This algorithm is used for clustering various states of the USA based on the total death toll in patients due to heart and cancer disease. This algorithm is also used to cluster the states based on the count of heart patients and cancer patients from each age group using the newly obtained k value from the elbow method. The K-means clustering algorithm is taken as it gives better result than rest of the clustering algorithms with minimum data preprocessing. A basic model of K-means algorithm is used in this work. The computational complexity of this algorithm is $O(n^2)$.

4.3 Decision Tree

The decision tree is basically defined as a flowchart, where each internal node signifies an experiment for the attribute, every part shows the output of the experiment, and every leaf node assigns a class. Decision trees can manage multidimensional data. The exhibition of trees helps us in gaining information. Generally, decision trees will have excellent accuracy [26]. Even though, validity and accuracy can be based on the data.

Using the various clusters that have been obtained using K-means clustering, along with the 'Risk factor and access to care' file, decision tree classifier was implemented to further find the ranking of risk factors that contribute to heart and cancer diseases. Decision tree is built in python. Feature importance score in python is implemented to score each attribute. The feature with the highest score is ranked the most important and so on. Based on this, correlation between the causes of both the diseases was found. A basic model of decision tree algorithm is used in this work. So, the computational complexity of the algorithm, as already defined, is $O(n \log n)$.

4.4 Data Visualization

Tableau tool is used for graphical visualization of data. Visualization of various states of the USA is done which are clustered based on the total death toll due to heart and cancer patients, respectively. A bar graph representing the total death toll due to heart disease in each age group in each state is shown in Fig. 2, and a scatter plot representing the total death toll due to cancer disease in each age group in each state is shown in Fig. 3. Different states in USA are clustered based on the number of heart patients in the three age groups that are 25–44, 45–64, and 65+. Similarly, visualization is done for cancer patients in five different age groups that are 1–14, 15–24, 25–44, 45–64, and 65+ .

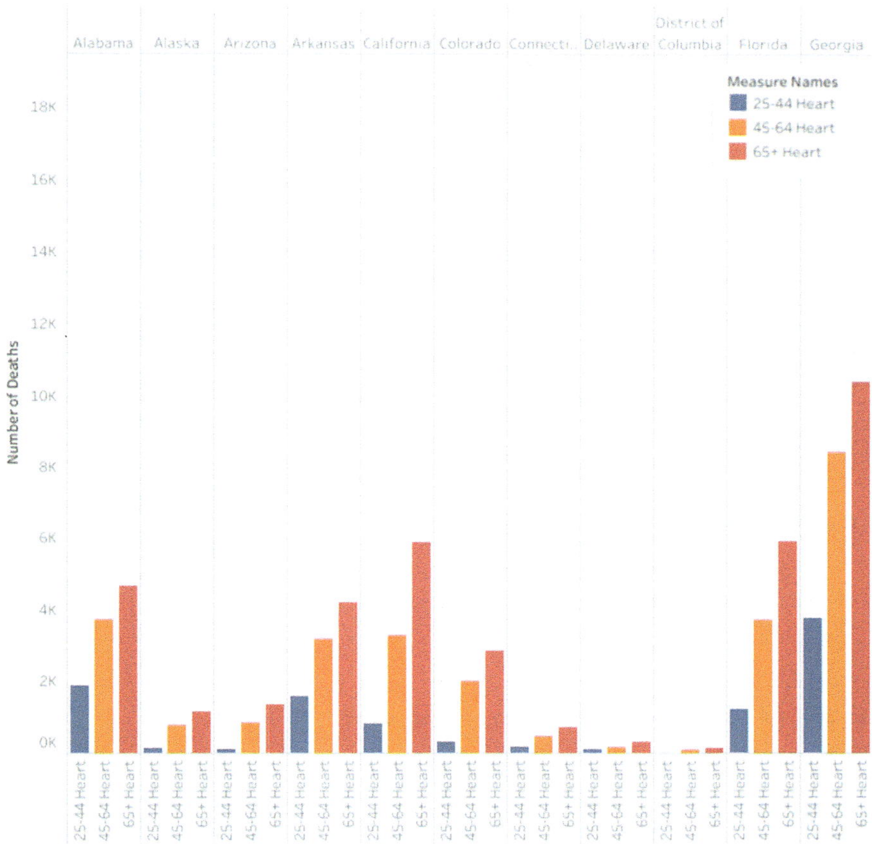

Fig. 2 Number of heart patients in each age group in each state

States of the USA are visualized based on the risk factors involved in these diseases. Visualization of cluster of states, which is shown in Figs. 5, 6, and 7, is done based on six criteria, i.e., the percentage of diabetic patients, high blood pressure patients, obese people, count of smokers, population who do less or no exercise, and people who have less access to fresh fruits and vegetables. Also, in Figs. 8 and 9, decision tree showing the risk factors involved in heart disease and cancer is visualized [27–30].

5 Result and Analysis

All the visualization of the clusters is done using tableau, and decision tree is obtained using python.

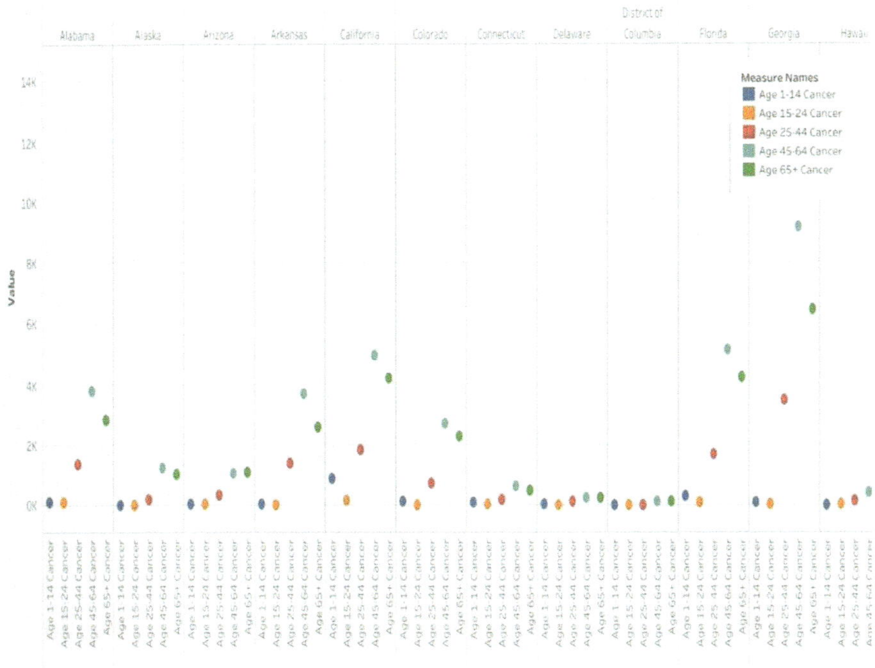

Fig. 3 Number of cancer patients in each age group in each state

A snapshot of a bar graph representation is exhibited in Fig. 2, showing the number of deaths caused by heart patients in each age group in each state. From the plot, it is clear that the minimum number of deaths in every age group is in the District of Columbia. The maximum number of deaths in the age group 25–44 is in Georgia, whereas in the age category 45 and above is in Texas.

A snapshot of a scatter plot representation is exhibited in Fig. 3, showing the number of deaths of cancer patients in each age group in each state. From the plot, it is clear that the minimum number of deaths in an average is in the District of Columbia. The maximum death toll in the age group 1–14 is in California, whereas in the age category 15 and above is Texas.

Figure 4 shows the states of the USA clustered based on the total number of deaths due to heart disease overlapped with the deaths in the age group 25–44. The transition from light red to dark red shows the various clusters ranging from 1 to 4 based on the total number deaths. Colored dots represent various clusters based on the age group 25–44 ranging from 1 to 4 where 1 being the cluster with states having the highest number of deaths and 4 being the one with least number of deaths recorded. Cluster 1 is represented using dark red dot, cluster 2 using light red, cluster 3 using light grey, and cluster 4 using dark grey. Table 3 shows the states of the USA clustered into four clusters based on the total number of deaths caused by heart disease.

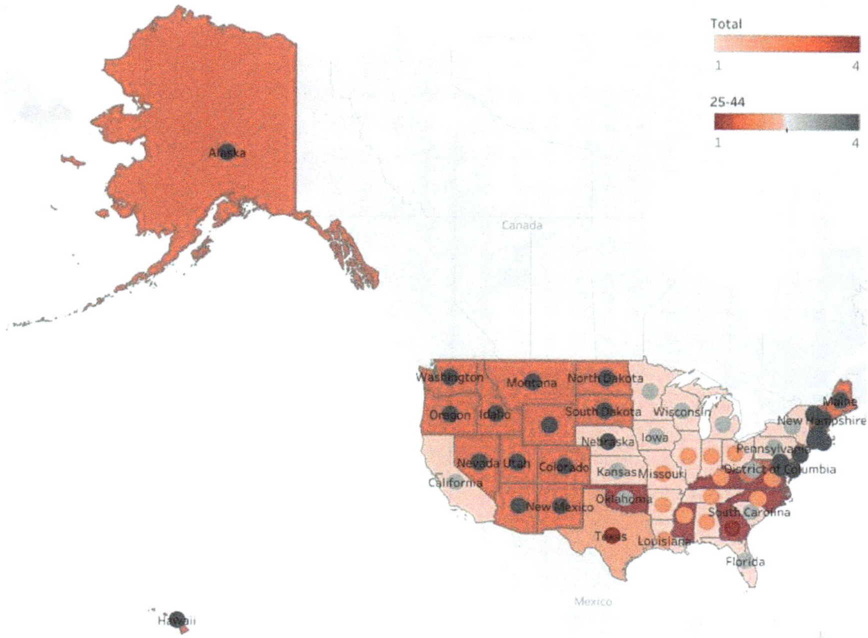

Fig. 4 Clustering of total number of heart patients and also patients in the age group 25–44

Table 3 Clusters based on deaths due to heart disease

Cluster	States	Color	Description
1	Alabama, Arkansas, California, Florida, Illinois, Indiana, Iowa, Kansas, Louisiana, Michigan, Minnesota, Missouri, Nebraska, New York, Ohio, Pennsylvania, South Carolina, Tennessee, Wisconsin	Light pink	Second lowest recorded number of deaths
2	Texas	Orange	Highest number of deaths
3	Alaska, Arizona, Colorado, Delaware, District of Columbia, Connecticut, Hawaii, Idaho, Maine, Maryland, Massachusetts, Montana, Nevada, New Hampshire, New Jersey, New Mexico, North Dakota, Oregon, Rhode Island, South Carolina, South Dakota, Utah, Vermont, Washington, West Virginia, Wyoming	Red	Lowest number of deaths
4	Georgia, Kentucky, Oklahoma, Virginia, Mississippi, North Carolina	Dark red	Second lowest number of deaths

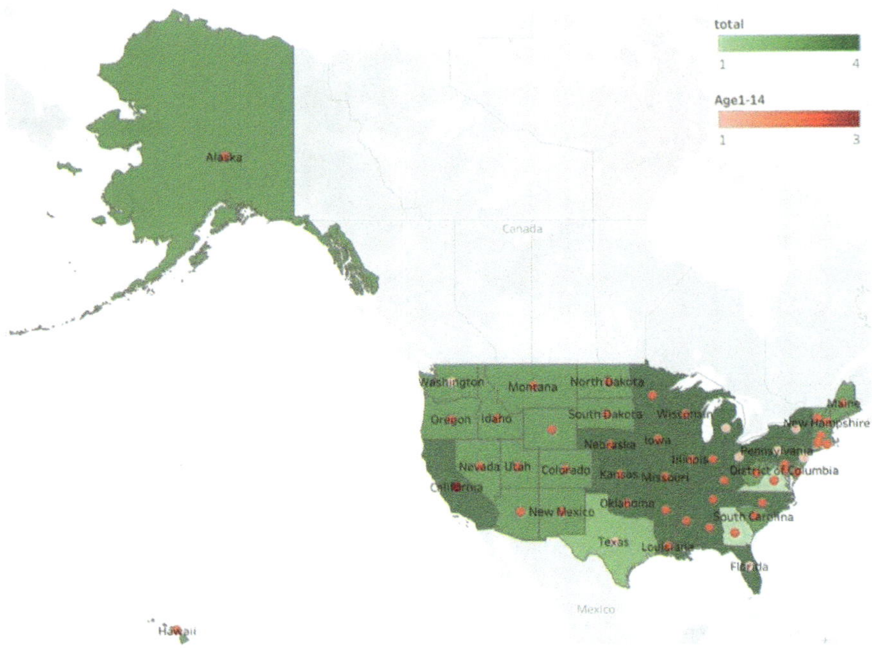

Fig. 5 Clustering of cancer patients and also patients in the age group of 1–14

Similarly, as represented in Fig. 4, clustering is done based on the number of deaths due to heart disease in the age group 45–64 and also above the age of 65, and the same has been visualized.

Figure 5 shows the states of the USA clustered based on the total number of deaths due to cancer overlapped with the deaths in the age group 1–14. The transition from light green to dark green shows the various clusters ranging from 1 to 4 based on the total number deaths. Colored dots represent various clusters based on the age group 1–14 ranging from 1 to 3 where 3 being the cluster with states having the highest number of deaths and 2 being the one with least number of deaths recorded. Cluster 1 is shown using light pink, 2 using orange, and 3 using dark red. Table 4 shows the states of the USA clustered into four clusters based on the total number of deaths caused by Cancer.

Similarly, as represented in Fig. 5, clustering is done based on the number of deaths due to cancer disease in the age groups 15–24, 25–44, 45–64, and also above the age of 65, and the same has been visualized.

Figure 6 shows the visualization of the states of the USA based on the percentage of diabetic people existing and also availability of fruits and vegetables. Light-yellow shaded state represents fewer diabetic populations. The transition from yellow to green shows an increase in population affected by diabetes where dark green shows the population most affected and light-yellow least. Colored dots represent the availability of fruits and vegetables in each state of the USA. Red dots mean that the

Table 4 Clusters based on cancer disease

Cluster	States	Color	Description
1	Georgia, Virginia	Light green	Second highest recorded number of deaths
2	Texas	Mid light green	Highest number of deaths
3	Alaska, Arizona, Colorado, Delaware, District of Columbia, Connecticut, Hawaii, Idaho, Maine, Maryland, Massachusetts, Montana, Nevada, New Hampshire, New Jersey, New Mexico, North Dakota, Oregon, Rhode Island, South Carolina, South Dakota, Utah, Vermont, Washington, West Virginia, Wyoming	Mid dark green	Lowest number of deaths
4	Alabama, Arkansas, California, Florida, Illinois, Indiana, Iowa, Kansas, Kentucky, Louisiana, Michigan, Minnesota, Missouri, Mississippi, Nebraska, New York, North Carolina, Oklahoma, Ohio, Pennsylvania, Tennessee, Wisconsin	Dark green	Second lowest number of deaths

availability of fruits and vegetables is scarce. Transition from red to blue indicates the increase in availability of fruits and vegetables.

Figure 7 shows the visualization of the states of the USA based on the percentage of people having high levels of blood pressure and also who exercise frequently. Light brown shaded state represents fewer people with high blood pressure. The transition from light brown to dark brown shows an increase in population suffering from high blood pressure. Colored triangles represent the percentage of the population who exercise on a daily basis in each state of the USA. The presence of green triangles means that the population there exercises less. The states with blue triangles represent that most of the people in that region exercise.

Similarly, as represented in Figs. 6 and 7, clustering is done based on the number of deaths due to obesity and smoking, and the same has been visualized.

By analyzing the data collected from the images above, preventive measures to reduce the occurrence of heart and cancer diseases can be taken. Various healthcare sectors can be opened or other measures can be taken based on the visualizations.

Next decision tree are constructed to find the ranking of various risk factors that lead to heart disease and cancer diseases, respectively, as shown in Figs. 8 and 9. Feature importance score is used to score each attribute. The feature with the highest score is ranked the most important. Figures 8 and 9 also contain entropy score of each attribute which helps in building the decision tree. Figure 10 shows the bar graph that shows the scores of each attribute for both heart and cancer. Ranking of

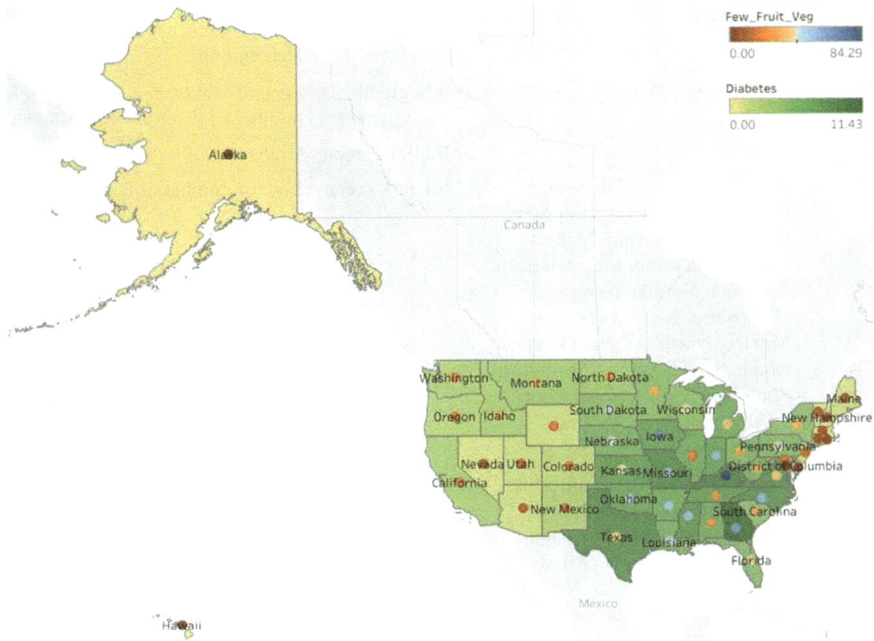

Fig. 6 Visualization of diabetic people and availability of fruits and vegetables in USA

the risk factors from highest to least for heart are as follows—diabetes followed by High_Blood_Pres, No_Exercise, obesity, smoker, and Few_Fruit_Veg. Ranking of the risk factors from highest to least for cancer are as follows—diabetes followed by High_Blood_Pres, Few_Fruit_Veg, and smoker.

Looking at both the feature rankings, it can be concluded that the two most important risk factors leading to both the disease which have to be dealt with is diabetes and high blood pressure. Handling these two risk factors can help reduce the number of deaths in the USA. The second thing to be noticed is that the amount of exercise and being obese is not a contributing factor to cancer but is a risk for heart disease.

6 Conclusion

This work has explored the CHSI dataset based on the patients in the states of the USA. The data was brought into a desired structure and extracted based on the chronic diseases heart and cancer along with the risk factors leading to these diseases. The need to understand whether there exists a relation between heart and cancer was understood based on the clusters and classification (ranking) results. Visualizations are done based on the number of patient's age wise (Age groups for Heart: 25–44, 45–64, and 65+. Age groups for Cancer: 1–14, 15–24, 25–44, 45–64, and 65+) and

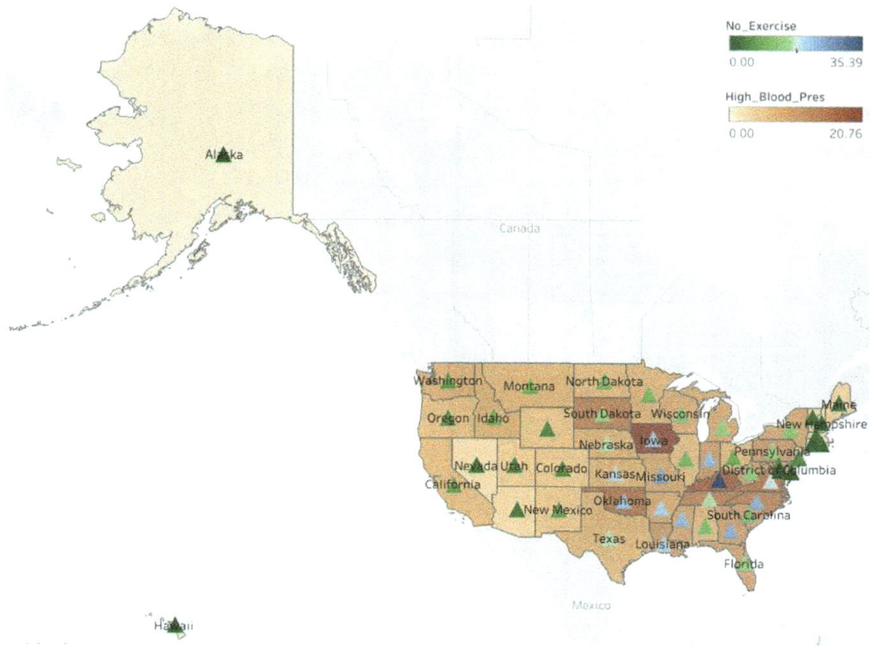

Fig. 7 Visualization of people with high blood pressure and people who exercise on a daily basis in USA

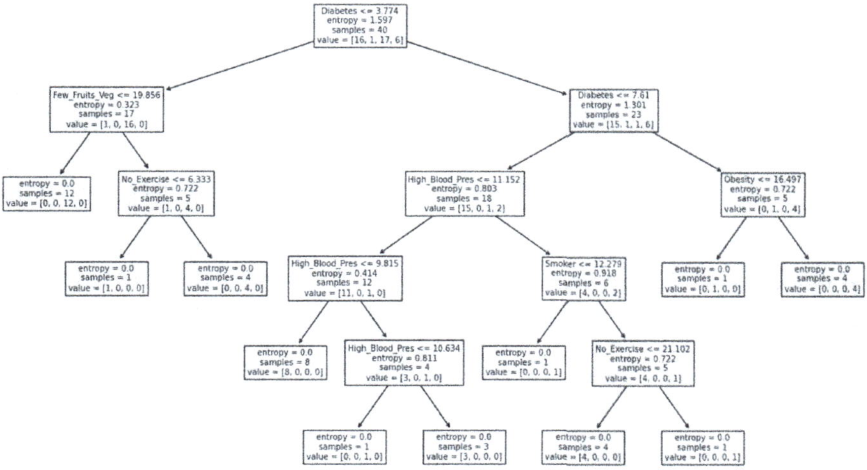

Fig. 8 Decision tree using risk factors that lead to heart disease

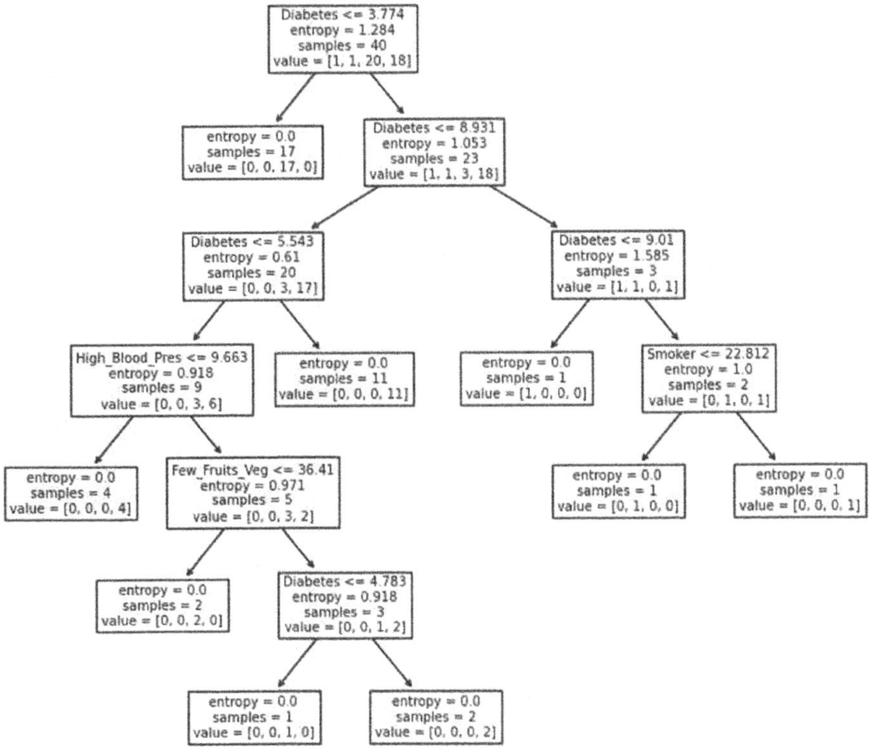

Fig. 9 Decision tree using risk factors that lead to cancer

based on the risk factors. This study can help understand which states of the USA are more prone to the disease and what are the leading cause of these diseases. After conducting the experiment, the derived conclusion is that the two major risk factors leading to both these diseases, diabetes and high blood pressure, in the states of the USA are similar, which matches closely to a blog [31] which mentions heart disease and cancer as the top two killers in USA, risk factors of these diseases and figures representing the average death annually due to these diseases. This work has also focused on analyzing and visualizing the number of deaths due to both the diseases, age wise, in different states, which would help build a diagnosis and preventive mechanism accordingly.

7 Future Scope

In future, other files of the CHSI dataset can be explored and visualized. Many more diseases or causes of death can be taken into consideration while analyzing the risk

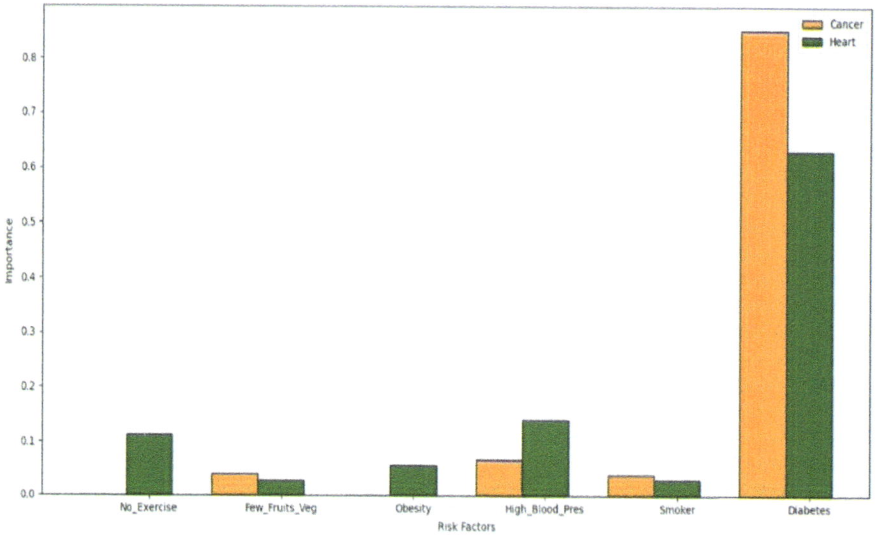

Fig. 10 Ranking of risk factors that lead to heart and cancer disease

factors. Many different algorithms can be implemented on the dataset. As the CHSI dataset is a huge dataset, a lot of exploration, analysis, and visualization can be done which would help in the medical sector of the USA. The future work would help track down various other diseases which are caused due to the risk factors mentioned in this dataset.

References

1. D. Gupta, S. Khare, A. Aggarwal, A method to predict diagnostic codes for chronic diseases using machine learning techniques. in *2016 International Conference on Computing, Communication and Automation (ICCCA)* (Noida, 2016), pp. 281–287. https://doi.org/10.1109/CCAA.2016.7813730
2. P. Ramachandran, N. Girija, T. Bhuvaneswari, Early detection and prevention of cancer using data mining techniques. Int. J. Comput. Appl. **97**(13), 48–53 (2014)
3. S. Khare, D. Gupta, Association rule analysis in cardiovascular disease. in *2016 Second International Conference on Cognitive Computing and Information Processing (CCIP)* (Mysore, 2016), pp. 1–6. https://doi.org/10.1109/CCIP.2016.7802881
4. N. Jothi, N.A.A. Rashid, W. Husain, Data mining in healthcare—a review. Procedia Comput. Sci. **72**, 306–313 (2015)
5. I. Yoo, P. Alafaireet, M. Marinov, K. Pena-Hernandez, R. Gopidi, J.F. Chang, H. Lar, Data mining in healthcare and biomedicine: a survey of the literature. J. Med. Syst. **36**, 2431–2448 (2011)
6. G. Ogbuabor, F.N. Ugwoke, Clustering algorithm for a healthcare dataset using Silhouette score value. Int. J. Comput. Sci. Inf. Technol. 10 (2018)

7. M. Deepika, K. Kalaiselvi, An empirical study on disease diagnosis using data mining techniques. in *2018 Second International Conference on Inventive Communication and Computational Technologies* (2018), pp. 615–620
8. A. Das, A. Halder, S. Samaddar, Analysis of behavioral patterns of different intelligent agents, k-nearest neighbour and probabilistic neural network. in *2017 International Conference on Energy, Communication, Data Analytics and Soft Computing (ICECDS)* (Chennai, 2017), pp. 439–442. https://doi.org/10.1109/ICECDS.2017.8390204
9. S. Babu, E.M. Vivek, K.P. Famina, K. Fida, P. Asawathi, M. Shanid, M. Hena, Heart disease diagnosis using data mining technique. in 2017 International conference of Electronics, Communication and Aerospace Technology (ICECA), Coimbatore, 2017, pp. 750-753, doi: https://doi.org/10.1109/ICECA.2017.8203643
10. R.A. Haraty, M. Dimishkieh, M. Masud, An enhanced K-means clustering algorithm for pattern discovery in healthcare data. Int. J. Distrib. Sens. Netw. **2015**, 1–11 (2015)
11. A. Golande, P. Kumar, Heart disease prediction using effective machine learning techniques. Int. J. Recent Technol. Eng. **8**, 2277–3878 (2019)
12. P. Ramachandran, N. Girija, T. Bhuvaneswari, Early detection and prevention of cancer using data mining techniques. Int. J. Comput. Appl. **97**(13), 0975–8887 (2014)
13. T. Barker, J. Mallon, T. Sethi, D. Walk, Visualizing Health Indicators. https://cluster.ischool.drexel.edu/~cchen/courses/INFO633/11-12/g5_Barker.pdf. Accessed on 14 Jul 2020
14. H. Wu, S. Yang, Z. Huang, J. He, X. Wang, Type 2 Diabetes Mellitus prediction model based on data mining. Inf. Med. Unlocked **10**, 100–107 (2018)
15. N.A. Anilkumar, D. Gupta, S. Khare, D.M. Gopalkrishna, A. Jyotishi, Characteristics and causes of malnutrition across Indian states: a cluster analysis based on indian demographic and health survey data. in *2017 International Conference on Advances in Computing, Communications and Informatics (ICACCI)* (Udupi, 2017), pp. 2115–2123. https://doi.org/10.1109/ICACCI.2017.8126158
16. I. Ko, H. Chang, Interactive Visualization of Healthcare Data Using Tableau. Healthc. Inf. Res. **23**(4), 349 (2017)
17. C. Kalaiselvi, G.M. Nasira, Prediction of heart diseases and cancer in diabetic patients using data mining techniques. Indian J. Sci. Technol. **8**(14), 1 (2015)
18. B.H. Frank, Globalization of diabetes: the role of diet, lifestyle, and genes. Diabetes Care **34**(6), 1249–1257 (2011)
19. S. Vijiyarani, S. Sudha, Disease prediction in data mining technique–a survey. Int. J. Comput. Appl. Inf. Technol. **2**, 2278–7720 (2013)
20. Dataset. https://healthdata.gov/dataset/community-health-status-indicators-chsi-combat-obesity-heart-disease-and-cancer. Accessed on 14 Jul 2020
21. J. Han, M. Kamber, J. Pei, Data mining concepts and techniques
22. M.A. Syakur, B.K. Khotimah, E.M. Rohman, B.D. Satoto, Integration K-means clustering method and elbow method for identification of the best customer profile cluster. IOP Conf. Ser.: Mater. Sci. Eng. **336**, 012–017 (2018)
23. M. Shukla, S. Agarwal, Hybrid approach for tuberculosis data classification on using optimal centroid selection based clustering. in *2014 Students Conference on Engineering and Systems* (Allahabad, 2014), pp. 1–5. https://doi.org/10.1109/SCES.2014.6880115
24. B. Zheng, S.W. Yoon, S.S. Lam, Breast cancer diagnosis based on feature extraction using a hybrid of K-means and support vector machine algorithms. Expert Syst. Appl. **41**(4), 1476–1482 (2013)
25. B.M. Patil, R.C. Joshi, D. Toshniwal, Hybrid prediction model for type-2 diabetic patients. Expert Syst. Appl. **37**, 8102–8108 (2010)
26. S.B. Patel, P.K. Yadav, D.P. Shukla, Predict the diagnostic of heart disease patients using classification mining techniques. J. Agric. Vet. Sci. **4**(2), 61–64 (2013)
27. S. Drazin, M. Montag, Decision tree analysis using Weka. in *Machine Learning Project II* (University of Miami, 2012), pp. 1–3
28. Z. Markos, F. Doyore, M. Yifiru, J. Haidar, Predicting undernutrition status of under-five children using data mining techniques: the case of 2011 Ethiopian demographic and health survey. J. Health Med. Inf. 5 (2014)

29. V. Dominic, D. Gupta, S. Khare, A. Aggarwal, Investigation of chronic disease correlation using data mining techniques. in *2015 2nd International Conference on Recent Advances in Engineering and Computational Sciences (RAECS)* (Chandigarh, 2015), pp. 1–6. https://doi.org/10.1109/RAECS.2015.7453329
30. T.C. Sharma, M. Jain, WEKA approach for comparative study of classification algorithm. Int. J. Adv. Res. Comput. Commun. Eng. **2**(4), 1925–1931 (2013)
31. https://www.air-worldwide.com/blog/posts/2019/3/heart-disease-and-cancer-the-top-2-killers-in-the-u-s/. Accessed on 20 Jul 2020

A Hybrid Deep Learning Approach for Predicting the Spread of COVID-19

V. K. Manojkumar, N. M. Dhanya, and P. Prakash

Abstract COVID-19 pandemic is creating a devastating effect on the global population in terms of health and wealth. The number of patients detected with COVID is increasing day by day and touched a total case count of around 11,000,000 and death count of 5 lakhs by July 2020. Along with other nations, India is also witnessing an exponential growth in the number of COVID cases. Providing proper treatment to infected patients seeks very close attention. There should be an efficient mechanism to predict the expected number of patients daily to prepare our medical system to accommodate them. This paper applies various deep learning models to predict the number of COVID cases. The paper also compares different deep learning techniques and traditional time-series analysis, such as ARIMA, and concludes that the deep learning hybrid model of CNN-LSTM predicts better compared to other models.

Keywords COVID prediction · Deep learning · Hybrid model · Predictive analytics

V. K. Manojkumar (✉) · N. M. Dhanya
Department of Computer Science and Engineering, Amrita School of Engineering, Amrita Vishwa Vidyapeetham, Coimbatore, India
e-mail: vkmanojk@gmail.com

N. M. Dhanya
e-mail: nm_dhanya@cb.amrita.edu

P. Prakash
School of Computer Science and Engineering, Vellore Institute of Technology, Chennai, India
e-mail: prakash.p@vit.ac.in

M. Paprzycki et al. (eds.), *Intelligent Systems, Technologies and Applications*,
Advances in Intelligent Systems and Computing 1353,
https://doi.org/10.1007/978-981-16-0730-1_13

1 Introduction

The entire world is in a health emergency due to the COVID-19 pandemic. This contagious disease started in the city of Wuhan, Hubei, China, in early December 2019. The World Health Organization (WHO) named the novel coronavirus SARS-CoV-2 disease as "COVID-19" on 30th January 2020 [1]. In March 2020, WHO declared COVID-19 as a global pandemic. The new virus has caused a significant threat to people's health due to potential harm and its nature of spreading.

The total number of people infected with COVID-19 worldwide is **11,468,979** as on 7 July 2020, whereas the numbers of reported deaths and recoveries are **535,181** and **6,036,634**, respectively. Most of the cases are recorded in the USA, Brazil, India, Russia, Peru, Chile, UK, Mexico, Spain and Iran [2], and these are the worst-affected countries. Figure 1 shows the worst-hit countries of COVID-19 disease. The number of cases in India is 719,661. Since February 2020, information technology services, such as mobile apps, have been used to track the potential risk of infection. The mobile applications suggest users to self-quarantine and alert the concerned health authorities when the virus infects someone. They also monitor infected people and the persons that they had contact with [3] in the recent past. Arogya Sethu is developed in India for these purposes. It is used for "Contact tracing, Syndromic mapping and Self-assessment" developed by the National Informatics Centre under the Ministry of Electronics and Information Technology.

Amidst such a very critical ongoing global pandemic crisis which has also affected the global economy very badly, this paper proposes a hybrid model for predicting the number of patients. Figure 2 shows the confirmed, death and recovery data.

Since the COVID data is a time series, time-series data analysis [4] techniques can be used for the analysis of this data. In this paper, we are using few time-series analysis and a hybrid model to predict the spread of COVID-19.

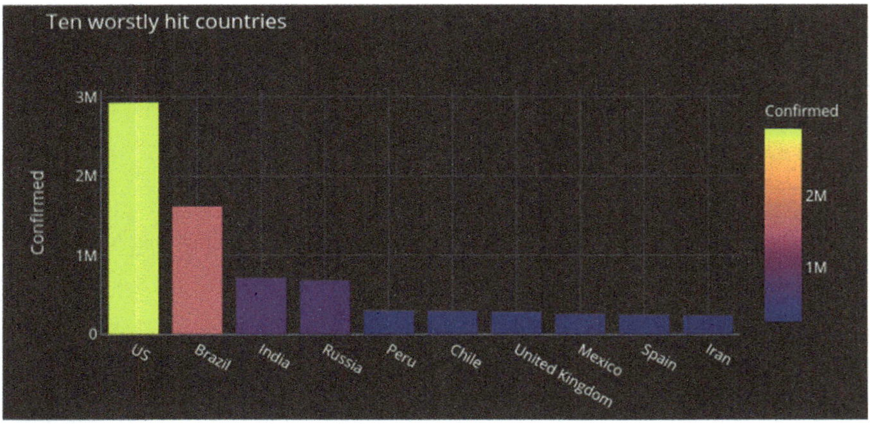

Fig. 1 Worst-hit countries

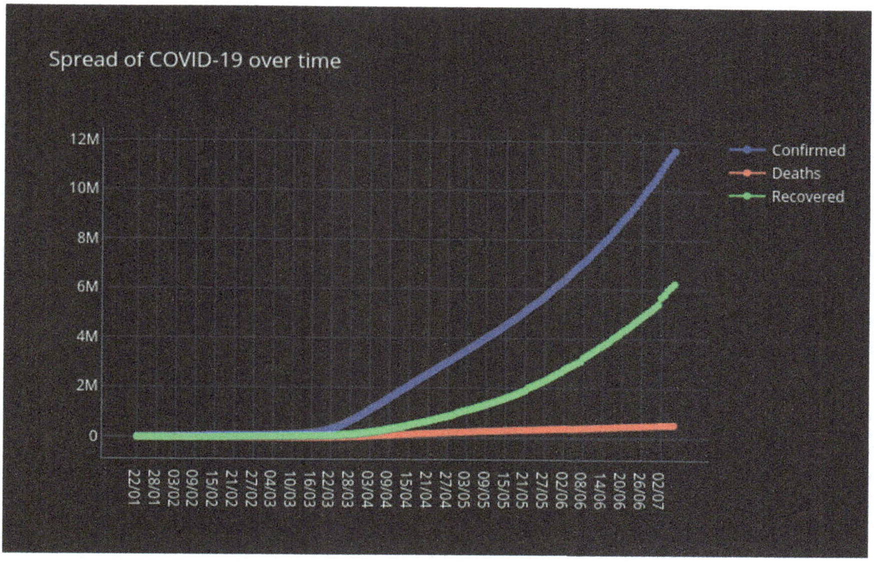

Fig. 2 The confirmed, death and recovered statistics

The organization of this paper is as follows, and Sect. 2 starts with surveys on the existing research whose inferences are aligned with our work. Section 3 explains the dataset features and pre-processing techniques employed. Section 4 describes the methodology used for the comparative study and evaluation for different methods. The results obtained from the deep learning model have been furnished in Sect. 5. The concluding remarks are presented in the last section.

2 Contributions of This Study

The following are the core contributions of this study:

- An AI-based intelligent predicting system to predict next day's number of confirmed patients using five different mechanisms, namely traditional ARIMA, MLP, CNN, LSTM and hybrid CNN and LSTM model.
- The proposed system has been trained on real-time data from John Hopkins University, which is updated on a daily basis.
- The dataset is pre-processed, and all models are applied to these pre-processed data.
- Comparative study of these models are generated based on MSE, RMSE, MAE and MAPE.
- The hybrid model gives the best results for predicting the future number of confirmed cases.

3 Literature Survey

Wang et al. [5] proposed a mechanism to detect this respiratory disease by investigating the radiographic changes in the CT scan images. A deep learning-based image analysis is done to extract the COVID specific features and clinical conclusion is given before the pathogenic test which in turn reduces the diagnostic time. Out of 1065 images of both COVID and non-COVID cases, the model shows an accuracy of 89.5%. Li et al. [6] created a simulation model to predict the incubation period, basic generations, average number of days to cure and average number of infected people. The data of Iran, Italy, Hubei and Korea is compared along with the simulated data.

Verity et al. [7] collected the COVID case data from the Hubei region, China and outside mainland China data. This data is used to analyse the time between the symptoms and outcome of the disease. Age stratified estimate of the fatality rate is also evaluated. An analysis is also done on the proportion of the population who needs hospitalization.

Sengupta and Mugde [8] analysed the COVID data with features like gender, location and age. They used machine learning algorithms for predicting the near-future scenario. A sigmoid model is used to predict the day in which peak cases are going to be reported. The data cleaning and feature engineering are done for better prediction.

Sujath et al. [9] created three models such as linear regression, MLP and vector autoregression for predicting the spread of COVID. Common data such as confirmed, received and death case dataset across India over time is taken and this is used for prediction. Roy et al. [10] investigated the spread of coronavirus and created a series of visualizations to analyse the data. They proposed an additive regression model which is taking into consideration the time-series nature of the data.

Anastassopoulou et al. [11] estimated the case lethality and case redemption ratios, accompanying 90% of confidence intervals as the outbreak emerges. The authors provided an estimation of the reproduction number (R0), which tells us the disease transmission growth and daily infection and recovery rates with the help of the Susceptible–Infectious–Recovered–Dead (SIDR) model. By adjusting the parameters in the SIDR model, they also forecasted the outbreak three weeks ahead in the epic centre at Hubei, China. SIRD modelling not considered critical factors such as transmission due to the contact, preventive action measures and the characteristics of the people's culture.

Corona Tracker Community Research group used the Susceptible–Exposed–Infectious–Recovered (SEIR) [12] model to understand the public sentiment patterns related to the health information provided by the government and also assess the legislative impact and economic significance over the expanse of the virus. The group found that more number of negativity articles compared to the positivity articles. They also mentioned that the transparency of the data also makes the SEIR model more accurate. The research helped on how to disseminate the information across the people to make them more proactive. The model has not predicted how many people will have COVID-19 soon, also not considered any COVID-19 disease dataset to perform an analysis.

Roy et al. [10] proposed an additive regression model to analyse the COVID time-series data and visualized the data over the country map. End of the research found that the lack of awareness regarding the social distance played a vital role in the virus spread. They predicted the number of cases with 93% of accuracy from the dataset during the period from 22nd January 2020 till 26th April 2020 across the globe.

Tuli et al. [13] came with the ML-based model to predict the possible threat of COVID-19 over all the countries. They referred to the data from the "Our World in Data" dataset, which is updated frequently. The robust curve fitting using the iterative weighting algorithm introduced, which is a step-by-step algorithm for iterative weighted curve fitting using *tanhshrink* function. The Robust Weibull iterative model outperforms the baseline Gaussian model, which is unfortunate in fitting the data, which may lead the situation to worsen.

From the above literature, there are some attempts which explore COVID data in various ways such as image analysis to detect COVID, data visualization and machine learning-based approaches. There is a lack of methods which will predict the number of COVID patients. Hence, this paper tries to predict the spread of COVID using a hybrid deep learning approach and this method gives better results in terms of prediction than other traditional time-series analysis and deep learning approaches.

4 Methodology and Results

The following section explains various steps involved in the proposed system. Figure 3 depicts the flow of the system. The system takes the raw data and preprocess data, which may feed into the learning model to predict the COVID cases daily. The various models have been applied to the dataset, such as CNN, LSTM, ARIMA and so on. The performance of the models has been evaluated based on metrics like root mean square error (RMSE), mean absolute error (MAE) and so on. Finally, the model predicts the number of COIVD cases on each day.

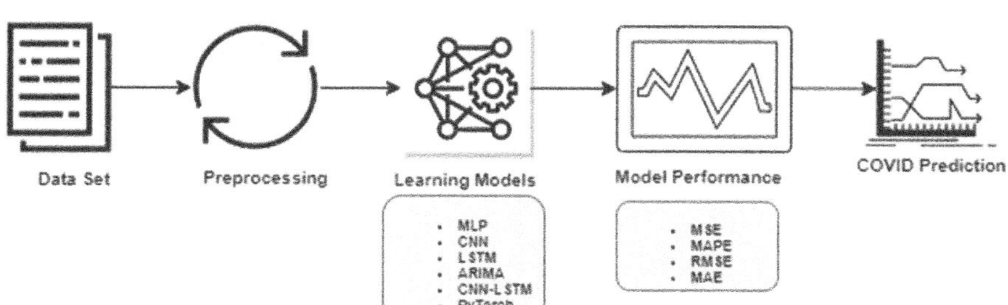

Fig. 3 System flow

4.1 Dataset

The dataset is in the form of a CSV file and is from the John Hopkins University GitHub page. On a daily basis, they are updating the data and are in separate CSV files for confirmed, deaths and recovered cases. Necessary data cleaning such as normalization, missing data filling and consolidation is done as pre-processing. The dataset is licensed under the Open Data Commons Public Domain and Dedication License. The data and its structure can be viewed as well as download it in alternative formats (e.g. JSON) from the DataHub [14]. The dataset contains the province/state, Country/Region, latitude, longitude and dates from 22nd January till date of confirmed, death and recovered number of patients. Day-wise information is available in the dataset.

4.2 Pre-processing

The data, in its raw form, contains an Unnamed column and few NaNs in the Recovered and the Province/State column. The unnamed column was dropped, and the NaNs were filled with zeros in the Recovered column and "Not specified" in the Province/State column. The country-wise and date-wise data was then extracted and stored separately as CSV files. The number of confirmed cases, fatalities and recovered cases were scaled using MinMaxScaler. The names of the countries were encoded using LabelEncoder. The data was then transformed into series and sequences so as to cater to the needs of the input requirements of different models.

The parameters which are finalized for all deep learning models are given in Table 1.

Table 1 Model parameters

Parameter	Value
Epochs	20
Optimizer	Adam
Learning rate	0.0003
Loss function	Mean square error
Parameter	Value

5 Models

5.1 *Multi-layer Perceptron*

The MLP model has input features equal to the window size (30). MLP models do not take the input as sequenced data, so for the model, it is just receiving inputs and does not treat them as sequenced data. In our case that may be a problem since the model won't see the data with the sequence pattern that it has.

5.2 *Convolutional Neural Network*

For the CNN mode configured with three convolutional hidden layers with a max-pooling layer. The filter maps are then flattened before being interpreted by six dense layers and outputting a prediction. The convolutional layer should be able to identify patterns between the timesteps.

5.3 *Long Short-Term Memory*

The LSTM model actually sees the input data as a sequence, so it is able to learn patterns from sequenced data (assuming it exists) better than the other ones, especially patterns from long sequences.

5.4 *Autoregressive Integrated Moving Average*

ARIMA is one of the traditional models used in time-series analysis either to understand the data or predict the future points in the series. ARIMA models are denoted as ARIMA (p, d, q) where p denotes the number of time lags (AR part), d denotes the degree of difference between timesteps and q denotes the order of MA model. It predicts the future point as a function of the data points present in the dataset. The best configuration for the dataset is ARIMA (4, 2, 0).

5.5 *CNN–LSTM Hybrid Model*

The CNN–LSTM hybrid model combines the benefits of both CNN and LSTM to output a prediction by dividing each sample into further subsequences. The model uses three time-distributed layers wrapped around convolutional layers which are

```
CoronaVirusPredictor(
  (lstm): LSTM(1, 128, num_layers=4, dropout=0.0029999999999999996)
  (linear): Linear(in_features=128, out_features=1, bias=True)
)
```

Fig. 4 Architecture of all models

then flattened and fed to three LSTM layers to interpret the results before making a prediction.

5.6 PyTorch

PyTorch is a Python-based library that is widely used in deep learning. PyTorch involves python-like coding which makes it easier to work with. The proposed model inherits the nn.Module class (a wrapper around lower-level PyTorch code) to implement a four-layered LSTM model to output a prediction (Fig. 4).

6 Comparing the Models

The comparative results of the models explained in the above section is given in the graph below. The MSE is decreasing as the number of epochs increases. The PyTorch and ARIMA model work well with training data but fail miserably with validation data. This is due to the overfitting of the model with training data and the size of the dataset.

The model such as MLP, CNN, LSTM and the hybrid model works well on the training, and validation data makes the model better than ARIMA. Among these four models, the hybrid model performs well with this dataset. It gives the advantage of feature selection by CNN and the power of time-series analysis of LSTM models.

The MSE plots for various models are given in Fig. 5.

6.1 Metrics

$$\text{Mean Squared Error (MSE)} = \frac{1}{2}\sum (y - \hat{y})^2 \tag{1}$$

$$\text{Root Mean Squared Error (RMSE)} = \sqrt{\text{MSE}} \tag{2}$$

$$\text{Mean Absolute Error (MAE)} = \frac{1}{n}\sum (y - \hat{y}) \tag{3}$$

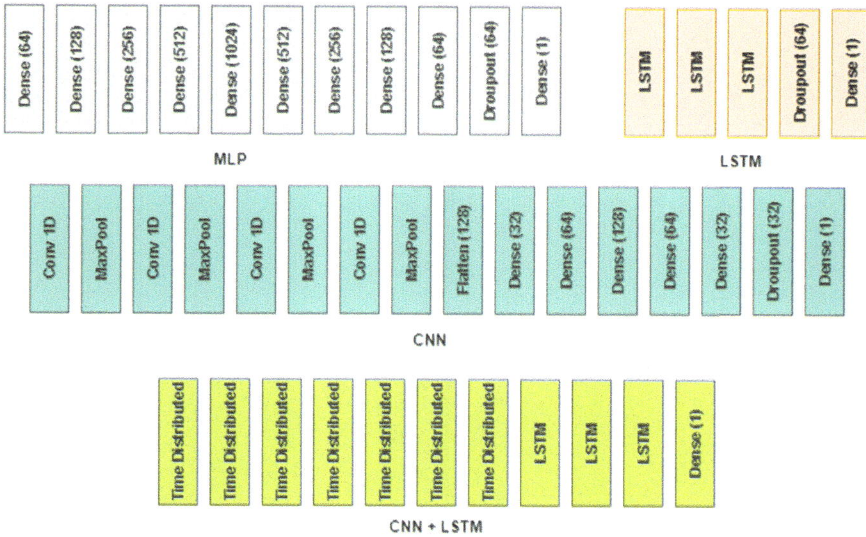

Fig. 5 Model architectures

$$\text{Mean Absolute Percentage Error (MAPE)} = \frac{100}{n}\sum\left|\frac{y-\hat{y}}{y}\right| \tag{4}$$

The MSE, RMSE and MAE are the most popular metrics for time-series analysis (Fig. 6).

The results are shown in Table 2. Both training and validation errors are shown in Table 2. In terms of MSE, RMSE and MAE, the hybrid model of CNN-LSTM shows the best results. Simple LSTM and CNN models are closely following the hybrid approach.

7 Conclusion and Future Scope

A comparative study on the number of COVID patients with different models is established in this paper. The results show that the hybrid model of CNN-LSTM gives the best results. The results are also compared with the traditional time-series prediction approach ARIMA. A promising result is obtained from the hybrid model, and this can be used for predicting the number of patients in the worldwide scenario. This will be helpful in predicting future cases and which will make the governments prepare and plan to face this pandemic in an effective manner. More deep learning models can be applied to make the prediction more accurate.

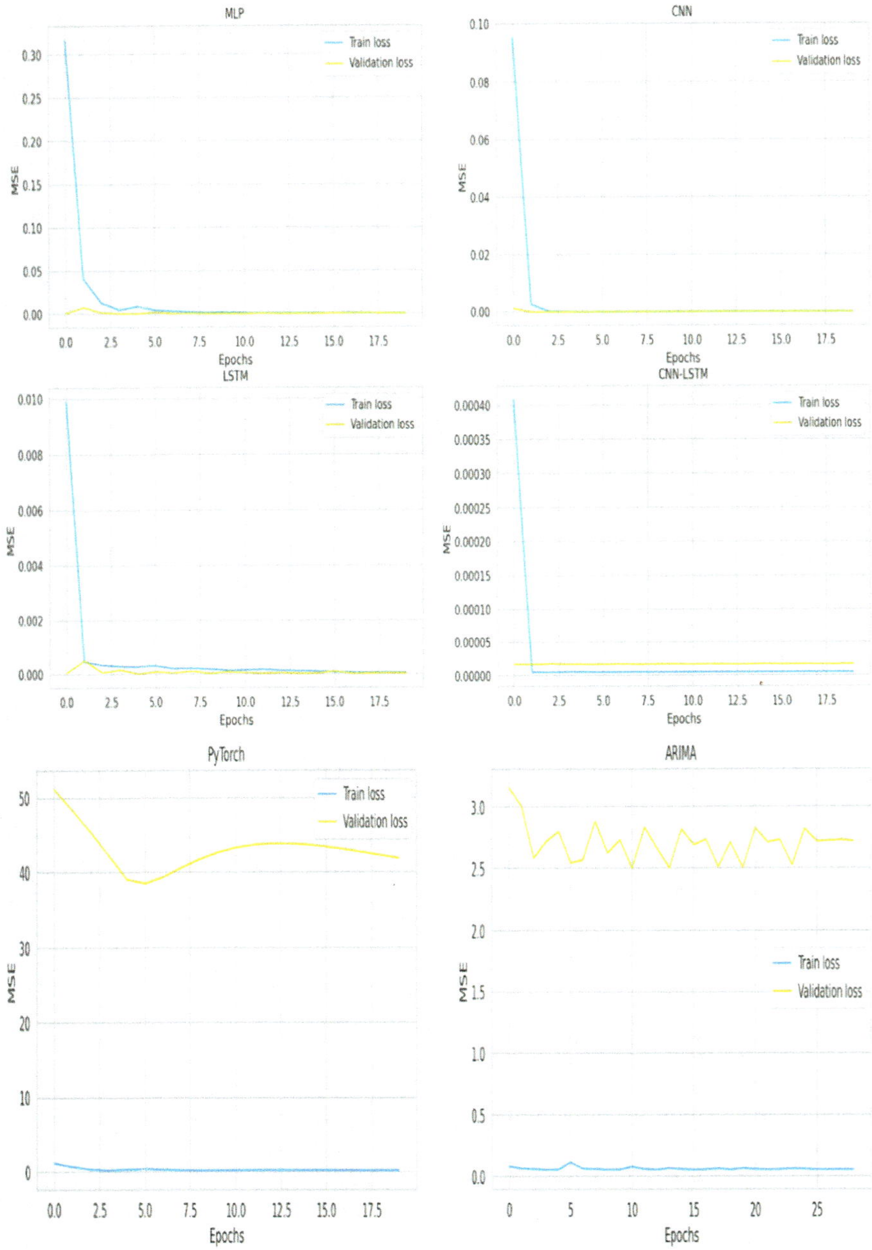

Fig. 6 Plots of MSE versus Epochs for all the six models

Table 2 Comparative results of various models

	MSE	RMSE	MAE	MAPE
MLP	Train: 0.000221 Validation: 0.000045	Train: 0.014889 Validation: 0.006744	Train: 0.002376 Validation: 0.002344	Train: 252.2784 Validation: 258.5793
CNN	Train: 0.000214 Validation: 0.000038	Train: 0.014652 Validation:0.006236	Train: 0.001795 Validation: 0.001673	Train: 251.9943 Validation: 237.9714
LSTM	Train: 0.000268 Validation: 0.000038	Train: 0.016371 Validation:0.006236	Train: 0.007501 Validation: 0.001673	Train: 1211.632 Validation: 237.9714
CNN–LSTM	**Train: 0.000211 Validation: 0.000035**	**Train: 0.014526 Validation:0.005996**	**Train: 0.001724 Validation: 0.001607**	Train: 301.5291 Validation: 283.3080
PyTorch	Train: 0.007188 Validation: 1.676518	Train: 0.084785 Validation: 1.294804	Train: 0.049153 Validation: 1.014143	**Train: 91.29479 Validation: 111.6819**

The bold values in the table signify the best results obtained

References

1. World Health Organization, WHO statement regarding cluster of pneumonia cases in Wuhan, China (2020). Available from: WHO Statement regarding cluster of pneumonia cases in Wuhan, China. Last accessed 01 Jul 2020
2. Worldometers, Coronavirus Cases (2020, April 6). Available: www.worldometers.info/coronavirus/. Last accessed 01 Jul 2020
3. A. Chen, China's coronavirus app could have unintended consequences. MIT Technology Review (2020). Available: https://www.technologyreview.com/2020/02/13/844805/coronavirus-china-app-close-contact-surveillance-covid-19-technology/. Last accessed 01 Jul 2020
4. S. Veerakumar, N.M. Dhanya, Performance analysis of various regression algorithms for time series temperature prediction. J. Adv. Res. Dyn. Control Syst. **10**(3), 996–1000, (2018). (Special Issue)
5. S. Wang, B. Kang, J. Ma, X. Zeng, M. Xiao, J. Guo, B. Xu, A deep learning algorithm using CT images to screen for Corona Virus Disease (COVID-19). medRxiv (2020). https://doi.org/10.1101/2020.02.14.20023028
6. L. Li et al., Propagation analysis and prediction of the COVID-19. Infect. Dis. Model. **5**, 282–292 (2020)
7. R. Verity et al., Estimates of the severity of coronavirus disease 2019: a model-based analysis. Lancet. Infect. Dis. (2020)
8. S. Sengupta, S. Mugde, *Covid-19 Pandemic Data Analysis and Forecasting using Machine Learning Algorithms* (2020). https://doi.org/10.1101/2020.06.25.20140004
9. R. Sujath, J.M. Chatterjee, A.E. Hassanien, A machine learning forecasting model for COVID-19 pandemic in India. Stochast. Environ. Res. Risk Assess. **34**(7), 959–972 (2020). https://doi.org/10.1007/s00477-020-01827-8

10. A.N. Roy, J. Jose, A. Sunil, N. Gautam, D. Nathalia, A. Suresh, Prediction and spread visualization of Covid-19 pandemic using machine learning (2020). https://doi.org/10.20944/preprints202005.0147.v1
11. C. Anastassopoulou, L. Russo, A. Tsakris, C. Siettos, Data-based analysis, modelling and forecasting of the COVID-19 outbreak. PLoS ONE **15**(3), e0230405 (2020)
12. F.A. Binti Hamzah, C. Lau, H. Nazri, D.V. Ligot, G. Lee, C.L. Tan et al., Corona Tracker: Worldwide COVID-19 outbreak data analysis and prediction. [Submitted]. Bull World Health Organ. E-pub: 19 Mar 2020. https://doi.org/10.2471/BLT.20.255695
13. S. Tuli, S. Tuli, R. Tuli, S.S. Gill, Predicting the growth and trend of COVID-19 pandemic using machine learning and cloud computing. Internet of Things (2020). https://doi.org/10.1101/2020.05.06.20091900
14. Novel Coronavirus 2019—Dataset—Kaggle Dataset. Last accessed 01 Jul 2020

Diabetes Prediction Using Machine Learning Techniques

Sriraj Vuppala, Nuha S. Varier, and Sangita Khare

Abstract One of the most common and serious diseases worldwide is Diabetes Mellitus, which is observed in people of all age groups ranging from children to older population. Annually it costs a lot of money for the cure and diagnosis of people suffering from Diabetes. The customary strategies for diagnosis are tedious. Therefore, accurate prediction and usage of reliable methods become the most important and top priority concern. Diabetes Type-1 is rare, and the only cure is insulin. Whereas Type-2 Diabetes is more common, and there are many factors leading to it that need to be predicted accurately. Most of the diabetics are unaware of their health condition or are in dark regarding the risk factors. Nowadays, there is availability of a huge range of tools and computational methods for data analysis. The availability of advancements in technology for developing the classification models can very well be implemented to detect the presence of Type-2 Diabetes in a patient. The goal of this work is to detect the existence of Type-2 Diabetes Mellitus in a person. If the existence of Diabetes in a patient is known beforehand, it would be easy for the doctors to understand the complexity of the disease. In this era of technological advancement, where the diagnosis is done using various sophisticated technologies, it is easy to identify a person with Diabetes using technology.

Keywords Diabetes Mellitus · Machine learning algorithms · Prediction · Classification models · Support vector machine classifier · Linear discriminant analysis classifier · Decision tree classifier · Neural networks · Random forest · Bagging · AdaBoost

1 Introduction

Diabetes is a persistent illness that is caused when the glucose or sugar in our blood is high [1]. The main source of energy for our body is glucose. Insulin, a hormone

S. Vuppala · N. S. Varier · S. Khare (✉)
Department of Computer Science and Engineering, Amrita School of Engineering, Amrita Vishwa Vidyapeetham, Bengaluru, India
e-mail: k_sangita@blr.amrita.edu

M. Paprzycki et al. (eds.), *Intelligent Systems, Technologies and Applications*, Advances in Intelligent Systems and Computing 1353,
https://doi.org/10.1007/978-981-16-0730-1_14

secreted by the pancreas, helps in converting glucose in our blood to energy. But sometimes, due to insufficient secretion of insulin or due to the failure of the pancreas or ineffective use of insulin, a lot of glucose remains in our blood and causes Diabetes. Heredity is also another major cause of Diabetes. Even though Diabetes does not have a cure, it can be prevented or can be dealt with in the initial stages. It is estimated that 415 million diabetics are present currently and around 46% of people with Diabetes are undiagnosed [2]. China holds the record of highest number of diabetics in the whole world with around 116 million diabetic patients which is estimated to rise to a massive amount of 134 million by the year 2045 [3].

Diabetes, also very commonly known as Diabetes Mellitus (DM), is of two types. Insulin-dependent Diabetes which is another name for Type 1 Diabetes Mellitus (T1DM) is caused because of lack of secretion of insulin. T1DM is less seen but is deadly serious and can result in loss of life. Even though the cause might not be known, it can be controlled by regular administration of insulin. It is mostly observed in children. Insulin-independent Diabetes is another name for Type 2 Diabetes Mellitus (T2DM) that is caused due to inefficient use of insulin. It is commonly seen in adults, mostly in elderly people. T2DM is not as dangerous as T1DM in the early stages. But, T2DM can mostly be detected only after it gets serious. Until recently, it was mostly observed in adults. But now it can be seen in children as well. Apart from heredity, there are various factors that lead to Diabetes. Today's lifestyle is a major contributing factor to the increase in diabetic patients.

Diabetes has hugely affected almost everyone's life, and there is a rapid growth in the number of people suffering with Diabetes. Approximately half of the deaths occurring before the age of 70 are due to high blood glucose levels [3]. Diabetes is also a cause for major issues like blindness, kidney failure, and heart attack. A patient suffering from Diabetes has to go through various complexities during his/her diagnosis of a disease. According to the WHO, the number of individuals with Diabetes has ascended from 108 million in 1980 to 422 million in 2014 [4]. Extensiveness of Diabetes has been growing quickly in most of the average and below average income countries. In 2012, about 2.2 million deaths were due to the presence of high glucose levels in the blood, and in 2016, another 1.6 million deaths were reported due to Diabetes. Diabetes has been recognized as the seventh most leading cause of death as of 2016 by the World Health Organization (WHO). So, increasing the awareness and predicting the onset of Diabetes in early stages can help people to take necessary precautions and thus decreasing the death caused by Diabetes.

Technology like machine learning (ML) which is a rapidly growing area of artificial intelligence (AI) is used to classify whether a person is diabetic or not based on various factors. Machine learning can be used for classifying, predicting, and even clustering. ML is basically the training of a model which is then used to perform some operation. In ML, the model learns something new when trained and executes the learned experience on a test data. Learning can be supervised, unsupervised, or reinforcement. Supervised learning is the type where a data set with a specific class label/dependent variable is used to train the model, whereas unsupervised learning does not require any output variable in its data. Unsupervised learning discovers information by itself. Reinforcement learning is learning using experience. This type of learning does not require a dataset. Various ML algorithms are—decision tree, support vector machine (SVM), random forest, K-means clustering, regression, etc. This work aims on improving the accuracy of the prediction of Diabetes, that can help in saving many lives.

Few existing studies that are similar to the work done have been explored and explained in Sect. 2. Dataset used for this work is explained in Sect. 3, followed by the experimental setup in Sect. 4 and proposed diagram in Sect. 5. The explanation and working of each model used for this work are explained next, which is followed by the result obtained by each of the models and its recall, precision, and F score value along with its comparison in Sect. 6. The future scope of this work is explained in Sect. 7.

2 Literature Survey

The primary point of this section is to comprehend the type of research that is being done on the clinical information consisting of factors contributing to Diabetes Mellitus. Previous work done on medical data has been explored to figure out various techniques that can be used for the dataset that is chosen.

Artificial neural network (ANN), one of the AI-driven systems, is used in a methodical manner in this paper [5], to identify if a person has Diabetes or not. The model constructed to train the dataset is a feed forward back propagation artificial neural network system. The paper concludes with the note that ANN reduces the error rate and tries to increase the accuracy. Few models like anomaly detection, clustering, classification, discriminant, decision tree, swarm intelligence (SI), K-nearest neighbor (KNN), logistic regression, and SVM [6] have been used on various healthcare datasets, and it is found that there is no one model in data mining for finding the best accuracy. So, a hybrid model has to be built for better accuracy [7, 8].

In a similar paper [9], a combination of techniques called as hybrid model is used to improve the accuracy of prediction. The various techniques that are used for classification of the data are: a simple K-means clustering algorithm to corroborate the given class labels and decision tree for classification. This paper again proves the statement that hybrid model gives better accuracy compared to individual simple data mining and machine learning techniques. Various preprocessing techniques, classification algorithms, and WEKA toolkit have been used on a related dataset [10]. Basically, the paper [11] takes up the experiments done by other researchers and then boosts up the accuracy by using few preprocessing techniques given in WEKA, improved K-means clustering, and logistic regression. This experiment focused on improving the accuracy and to make the model adapt to various other datasets.

Techniques like ANN, decision tree, support vector machine, and Bayesian networks have been used for cancer prediction. Based on the research done [12], it is concluded that the combination of applications of various classification and feature selection techniques along with multi-dimensional non-homogeneous data can provide useful tools that help in coming to a conclusion in the cancer diagnosis. Another very similar work [13] summarizes various algorithms used in the data mining domain for classification, clustering, and association along with their drawbacks, advantage, and three examples of each. An insight on what data mining is also given. The paper concludes that data mining can be a crucial technology in various fields. There is also a work [14] that describes about 10 algorithms, effects of those algorithms, and checks current and future research that can be done on those algorithms. They have focused on C4.5, K-means clustering, expectation maximization, naive Bayes, PageRank, AdaBoost, apriori algorithm, KNN, classification and regression trees (CART) and support vector machine.

This research publication [15] focuses on a dataset related to heart disease and the way few machine learning and data mining techniques can help in better learning of the intensity of risks involved in it. Feature selection, classifiers, and validations can be learnt from this paper. Decision tree using J48 algorithm, multilayer perceptron, support vector machine, linear regression, naïve Bayes, and AdaBoost [16] were used to classify and validate the reduced set of data. A study which deals with Diabetes using machine learning models has been come across. This research [17] is conducted using the study of genome-wide and phenome-wide association study. A semiautomated framework has been developed to support expert algorithms that have been designed using machine learning techniques. This framework successfully supported the expert algorithms by predicting the missing values and filtering out the outliers.

Oracle Data Miner has been used as a software support, and SVM is used as an experimental tool for predicting various treatments and their effectiveness on various age groups of people. It is concluded that treatment can be delayed for the young age group to prevent side effects, whereas treatment should be started right away for the old-age group using certain efficient methods [18].

A study [19] is conducted on the working of ensemble methods. The working of ensemble methods and how to improve ensemble learning are discussed in this study. Two algorithms, namely bagging and AdaBoost, are considered as an example and their working, and superiority in accuracy is illustrated using various datasets. A similar study [20] also focuses on ensemble methods such as bagging and boosting. The working of the techniques along with their pros and cons are discussed which helps in better understanding of the techniques.

Along with the rise in new technologies, the need for better accuracy is also rising. So, the main aim of this work is to predict the outcome with better accuracy which will prove to be very helpful in the real-life situations. After going through all the various models used in various datasets, this work can add on by using some of these algorithms in a dataset that has been obtained. The machine learning techniques that have been used are: SVM classifier, linear discriminant analysis (LDA) classifier, decision tree classifier, neural networks and ensemble methods such as bagging and AdaBoost to identify the existence of Diabetes Mellitus in a patient. After applying preprocessing techniques and implementation of the above-mentioned machine learning algorithms, the best suitable model that gets the highest recall for the dataset used has been finalized.

3 Data Description

Dataset that is used in this work is publicly available and is obtained from Kaggle [21]. Dataset has been updated in 2018 and has 10 features and 15K records. The file contains data explaining whether a person is diabetic or not using the output attribute Diabetic. The attribute contains the value 1 if a person is diabetic and 0 otherwise. The features in the dataset are:

Patient ID—unique identification number assigned to each patient
Pregnancies—number of pregnancies that a patient has gone through
Plasma Glucose—the plasma glucose level in the body of a patient in every two hours duration (represented in the units of mM—millimolar)
Diastolic Blood Pressure—(represented in the units of mm Hg)
Triceps Thickness—Triceps skinfold thickness (represented in the units of mm)
Serum Insulin—insulin level in every two hours gap (represented in the units of mu U/ml)
Body Mass Index (BMI)—The ratio of weight is to height square of the person (represented in the units of kg/m^2)
Diabetes Pedigree—Depends on the genetic conditions of the family of the person
Age—(represented in the units of years)
Diabetic (Output)—whether a person is diabetic or not

4 Experimental Setup

The experiment consists of various models that are used for binary classification. It is done using python as the programming language and is implemented with the help of various python libraries. Python is a programming language which offers various packages and libraries to the coder for an easy implementation. This work imports inbuilt libraries like numpy, pandas, scikit learn, and keras. Numpy is used for various mathematical calculations, whereas pandas are used for working with the dataset. Scikit learn is used for the entire machine learning coding while keras is used for implementing neural networks. Various methods constitute these libraries, and this experiment has taken complete advantage of that.

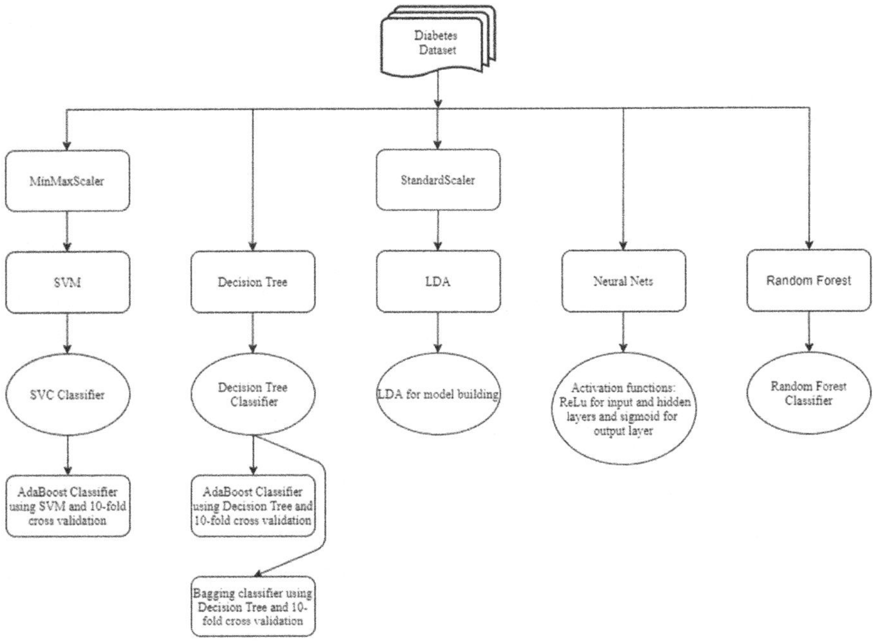

Fig. 1 General flowchart

5 Proposed Design

The architecture shown in Fig. 1 follows a top–down method in executing the algorithms [22]. Each different branch originating from a same given node represents a different method of execution. Different methods [23] involved are support vector machine, decision tree, neural networks, linear discriminant analysis, and random forest. Ensemble methods like bagging and AdaBoost have been used to enhance the accuracy of the models. Various data preprocessing tools like MinMaxScaler and StandardScaler (explained in Sect. 5.1) have been used on the data that has been extracted from the dataset to suit the classification algorithms. Each step in Fig. 1 is explained from Sects. 5.1–5.7.

5.1 Data Preprocessing

Data preprocessing is a technique which transforms the raw data into meaningful and usable data. This includes three steps which are—data cleaning, data transformation, and data reduction. In this work, data preprocessing has been applied to various algorithms based on the requirement.

MinMaxScaler()—This function converts the data in the dataset into a value between 0 and 1 to make the computations easier. It takes the least available value in the set of features and then subtracts every value in that feature set with that minimum value. It then calculates the range by subtracting the original maximum and original minimum value of that feature. Every subtracted value is then divided by the range calculated. Formula used by this technique is shown in Eq. (1).

$$X_{\text{new}} = (X_i - (X))/((X) - \min(X)) \tag{1}$$

Standardscaler()—Standardscaler is a tool for data preprocessing which is used for data transformation. It converts the value of a feature such that it gives a distribution with a standard deviation equals 1, i.e., variance is equal to 1. This is achieved by subtracting the average from all the values in the feature and then scaling it to unit variance. The formula used is shown in Eq. (2).

$$z = (x - u)/s \tag{2}$$

where x is the value, u is the average of the feature, and s represents the standard deviation of the training sample.

5.2 SVM

Support vector machine (SVM) [24] is a supervised classification algorithm which is best suited for binary class classification. This algorithm groups data into two classes by forming a hyperplane by using the data points in the dataset. One side of the hyperplane represents class A, and the other side represents class B. Points that are close to the hyperplane are called support vectors. Margins are formed using these support vector points. The model gives maximum accuracy when the distance between the margin and hyperplane is the maximum. In this work, min–max scalar is used for preprocessing.

To increase the accuracy, AdaBoost classifier is implemented using svc as a base classifier and tenfold cross-validation. The accuracy is calculated by weighted average of all the accuracies of the entire model. The weights are changed after every iteration. To boost up the accuracy, this model combines various SVM classifier models with different train and test data for each model.

5.3 Decision Tree

Decision tree [25, 26] is a supervised learning which is a structure similar to flowcharts and is used for classification and regression where every node illustrates a test condition on a feature, every branch illustrates a result of the test, and each leaf node represents the output. There are various criteria that can be used to determine the split of a node like—entropy, gain ratio, Gini index, and information gain. In this paper, the criteria used for split is entropy. Entropy is the degree of uncertainty that is inked with an arbitrary variable. Entropy is directly proportional to uncertainty. Entropy is measured between 0 and 1. Formula to calculate entropy is shown in Eq. (3).

$$E(S) = \sum_{i=1}^{c} -p_i p_i \tag{3}$$

where p_i represents the probability of an element in class 'i' in the dataset, H and E represents the entropy.

Based on the dataset used, the ranking of the features using correlation ranking filter from the greatest value to least value is as follows: Pregnancies, Age, Serum Insulin, BMI, Diabetes Pedigree, Tricep Thickness, Plasma Glucose, and Diastolic Blood Pressure. According to the work done, Diabetes mainly depends on a few factors that have been taken into consideration. A data mining process called ranking is used on the decision tree obtained to find the factors affecting the cause of Diabetes in a person. As per the results, the main cause of Diabetes is the number of pregnancies. Other major affecting factors are age, serum insulin, and BMI of a person.

5.4 Ensemble Methods

Ensemble methods [27] are machine learning techniques that incorporate various similar or hybrid of different classification models to produce a better accuracy than the single classification model. Average of all the models is taken which proves to be more accurate.

To increase the accuracy, ensemble methods like bagging and boosting technique called AdaBoost are used with base classifiers such as SVM [28], decision tree. These ensemble methods also use tenfold cross-validation. In this, the data is arbitrarily divided into 10 mutually exclusive subsets of similar sizes. In this work, 100 trees are built in all the ensemble methods.

Bootstrapping—This technique is used in ensemble methods to get the training data from the given dataset. This algorithm takes a random data sample and adds it to the training set. A particular data sample can be chosen more than once to be placed in a training set. So, a training set can contain redundant data. The remaining data that is not selected from the given dataset for the training data will be considered as the test data.

Bagging—Also called bootstrap aggregating, it is an ensemble method which is used for enhancing the accuracy of a model. It usually uses a decision tree, by default, to construct the model. For the given train–test percentage, bagging uses bootstrapping method to construct a train data which have given percent equivalent number of rows, and the remaining data is considered as test data. Then, it calculates the accuracy percentage for the given train–test data. This action is repeated for n times and gets the accuracy for different train–test combinations. After all the iterations are done, the resulting accuracies will be taken and derive an average out of all the obtained accuracies. The resultant accuracy will be the final accuracy of the model. This accuracy proves to be greater than the accuracy obtained by using just a simple model.

AdaBoost—This is another ensemble method that is used for improving the accuracy of a model. AdaBoost is a type of boosting classifier. AdaBoost usually uses bootstrapping method to get the test train data. It predicts the final accuracy by the process which is similar to bagging. But AdaBoost uses a weighted average. Initially, each data sample in the dataset is given equal importance; i.e., each sample has the same weight. The training set is formed by using the process of bootstrapping where the sample with more weight is most likely to be selected more often. AdaBoost uses a series of techniques to get the result. After modeling the data with a certain result, the weights of the sample that are wrongly classified will be increased. The process of bootstrapping happens again, and the cycle continues till all the techniques have been used for modeling the data. Then, each accuracy is multiplied with the weight associated with that particular technique and average is taken. The resultant accuracy will be considered as the final accuracy. This technique tends to give better accuracy, but often leads to overfitting.

5.5 Neural Networks

A neural network [29, 30] is a group of algorithms which is inspired and modeled by the human brain. These systems learn to perform tasks with the help of examples, without being programmed by a set of specific rules. In this experiment, a neural network of three layers is built, namely input, hidden, and output layers. The input layer consists of 8 nodes, hidden has 5 nodes, and output has 1 node. In the beginning, random weights which lie in between −0.5 and 0.5 are given. The input and hidden layers have ReLU as its activation function, and the output layer has sigmoid function. An activation function is a function which gives us a result which in turn provides information to a neuron whether it can fire the output or not. The value is calculated with the help of weighted sums that a neuron receives and adding bias to the result obtained. In this work, 100 epochs are considered. A general flowchart showing the number of nodes in each layer is shown in Fig. 2.

Sequential(): This function allows us to build a neural network, layer by layer and limits the neural network from sharing the layers, i.e., a layer in a neural network can maximum be connected to the layers next and previous to it.

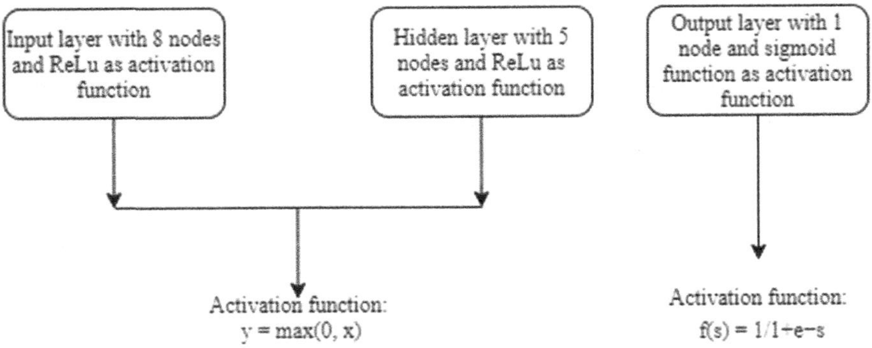

Fig. 2 Neural networks flowchart

Kernel_initializer = 'random_uniform': This function is used to get the initial weights of the neural network. It uses uniform distribution with default interval as (−0.5, 0.5) to update the weights.

Optimizer = 'Adam': This is used for updation of weights in the neural net. Adaptive moment (Adam) works in the following way.

$$m_t = \beta_1 m_{t-1} + (1 - \beta_1) g_t \tag{4}$$

$$v_t = \beta_2 v_{t-1} + (1 - \beta_2) g_t^2 \tag{5}$$

where the values of betas are 0.9 and 0.999, respectively, (default values) and g is the gradient value of the net. The process of obtaining m_t in Eq. (4) and v_t in Eq. (5) is shown from Eqs. (6–12).

$$m_0 = 0 \tag{6}$$

$$m_1 = \beta_1 m_0 + (1 - \beta_1) g_1 = (1 - \beta_1) g_1 \tag{7}$$

$$m_2 = \beta_1 m_1 + (1 - \beta_1) g_2 = \beta_1 (1 - \beta_1) g_1 + (1 - \beta_1) g_2 \tag{8}$$

$$m_3 = \beta_1 m_2 + (1 - \beta_1) g_3 = \beta^2 (1 - \beta_1) g_1 + \beta_1 (1 - \beta_1) g_2 + (1 - \beta_1) g_3 \tag{9}$$

$$m_t = (1 - \beta_1) \sum_{i=0}^{t} \beta_1^{t-i} g_i \tag{10}$$

$$\widehat{m_t} = \frac{m_t}{1 - \beta_1^t} \tag{11}$$

$$\widehat{v_t} = \frac{v_t}{1 - \beta_2^t} \quad (12)$$

$$wt = wt - 1 - \eta \frac{\widehat{m_t}}{\sqrt{\widehat{v_t}} + \in} \quad (13)$$

where η is the step value and can be changed with each iteration w_t in Eq. (13) is the updated weight, $\widehat{m_t}$ in Eq. (11) is the moving average of the gradient (mean), and $\widehat{v}_t$ in Eq. (12) is the moving average of squared gradient (variance).

The two activation functions used are:

ReLU: ReLU function [Eq. (14)] is one of the fastest and most used activation functions in neural networks. If the input is positive, then the output will be the same as the input else the output will be 0. It is a nonlinear function. This has better performance than most other activation functions and hence has been used for the input and hidden layers. It uses simple mathematical functions to compute the output.

$$y = \max(0, x) \quad (14)$$

Sigmoid: Sigmoid function is basically used when there is a probability as an output. The values range from 0–1. The sigmoid function [Eq. (15)] is monotonic and differentiable. Hence, slope of a sigmoid function can be found for any use. Its output is monotonically increasing from -1 to 1 or 0 to 1 as required.

$$f(s) = \frac{1}{1 + e - s} \quad (15)$$

5.6 Linear Discriminant Analysis (LDA)

Linear discriminant analysis [31, 32] is used for dimensionality reduction. LDA usually creates a new feature set from the original set and uses the newly obtained feature set for classification. It can also be used as a supervised classification technique. In this paper, linear discriminant analysis is used as a classification technique. LDA technique is used for separating two linearly separable classes by creating a linear line that acts as a separating plane. The main aim in LDA is to maximize the distance between 2 classes and minimize the variation within a class with a minimum overlapping as possible.

The above criterion is used to create a new separating line which reduces the overlapping as much as possible. After a new axis is created, all the data points are plotted on the new axis obtained.

The formula used to find LDA factor or function value is shown in Eq. (16)

$$f_i = \mu_i C^{-1} X_k^T - \frac{1}{2}\mu_i C^{-1} \mu_i^T + \ln(p_i) \quad (16)$$

where f_i is the LDA function value.

μ_i is average value for record x_i.

C^{-1} is inverse of pooled covariance matrix.

x_k is data of row k.

p_i is prior probability of vector.

5.7 *Random Forest*

Random forest is an algorithm which uses train and test data. The random forest algorithm can be used for both classification and regression problems. But it is mostly used for solving classification problems. A forest is a collection of many trees. In a similar way, random forest is a collection of multiple decision trees. Random forest follows a voting system. A dataset will be randomly split into multiple samples, and each sample will be assigned to a decision tree. After assigning each sample to a decision tree, all the decision trees will be executed and each result will be considered as a vote of that particular decision tree. Outcome which has the maximum number of votes is considered to be the final outcome of the random forest. Since the result of random forest depends on multiple decision trees, it is considered to be an ensemble method.

6 Result

The experiment is carried out using various data mining and machine learning techniques like support vector machine (SVM), decision tree, neural networks, linear discriminant analysis, random forest, AdaBoost using decision tree, AdaBoost using SVM and bagging using decision tree. To increase the accuracy and recall obtained using those methods, ensemble methods are also used. The experiment is carried to predict the best-fitting model for the given Diabetes dataset which gives the highest accuracy. The aim of conducting the experiment is to find the presence of T2DM in a person using various machine learning classification algorithms.

Figure 3 represents the bar graph of the models—SVM, decision tree, neural networks, LDA, random forest, AdaBoost using SVM, AdaBoost using decision tree, and bagging using decision tree. The models are plotted against their accuracies that are obtained when implemented on the dataset. Accuracy is shown in terms of percentage.

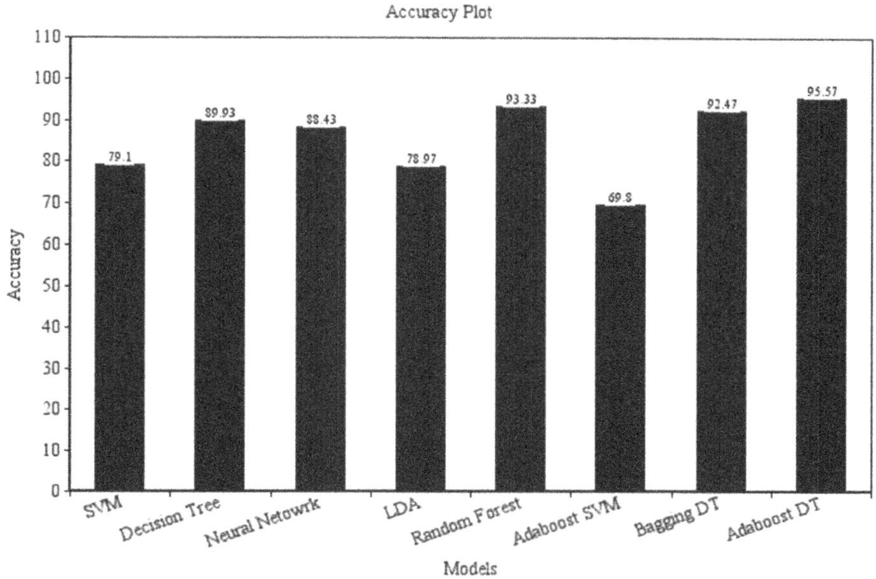

Fig. 3 Graphical representation and comparison of accuracy

Table 1 shows recall, precision, F score measure, accuracy percentages for all the models that have been applied on the dataset. The experiment proved that ensemble methods like boosting gives the best accuracy and hence is the best model to be used for Diabetes prediction with a similar dataset. The work concludes that AdaBoost using decision tree is the best suitable model for the prediction of T2DM in a patient. The experiment proves that models that use decision tree give the best result which is followed by the random forest model.

Table 1 Result obtained

Algorithm	Recall (%)	Precision (%)	*F* score measure (%)	Accuracy (%)
SVM	87.163	82.506	84.770	79.100
Decision tree	92.358	92.543	92.450	89.933
Neural networks	88.980	79.990	83.655	88.430
LDA	87.463	82.168	84.733	78.967
Random Forest	95.405	94.648	95.025	93.333
AdaBoost using SVM	98.751	69.174	81.358	69.800
Bagging using decision tree	94.655	94.091	94.373	92.466
AdaBoost using decision tree	96.404	96.936	96.669	95.567

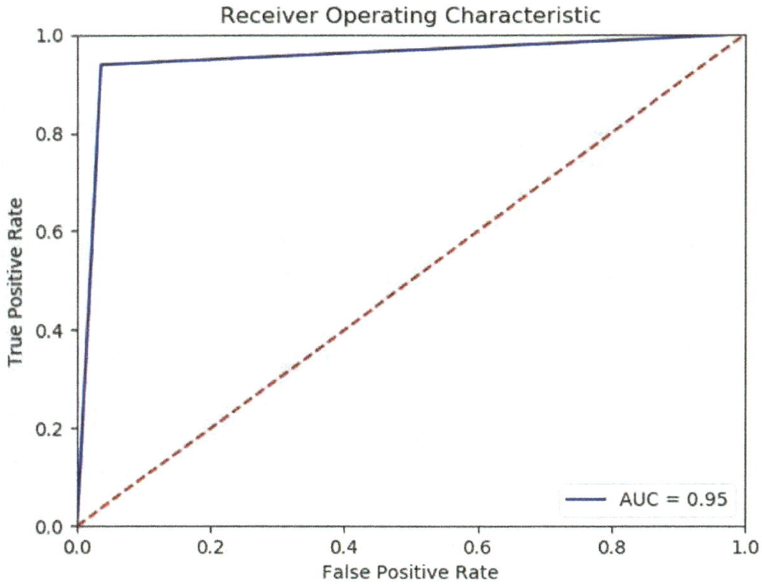

Fig. 4 ROC curve for AdaBoost using decision tree

ROC curve is a measure of performance which shows the probability of how accurate the prediction of a classification model can be. Area under the curve (AUC) is a degree of separability. Higher the value of AUC, better the predicting capability of the model. A ROC curve has been drawn for AdaBoost using the decision tree to check the predicting ability of the model. From Fig. 4, AUC value is shown as 0.95. This means that the model that has been trained on the dataset has the ability to classify the dataset precisely up to 95%.

A ROC curve has been drawn for random forest to check the predicting ability of the model. From Fig. 5, the model has the ability to classify the dataset precisely up to 92%.

This work has succeeded in achieving a higher accuracy of 95.567% using AdaBoost than the 94.86% accuracy obtained in previous work [33] done on this dataset using AdaBoost. It also gives a better precision, recall, and F score measure when compared to the work done before. Other experiment done on this dataset uses the extra trees method which gives an accuracy of 92.6% and GaussianNB method which gives an accuracy of 79%. From the above-mentioned results, it can be concluded that ensemble methods such as bagging and AdaBoost give us the highest accuracy and recall rate. This statement can be validated with the help of an article [34] wherein it states that higher accuracy can be obtained through ensemble methods than solely depending on a single technique.

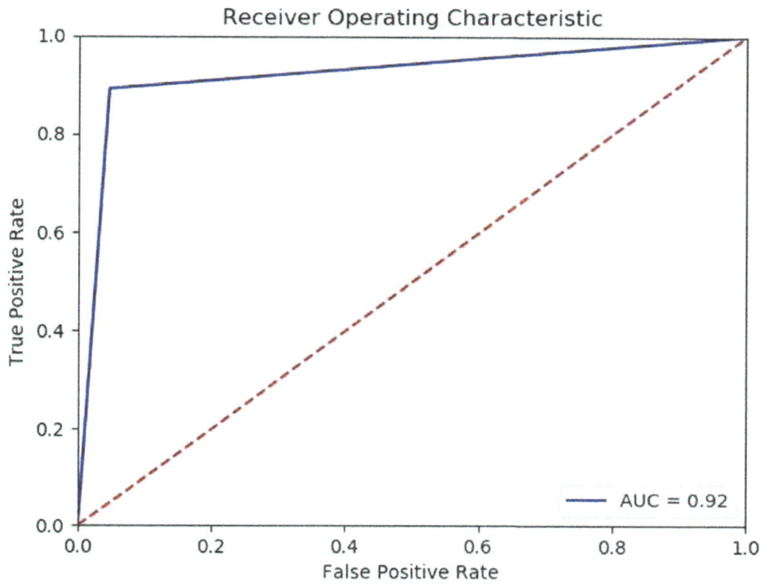

Fig. 5 ROC curve for random forest

7 Future Scope

There is scope of enhancement in the experiment by using various other models like Gaussian NB, KNN, logistic regression, Naïve Bayes and extra trees classifier. The accuracy might vary by applying various other preprocessing techniques and tools which can be explored further. Neural networks can be altered in numerous ways by—changing the number of layers, the number of neurons in each layer can be changed, different activation functions can be tried on different layers and density of the network can be altered. The optimizer and the batch size also can be altered and checked for better accuracy. Furthermore, the number of trees can be changed in decision tree, bagging and boosting algorithms and dimensionality reduction or feature selection methods can be tried on to get an increase in the accuracy.

References

1. The National Institute of Diabetes and Digestive and Kidney Diseases, https://www.niddk.nih.gov/health-information/diabetes. Accessed on 05 Jul 2020
2. https://www.diabetes.co.uk/diabetes-prevalence.html. Accessed on 29 Sept 2020
3. https://www.statista.com/statistics/281082/countries-with-highest-number-of-diabetics/. Accessed on 05 Jul 2020
4. World Health Organization, https://www.who.int/news-room/fact-sheets/detail/diabetes. Accessed on 05 Jul 2020

5. N.S. El_Jerjawi, S.S. Abu-Naser, Diabetes prediction using artificial neural network. Int J Adv Sci Technol **121**, 54–64 (2018)
6. N. Jothi, N.A.A. Rashid, W. Husain, Data mining in healthcare-a review. Procedia Comput. Sci. **72**, 306–313 (2015)
7. B. Zheng, S.W. Yoon, S.S. Lam, Breast cancer diagnosis based on feature extraction using a hybrid of K-means and support vector machine algorithms. Expert Syst. Appl. **41**(4), 1476–1482 (2013)
8. E. Corchado, M. Grana, M. Wozniak, A survey of multiple classifier systems as hybrid systems. Inf. Fusion **16**, 3–17 (2014)
9. B.M. Patil, R.C. Joshi, D. Toshniwal, Hybrid prediction model for type-2 diabetic patients. Expert Syst. Appl. **37**, 8102–8108 (2010)
10. V. Dominic, D. Gupta, S. Khare, A. Aggarwal, Investigation of chronic disease correlation using data mining techniques. in *2015 2nd International Conference on Recent Advances in Engineering and Computational Sciences (RAECS)* (Chandigarh, 2015), pp. 1–6. https://doi.org/10.1109/RAECS.2015.7453329
11. H. Wu, S. Yang, Z. Huang, J. He, X. Wang, Type 2 diabetes mellitus prediction model based on data mining. Inf. Med. Unlocked **10**, 100–107 (2018)
12. K. Kourou, T.P. Exarchos, K.P. Exarchos, M.V. Karamouzis, D.I. Fotiadis, "Machine Learning applications in cancer prognosis and prediction. Comput. Struct. Biotechnol. J. **13**, 8–17 (2015)
13. I. Yoo, P. Alafaireet, M. Marinov, K. Pena-Hernandez, R. Gopidi, J.F. Chang, H. Lar, Data mining in healthcare and biomedicine: a survey of the literature. J. Med. Syst. **36**, 2431–2448 (2011)
14. X. Wu, V. Kumar, J.R. Quinlan, J. Ghosh, Q. Yang, H. Motoda, G.J. McLachlan, A. Ng, B. Liu, P.S. Yu, Z. Zhou, M. Steinbach, D.J. Hand, D. Steinberg, Top 10 algorithms in data mining. Knowl. Inf. Syst. **14**, 1–37 (2008)
15. V. Dominic, D. Gupta, S. Khare, An effective performance analysis of machine learning techniques for cardiovascular disease. Appl. Med. Inf. **36**, 23–32 (2015)
16. D. Gupta, S. Khare, A. Aggarwal, A method to predict diagnostic codes for chronic diseases using machine learning techniques. in *2016 International Conference on Computing, Communication and Automation (ICCCA)* (Noida, 2016), pp. 281–287. https://doi.org/10.1109/CCAA.2016.7813730
17. T. Zheng, W. Xie, L. Xu, X. He, Y. Zhang, M. You, G. Yang, Y. Chen, A machine learning-based framework to identify type 2 diabetes through electronic health records. Int. J. Med. Inf. **97**, 120–127 (2017)
18. A.A. Aljumah, M.G. Ahamad, M.K. Siddiqui, Application of data mining: Diabetes health care in young and old patients. J. King Saud Univ. – Comput. Inf. Sci. **25**(2), 127–136 (2013)
19. H. Faliang, X. Guoqing, X. Ruliang, Research on ensemble learning. in *2009 International Conference on Artificial Intelligence and Computational Intelligence* (Shanghai, 2009), pp. 249–252. https://doi.org/10.1109/AICI.2009.235
20. O. David, M. Richard, Popular ensemble methods: an empirical study. J. Artif. Intell. Res. **11**, 169–198 (1999)
21. Diabetes Dataset, https://www.kaggle.com/fmendes/diabetes-from-dat263x-lab01. Accessed on 05 Jul 2020
22. J. Han, M. Kamber, J. Pei, *Data Mining Concepts and Techniques*
23. D.G. Amorim, S. Barro, E. Cernadas, M.F. Delgado, Do we need hundreds of classifiers to solve real world classification problems. J. Mach. Learn. Res. **15**(9), 3133–3181 (2014)
24. D.K. Srivastava, L. Bhambhu, Data classification using support vector machine. J. Theor. Appl. Inf. Technol. **12**, 1–7 (2010)
25. J.R. Quinlan, Induction of decision trees. Mach. Learn. **1**, 81–106 (1986)
26. J.D. Kelleher, B. Mac Namee, A. D'Arcy, *Machine Learning for Predictive Data Analytics*
27. Ensemble Learning to improve Machine Learning results. https://blog.statsbot.co/ensemble-learning-d1dcd548e936. Accessed on 26 Sept 2020
28. S. Kang, P. Kang, T. Ko, S. Cho, K.S. Yu, An efficient and effective ensemble of support vector machines for anti-diabetic drug failure prediction. Expert Syst. Appl. **42**(9), 4265–4273 (2015)

29. L. V. Fausett, *Fundamentals of Neural Networks: Architectures, Algorithms, and Applications*
30. Building neural network using keras for classification. https://medium.com/datadriveninvestor/building-neural-network-using-keras-for-classification-3a3656c726c1. Accessed on 05 Jul 2020
31. N.A.A. Shashoa, N.A. Salem, I.N. Jteta, O. Abusaeeda, Classification depend on linear discriminant analysis using desired outputs. in *2016 17th International Conference on Sciences and Techniques of Automatic Control and Computer Engineering (STA)* (Sousse, 2016), pp. 328–332. https://doi.org/10.1109/STA.2016.7952041
32. Linear Discriminant Analysis Bit by Bit explanation. https://sebastianraschka.com/Articles/2014_python_lda.html. Accessed on 26 Sep 2020
33. Kaggle Dataset Activity, https://www.kaggle.com/plarmuseau/100-classification. Accessed on 05 Jul 2020
34. https://www.sciencedirect.com/science/article/pii/S1877050916300308. Accessed on 26 Jul 2020

Deep Neural Network Based Multi-class Arrhythmia Classification

K. A. Akhila Naz, R. S. Jeena, and P. Niyas

Abstract An arrhythmia is a condition which represents irregular beating of the heart, beating of the heart too fast, too slow, or too early compared to a normal heartbeat. Diagnosis of various cardiac conditions can be done by the proper analysis, detection, and classification of life-threatening arrhythmia. Computer-aided automatic detection can provide accurate and fast results when compared with manual processing. This paper proposes a reliable and novel arrhythmia classification approach using deep learning. A deep neural network (DNN) with three hidden layers has been developed for arrhythmia classification using MIT–BIH arrhythmia database. The network classifies the input ECG signals into six groups: normal heartbeat and five arrhythmia classes. The proposed model was found to be very promising with an accuracy of 99.45%.

Keywords Electrocardiogram · Arrhythmia · Deep neural network

1 Introduction

The electrical activities of the human cardiac system are obtained from the electrocardiogram (ECG) signals using surface electrodes. A typical ECG signal consists of peaks and waves of different amplitudes and frequencies, and the analysis of these signals plays an important role in heart disease diagnosis such as arrhythmia. In general, ECG signal contains six different frequency components. Careful analysis of these components is to be carried out for qualified diagnosis and treatment. According to the Association for the Advancement of Medical Instrumentation (AAMI) [1], five kinds of cardiac diseases can be defined from the ECG signals, viz. unknown beats (Q), normal (N), ventricular (V), supra-ventricular (S), and fusion of normal and ventricular beats (F). An irregularity in the rate or rhythm of the heartbeat is known as arrhythmia. During this condition, the heart can beat too fast (tachycardia) or too

K. A. Akhila Naz (✉) · R. S. Jeena · P. Niyas
Department of Electronics and Communication Engineering, College of Engineering Trivandrum, Thiruvananthapuram, Kerala 695017, India

M. Paprzycki et al. (eds.), *Intelligent Systems, Technologies and Applications*,
Advances in Intelligent Systems and Computing 1353,
https://doi.org/10.1007/978-981-16-0730-1_15

slow (bradycardia). Most arrhythmia are harmless, but some can be serious or even life-threatening.

2 Literature Review

During an arrhythmia, inconsistent blood flow can cause damaging effects to various organs of the human body especially, the brain and heart. In the daily life of a patient, arrhythmia is found to occur intermittently. If the patient is undergoing bed-side monitoring or wearable healthcare monitoring, the corresponding manual inspection and analysis of the ECG signal recordings for a long period are tedious and time-consuming. Therefore, the ECG signals can be analyzed and interpreted using computer-assisted methods in medical applications. So, a Holter device is employed to capture these infrequent events and to record long term ECG data. Therefore, automatic recognition of abnormal heartbeats from a huge amount of ECG data is a salient and crucial task.

A reasonable amount of research has been dedicated to the advancement of automatic heartbeat classification by Martis et al. [1] and Llamedo et al. [2]. The works can be broadly classified into two categories, namely 'intra-patient' and 'inter-patient,' also called 'class-oriented' and 'subject-oriented,' respectively [3]. According to Ye et al. [3] and Arif et al. [4], both subsets of data can have part of the ECG recordings when the intra-patient paradigm separates the entire data-set into training and testing subsets, considering only the beat labels. With this strategy, the classifiers usually produce bleak results. In clinical practice, the classification efficiency declines due to the variations among the individuals was shown by Garcia et al. [5].

Chazal et al. [6] proposed an inter-patient automatic heartbeat classification while both the sub-sets, i.e., training and testing subsets, were developed using different ECG recordings so that the variation from different individuals would be taken into account and the classifier would clearly manifest the generalization ability. It was found that one single lead classifier obtained an efficiency of 83% which was found to be the maximum accuracy. The specificity values were also found to be high with a value of 88.1%, but the sensitivity values were very low. Additionally, a patient-specific automatic heartbeat classification was also proposed by Chazal et al. wherein training a global classifier was done initially and to attune this global classifier, a local classifier was used [7].

Another approach proposed by Hu et al. [8] involved a fully mechanized configuration approach known as GE classifier. The accuracy, sensitivity, and specificity of this approach were 75.3%, 69.6%, and 76.6%, respectively. A partially automatic mixture of expert's approach was utilized to arrive at these results. Apart from frequency domain, feature analysis of the ECG signals also provides appreciable perception of the signals. These analyses can be done by various signal processing methods like wavelet decomposition (WT) [1, 4, 9], principal component analysis (PCA), and independent component analysis (ICA) [4]. Various other classifier methods based on feature extraction methods are also carried out by de Chazal et al. [7], Hu et al. [8], and Hanbay [10].

In DNN, the network is feed forward, which means there is no looping back of the data from the output layer. Initially, the DNN creates a cluster of virtual neural networks and assigns random numerical values, or 'weights,' to connections (nodes) between them. The weights and inputs are multiplied and return an output whose value is between 0 and 1.

Al Rahhal et al. [11] proposed a novel approach based on deep learning for active classification of ECG signals were a softmax regression layer is added on the top of the resulting hidden representation layer yielding the so-called DNN. Here method for ranking relies on the DNN posterior probabilities to associate confidence measures such as entropy and Breaking-Ties (BT) to each test beat in the ECG record under analysis.

Joshi et al. [12] showed that the problem of class imbalance which is rife in biomedical signals have been overcome by a Holder-SVM detection algorithm using a novel hybrid arrangement of binary and multi-class SVMs with highest accuracy. By integrating a 16-layer convolutional neural network (CNN), recurrent cells and attention module, a DNN model obtaining both spatial and temporal fusion of information from ECG signals has been developed by Yao et al. [13]. Another algorithm having four weighted hidden layers with biases in MLP and a four-layer CNN has been developed by Shalin Savalia et al., which map ECG signals to different arrhythmia classes with a considerable accuracy has been proposed [14]. In this model, bigeminy features were easily mistaken for normal, FAV, VT, AF, and AFIB signals, which would lead to false positives. Therefore, deep learning becomes the most promising direction for cardiac abnormality detection.

A DNN for the heartbeat classification composed of seven hidden layers and has been built empirically by Saninano et al. [15]. Along with the MITBIH dataset, certain features has also been also extracted and given as the input to the classifier model. The model classifies the heartbeats into normal and abnormal beats.

The proposed system has a deep neural network classifier of three hidden layers for classifying arrhythmia into normal and five abnormal classes with promising efficiency parameters.

3 Proposed System

This section gives an overview of MIT–BIH arrhythmia database, signal processing steps implemented, and the proposed DNN. Figure 1 shows the block diagram of the proposed system.

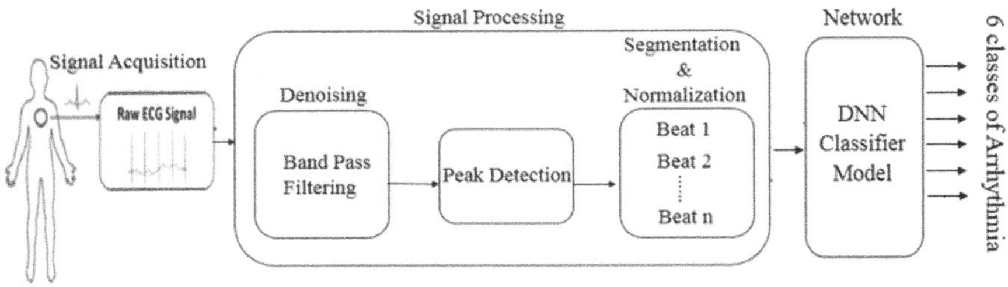

Fig. 1 Block diagram of proposed method

3.1 MIT–BIH Arrhythmia Dataset

MIT–BIH arrhythmia database is a dataset compiled from ECG recording done for 24 h carried out on 47 subjects. Subjects included 25 men of ages ranging from 32 to 89 years and 27 women of ages ranging from 23 to 89 years. The MIT–BIH arrhythmia database ECG records consist of two leads which were developed from different electrodes. The modified limb leads II (MLII) and VI, obtained by placing the electrodes on the chest, are the most common leads in the database.

Initially, ECG data used as the input for the network where randomly chosen from the MIT–BIH dataset itself. This data were used for training and testing of the proposed model.

3.2 Signal Processing

A new dataset, starting from the MIT–BIH arrhythmia database, has been created by applying the following processing steps to the ECG signals. The Python programming language has been used for all these processing, which were carried out on the VI lead of each subject of the MIT–BIH arrhythmia database.

Denoising Patient movements or respiration may cause possible interference in power line and baseline variations which in turn can result in errors in the detection of ECG peaks. Such interference's can be removed by denoising. The ECG signals are usually manipulated by various kinds of noise signals, namely artifacts [15]. These artifacts hinder the ready removal of the information of interest from the raw signal. The denoising section has been carried out using a band-pass filter of frequency range [0.8–48 Hz]. This frequency range has been set based on baseline wander for the lowest frequency and power line interference for the highest frequency.

Peak Detection Peak detection phase is useful in determining the ECG signal peaks that are of interest. Governing the positions of all peaks of interest found in the ECG signal is done by peak detection. The sectioning of the ECG signals is represented by P-wave and T-wave positions for the extraction of time-based features since the R-wave peak detection is used. The position of the ECG signal peaks which are of interest is obtained by applying a function wfdb () to the ECG signal. This is done by the waveform database toolbox. This technique provides an easier way of QRS detection especially the R-wave detection since the structural form and high amplitude of QRS wave can be easily distinguished from the other parts of the ECG signals. Each R-peak detection corresponds to a single heartbeat. The Pan and Tompkins algorithm is used for QRS detection. In the proposed method, only the R-peaks are detected, thereby the output annotation file contains annotations that indicate the R-wave peak.

Signal segmentation and Normalization Signal segmentation is done to categorize the ECG signal into single beats that will be classified as normal and different types of arrhythmia. After locating the R peaks, the ECG signals are then segmented into single heartbeats. A window of length 150 is used for segmentation. Then the signals are made more normal or regular by re-scaling into segments of same number of points and amplitude for easier mathematical computation. The signal is generally reduced into their canonical forms for easier mathematical processing.

3.3 *The Proposed DNN*

Generally for prediction, a DNN consists of an input layer for the raw descriptors $X(i)$, an output layer and L hidden layers, (three or more). An overview of the proposed DNN is given in Fig. 2.

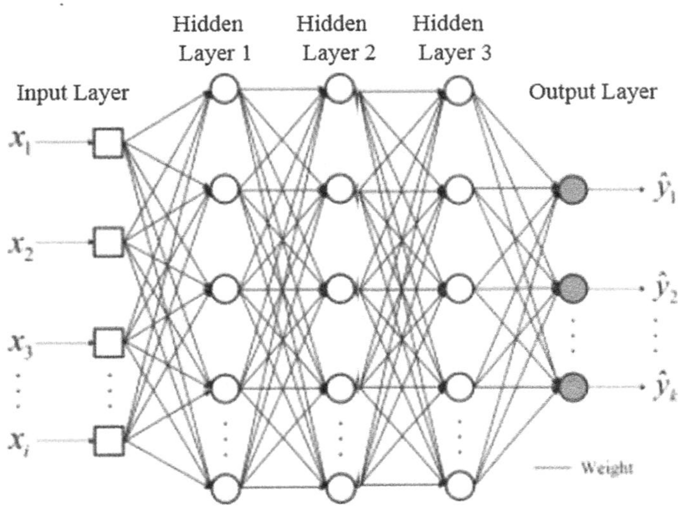

Fig. 2 An overview of the proposed DNN

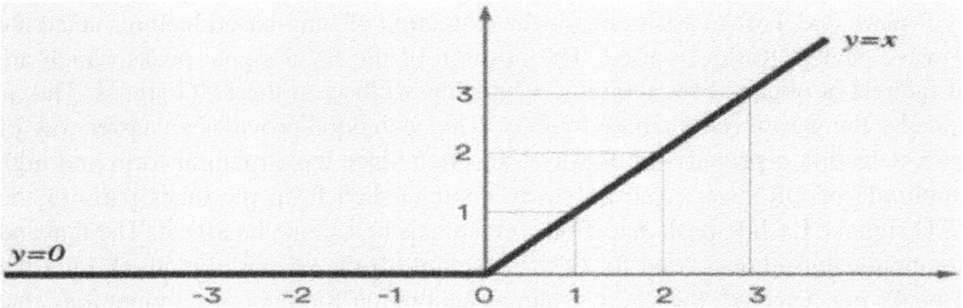

Fig. 3 Graphical representation of ReLU activation function

In detail, the proposed DNN has been developed using the Keras with TensorFlow back-end in the Python programming language. Performed in a wide set of trials, this DNN has been built empirically. The number of hidden layers, corresponding number of neurons that make up each layer, the activation function, the number of learning steps are the parameters that have been manually configured in each trial and the changes were noted. The classifier model accuracy is eradicated on the testing set for each manual configuration. After this arduous manual phase, the best classification performance was obtained with a DNN composed of three hidden layers, i.e., 300, 500, and 300 neurons, respectively. This DNN classifier makes all the neural layers based on the activation function called ReLU which is an abbreviation for 'Rectified Linear Unit.' ReLU stands for rectified linear unit and is a type of activation function. Mathematically, it is defined as Eq. (1), and visually, it looks like the following Fig. 3.

$$y = \max(0, x). \tag{1}$$

The aim of machine learning and deep learning is to make the predicted output as much as accurate as that of the actual output. This is also called as a cost function (C) or loss function. The cost function used is sparse categorical cross-entropy. This function is used when the classes are mutually exclusive (e.g., when each sample belongs exactly to one class) otherwise categorical cross-entropy is used, where one sample can have multiple classes or labels. The formula for sparse categorical cross-entropy (*s*—samples, *c*—classes, *s* in *c*—sample belongs to class *c*), *p*—probability is given in Eq. (2).

$$f = -\{\log p(s \backslash\text{in}\, c)\} \tag{2}$$

The goal is to find an optimal weight value to minimize the cost function and ensuring that the algorithms achieve accurate recognition of the model. To achieve this, different weight values are used by carrying out multiple iterations. This helps to finding the minimum cost. This process is called gradient descent. Gradient descent is mainly done to reduce the cost function. The direction of increase is along the gradient; hence, the minimum point can be obtained moving in the opposite direction of the gradient. To minimize the loss function, the weighted parameters need to be updated. This is done by optimizers. Loss function informs the optimizer about the direction of its movement along the terrain to reach a global minimum, which is at the bottom of the curve, and Eq. (3) defines the minimization of loss function.

$$\theta = \theta - \eta J(\theta; x, y) \tag{3}$$

where θ is the weight parameter, η is the learning rate and $\nabla J(\theta; x, y)$ is the gradient of weight parameter θ.

The optimizer used within the network is Adamax type. Adamax is a variant of the Adam optimizer that uses the infinity norm, while the Adam optimizer itself uses the L2 norm for optimization. The softmax function is often used in the final layer of a neural network-based classifier. The training of such networks is done under a log loss or cross-entropy domain which gives a nonlinear variant of multinomial logistic regression.

At the softmax regression layer, six nodes are employed for classification purpose; i.e., the ECG signals are classified into normal and five classes of arrhythmia. The labels denoting the six different classes are *N*—normal, *L*—left bundle block, *R*—right bundle block, *V*—ventricular block, *A*—atrial blocks, and *a*—other arrhythmia. All the other types of arrhythmia are categorized in the class '*a*.'

4 Results and Discussion

4.1 The Dataset

The MIT–BIH dataset consists of a series of ECG data's named as a number list, that is {100, 101, 103, 105, 106, 108, 109, 111, 112, 113, 114, 115, 116, 117, 118, 119, 121, 122, 123, 124, 200, 201, 202, 203, 205, 207, 208, 209, 210, 212, 213, 214, 215, 219, 220, 221, 222, 223, 228, 230, 231, 232, 233, 234}. This database consists of randomly chosen database of 23 records (numbered from 100 to 124 inclusive with some numbers missing) and 25 records of clinically important arrhythmias (numbered from 200 to 234 inclusive, again with some numbers missing). Each of these 48 records possess a duration of slightly more than half an hour [16, 17].

For each subject of the MIT–BIH database, the four preprocessing steps mentioned in the signal processing section have been performed. These data are then randomly shuffled and then 80% of the data is used for training of the model and the remaining 20% for testing of the model. The length of the whole data used in modeling the proposed system is 102,885 and thereby the length of the data used for training is 82,308 in which 65,846 beats are used for actual training and 16,462 beats are used for validation. The size of the remaining 20% of the actual data used for testing is 20,577.

Eventually, real-time raw ECG data were collected from randomly selected people of different ages from 22 to 60. The ECG electrodes and AD8232 ECG module were used for the raw ECG sample collection.

4.2 Dataset Preprocessing

The original signal of the subject 100 is taken as an example and is shown in Fig. 4. The input signal is then allowed to undergo denoising processes using band-pass filter followed by a baseline removal. The peaks in the filtered signal are detected using waveform database toolbox. These signals are then normalized and segmented into beats by setting windows of definite length to the both sides of the detected peaks. Figure 4 shows the results obtained at each of the signal processing stages, i.e., original signal, filtered signal, peak detected signal, segmented, and normalized signal.

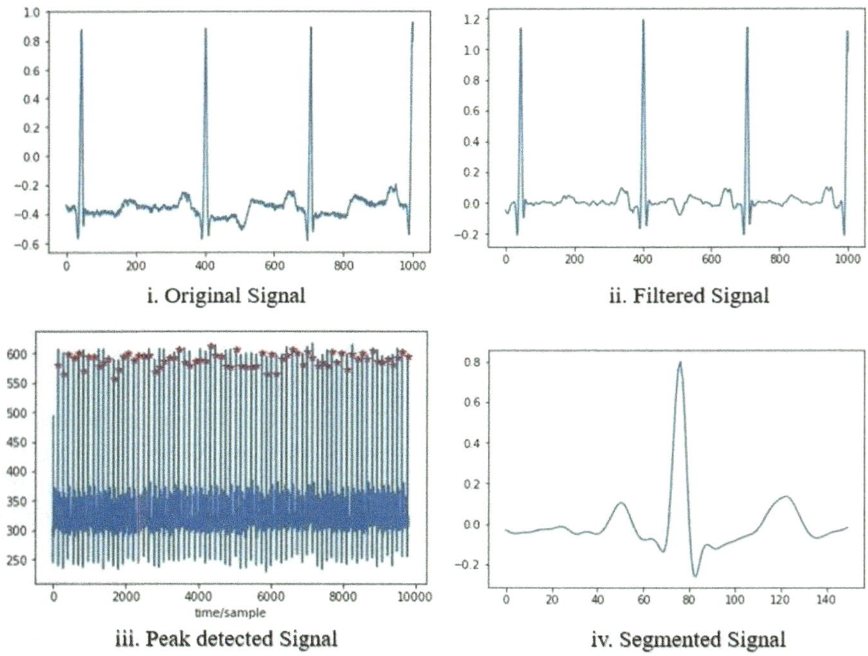

Fig. 4 Original signal of subject 100 of the MIT–BIH database and its filtered, peak detected, and segmented form

Table 1 Network layers, weights, and biases

Network layers	Weights	Biases
input_1 (input layer)	(None, 150)	0
dense_1 (dense)	(None, 300)	45,300
dropout_1 (dropout)	(None, 300)	0
dense_2 (dense)	(None, 500)	150,500
dropout_2 (dropout)	(None, 500)	0
dense_3 (dense)	(None, 300)	150,300
dropout_3 (dropout)	(None, 300)	0
dense_4 (dense)	(None, 6)	1806
Total parameters: 347,906	Trainable parameters: 347,906	Non-trainable parameters: 0

4.3 Deep Neural Network

A DNN model to classify six classes of arrhythmia has been developed. Classifier model designing was completed because of a number of trial and error methods. In each trial, the improvement in model has been manually configured by changing the number of hidden layers, nodes of hidden layers, and the learning steps. The developed classifier has three hidden layers, along with the neuron arrangement of [300, 500, and 300], the learning and decay rate as 0.01 and 0.001, respectively, with an Adamax optimizer. The output layer consists of softmax regression layer, and the losses are calculated using categorical cross-entropy function. Hundred number of epochs are used for the training and testing process above and below this range various anomalies like over-fitting, under-fitting, and other accuracy problems were encountered. The network layers and the parameters are described in Table 1. The accuracy and error graphs obtained after the 300 epochs, using the mentioned parameters for the proposed network is shown in Fig. 5.

In machine learning, especially in stratified classification, an error matrix known as confusion matrix is calculated, how an algorithm performs on a test dataset for which

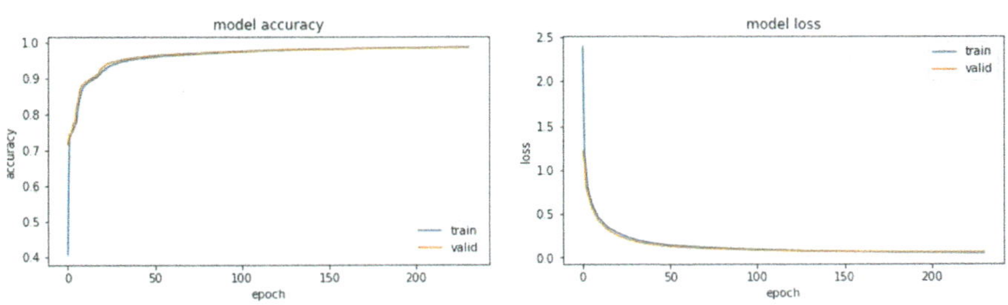

Fig. 5 Accuracy and loss of the network with dropout layer and Ada-max optimizer with a learning rate of 0.01

the true values are known is depicted by a table called confusion matrix. Confusion matrix helps in easy identification of one class being commonly mislabeled as another class. Confusion matrix is used for the computation of performance measures of classes. The confusion matrix for the developed model is shown in Fig. 6. From the obtained confusion matrix, the True Positive (TP), True Negative (TN), False Positive (FP), and False Negative (FN) values are calculated for each class. Table 2 gives the performance matrix. From the obtained values of TP, TN, FP, and FN, the accuracy, sensitivity, and specificity of each of the classes and from which these parameters of the whole model can be estimated from Eqs. (4), (5) and (6), respectively.

$$\text{Accuracy} = \frac{(\text{TP} + \text{TN})}{((\text{TP} + \text{TN}) + (\text{FP} + \text{FN}))} \tag{4}$$

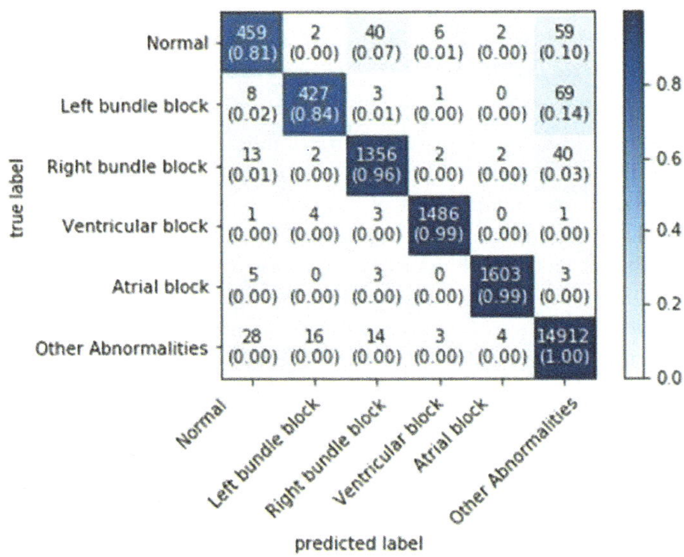

Fig. 6 Confusion matrix

Table 2 TP, TN, FP, FN of the classes *N*, *L*, *R*, *V*, *A*, *a*

Class	TP	TN	FP	FN
N	501	19,949	60	67
L	450	200,025	44	58
R	1375	19,104	58	40
V	1484	19,074	8	11
A	1603	18,955	8	11
a	14,879	5493	107	98

Table 3 Accuracy, sensitivity, and specificity of each of the classes

N	99.38	88.2	99.7
L	99.5	88.58	99.7
R	99.5	97.17	99.69
V	99.9	99.26	99.95
A	99.9	99.3	99.95
a	99	99.3	98

$$\text{Sensitivity} = \frac{(\text{TP})}{(\text{TP} + \text{FN})} \tag{5}$$

$$\text{Specificity} = \frac{(\text{TN})}{(\text{TN} + \text{FP})} \tag{6}$$

The performance matrix has been computed and the accuracy, sensitivity, and specificity of each of the classes obtained is shown in Table 3. From this data, the overall efficiency of the entire model is calculated by calculating the average of the accuracy, sensitivity, and specificity of all the classes, and the values obtained are 99.45%, 93.16%, and 99.35%, respectively.

To demonstrate the quality of the DNN in terms of accuracy, sensitivity, and specificity, comparative analysis was carried out on the same dataset using eleven other well-known classifiers [15].

The most representative algorithms of each category, like the AdaBoostM1 [18], Bagging [19], Naive Bayes [20], Multilayer Perceptron (MLP) [21], KStar [22], OneR [23], Part [24], J48 [25], Random Tree [26], JRip [27], support vector machine (SVM) [28–30], were chosen for the classification performance. The results obtained from this analysis depicts the fact that the accuracy of the DNN models was highest, with values greater than 99%.

Comparative analysis for multi-class arrhythmia classification denoted that the proposed DNN model achieved the accuracy, sensitivity, and specificity values at 99.45, 93.16, and 99.35%, respectively. To the best of our knowledge, no other works have reported a multi-class arrhythmia classification using deep neural network with the obtained accuracy level. Table 4 shows the results of the accuracy obtained for different algorithms used in multi-class arrhythmia classification.

Table 4 Accuracy for multi-class arrhythmia classifiers

Algorithm	Overall accuracy
CNN + RNN + LSTM [31]	81.5
Levenberg–Marquardt, Bayesian-Regularization [32]	93.8
Random Forest [33]	98.03
CNN [34]	98.30
Proposed DNN model	99.45

5 Conclusions and Future Work

In healthcare, biomedical, and bioinformatics fields, classification is currently one of the emerging research topics, especially for arrhythmia detection. In this proposed system, a novel approach based on a DNN for the automatic classification of multi-class arrhythmia has been developed. The comparative analysis with several other classification methods shows that DNN networks are having highest accuracy, sensitivity, and specificity for classifying the heartbeats into normal and five classes of arrhythmia. The proposed model can differentiate the normal ECG beats from five classes of arrhythmia with promising accuracy, sensitivity, and specificity values of 99.45, 93.16, and 99.35%, respectively, which indicates the efficiency of the proposed model is very promising.

Moreover, real-time ECG monitoring system and arrhythmia classification will be a great boon to healthcare industries. Increasingly, organizations in a variety of industries are using Internet of things (IoT) to operate more efficiently, better understand customers to deliver enhanced customer service, improve decision-making and increase the value of the business. The current research work involves incorporating IOT to the proposed model to enhance the availability of system to a wide range of users.

References

1. R.J. Martis, C. Chakraborty, A.K. Ray, A two-stage mechanism for registration and classification of ECG using Gaussian mixture model. Pattern Recogn. **42**(11), 2979–2988 (2009)
2. M. Llamedo, J.P. Martínez, Heartbeat classification using feature selection driven by database generalization criteria. IEEE Trans. Biomed. Eng. **58**(3), 616–625 (2011)
3. C. Ye, B.V. Kumar, M.T. Coimbra, Heartbeat classification using morphological and dynamic features of ECG signals. IEEE Trans. Biomed. Eng. **59**(10), 2930–2941 (2012)
4. M. Arif et al., Robust electrocardiogram (ECG) beat classification using discrete wavelet transform. Physiol. Meas. **29**(5), 555 (2008)
5. G. Garcia, G. Moreira, D. Menotti, E. Luz, Inter-patient ECG heartbeat classification with temporal VCG optimized by PSO. Sci. Rep. **7**(1), 10543 (2017)
6. P. de Chazal, M. O'Dwyer, R.B. Reilly, Automatic classification of heartbeat using ECG morphology and heartbeat interval features. IEEE Trans. Biomed. Eng. **51**(7), 1196–1206 (2004)
7. P. de Chazal, R.B. Reilly, A patient-adapting heartbeat classifier using ECG morphology and heartbeat interval features. IEEE Trans. Biomed. Eng. **53**(12), 2535–2543 (2006)
8. Y.H. Hu, S. Palreddy, W.J. Tompkins, A patient-adaptable ECG beat classifier using a mixture of experts approach. IEEE Trans. Biomed. Eng. **44**(9), 891–900 (1997)
9. D. Bianchi, P.D. Michele, C. Marchetti, B. Tirozzi, S. Cuomo, H. Marie, M. Migliore, Effects of increasing CREB-dependent transcription on the storage and recall processes in a hippocampal CA1 microcircuit. Hippo-campus **24**(2), 165–177 (2014)
10. K. Hanbay, Deep neural network based approach for ECG classification using hybrid differential features and active learning (Bingol University, Computer Engineering Department, Bingol 12000, Turkey), www.ietdl.org. https://doi.org/10.1049/iet-spr.2018.5103

11. M.M. Al Rahhal, Y. Bazi, H. Al Hichri, N. Alajlan, F. Melgani, R.R. Yager, Deep learning approach for active classification of electrocardiogram signals. Inf. Sci. **345**, 340–354 (2016)
12. A.J. Joshi, S. Chandran, V.K. Jayaraman, B.D. Kulkarni, Hybrid SVM for multiclass arrhythmia classification. in *2009 IEEE International Conference on Bioinformatics and Biomedicine* (IEEE, 2009), 287–290
13. Q. Yao, R. Wang, X. Fan, J. Liu, Y. Li, Multi-class arrhythmia detection from 12-lead varied-length ECG using attention-based time-incremental convolutional neural network. Inf. Fusion **53**, 174–182 (2020)
14. S. Savalia, V. Emamian, Cardiac arrhythmia classification by multi-layer perceptron and convolution neural networks. Bioengineering **5**(2), 35 (2018)
15. G. Sannino, G. De Pietro, A deep learning approach for ECG-based heartbeat classification for arrhythmia detection. Future Gener. Comput. Syst. **86**, 446–455 (2018)
16. G.B. Moody, R.G. Mark, The impact of the MIT-BIH arrhythmia database. IEEE Eng. Med. Biol. **20**(3), 45–50 (2001). (PMID: 11446209)
17. A. Goldberger, L. Amaral, L. Glass, J. Hausdorff, P.C. Ivanov, R. Mark, H.E. Stanley, PhysioBank, PhysioToolkit, and PhysioNet: components of a new research resource for complex physiologic signals. Circulation **101**(23), e215–e220 (2000)
18. Y. Freund, R.E. Schapire, et al., Experiments with a new boosting algorithm. in *ICML 96* (1996), pp. 148–156
19. L. Breiman, Bagging predictors. Mach. Learn. **24**(2), 123–140 (1996)
20. G.H. John, P. Langley, Estimating continuous distributions in Bayesian classifiers. in *Proceedings of the Eleventh Conference on Uncertainty in Artificial Intelligence* (Morgan Kaufmann Publishers Inc., 1995), pp. 338–345
21. J. Rumelhart, Learning representations by back-propagation error. Nature **323**(9), 533–536 (1986)
22. J.G. Cleary, L.E. Trigg, et al., K*: an instance-based learner using an entropic distance measure. in *Proceedings of the 12th International Conference on Machine Learning*, vol. 5, (1995), pp. 108–114
23. R.C. Holte, Very simple classification rules perform well on most commonly used datasets. Mach. Learn. **11**(1), 63–90 (1993)
24. E. Frank, I.H. Witten, Generating accurate rule sets without global optimization (1998), pp. 1–15
25. J.R. Quinlan, *C4.5: Programs for Machine Learning* (Elsevier, 2014). https://books.google.it/books?id=b3ujBQAAQBAJ
26. L. Breiman, Random forests. Mach. Learn. **45**(1), 5–32 (2001)
27. W.W. Cohen, Fast effective rule induction. in: *Twelfth International Conference on Machine Learning* (Morgan Kaufmann, 1995), pp. 115–123
28. J. Platt, Fast training of support vector machines using sequential minimal optimization. in: *Advances in Kernel Methods—Support Vector Learning* (MIT Press, 1998)
29. S. Keerthi, S. Shevade, C. Bhattacharyya, K. Murthy, Improvements to Platt's SMO algorithm for SVM classifier design. Neural Comput. **3**, 637–649 (2001)
30. T. Hastie, R. Tibshirani, Classification by pairwise coupling. in *Advances in Neural Information Processing Systems* (MIT Press, 1998), pp. 507–513.
31. T.-M. Chen, C.-H. Huang, E.S.C. Shih, Y.-F. Hu, M.-J. Hwang, Detection and classification of cardiac arrhythmias by a challenge-best deep learning neural network model. iScience **23**(3), 100886 (2020)
32. H. Lassoued, R. Ketata, ECG multi-class classification using neural network as machine learning model. in *2018 International Conference on Advanced Systems and Electric Technologies* (IEEE, 2018), pp. 473–478
33. F.I. Alarsan, M. Younes, Analysis and classification of heart diseases using heartbeat features and machine learning algorithms. J. Big Data **6**(1), 1–15 (2019)
34. S.K. Pandey, R.R. Janghel, Automatic detection of arrhythmia from imbalanced ECG database using CNN model with SMOTE. Australas. Phys. Eng. Sci. Med. **42**(4), 1129–1139 (2019)

Lung Nodule Detection from Computed Tomography Images Using Stacked Deep Convolutional Neural Network

Mahender Gopal Nakrani, Ganesh S. Sable, and Ulhas B. Shinde

Abstract The advances in deep convolutional neural networks (DCNNs) have made remarkable progress in image classification and object detection methods. DCNN also has been successfully implemented to medical images problem in the past few years. One of the major issues in medical imaging is the detection of lung cancer. Detection of lung nodules is a crucial step in lung cancer screening. To detect lung nodule from computed tomography (CT) images, we proposed a novel stacked DCNN-based computer-aided detection (CAD) system. Three DCNNs are stacked together to form a single stacked DCNN. In this work, we segment raw CT images to get lung region for the images using basic morphological operations. The segmented images are directly put into stacked DCNN to generate probabilities of probable lung nodules. The largest publicly available archive of lung cancer screening thoracic CT scan was used for experimentation. Our proposed method achieves a sensitivity of 96.81% at 8 false positives per scan and top-5 accuracy of 96.23%.

Keywords Lung nodules · Deep convolutional neural network · Computer-aided detection · Computed tomography · Segmentation · Morphological operations

1 Introduction

Lung cancer is one of the common cancer in the world, accounting for more than 2 million new cases and 1.7 million deaths in 2018 [1]. It is also estimated that one in every five deaths occurs due to lung cancer. The lung cancer patients five-year survival rate is just 17.7% which is the lowest among all other leading sites of cancer. The widely used medical imaging modality for lung cancer detection is computed tomography (CT). Lung cancer can be controlled and treated if it is detected at an

M. G. Nakrani (✉) · U. B. Shinde
Electronics and Tele-Communication Engineering, CSMSS's Chh. Shahu College of Engineering, Aurangabad, Maharastra 431011, India

G. S. Sable
Electronics and Tele-Communication Engineering, G.S. Mandal's Maharashtra Institute of Technology, Aurangabad, Maharastra 431010, India

M. Paprzycki et al. (eds.), *Intelligent Systems, Technologies and Applications*, Advances in Intelligent Systems and Computing 1353,
https://doi.org/10.1007/978-981-16-0730-1_16

early stage when the cancer is limited to lung. The detection of lung cancer at the early stage depends on the detection of lung nodules which are small lesion in the lungs. The lung nodule detection is not an easy task as the size of the nodules is very small and also due to the structure of parenchyma where nodules are formed. The experience of the radiologist also affects the diagnosis of the CT scans in front. Many CAD systems are proposed to assist radiologists with this task of nodule detection. Most of these CAD systems use machine learning algorithms and mainly consist of four pipelining stages [2]. The first stage is the pre-processing stage, the second stage is the feature extraction stage, the third stage is the nodule detection stage and the final stage is false-positive reduction stage. On the other hand, the deep learning-based CAD systems have only three pipelining stages as the convolutional neural networks (CNNs) used for nodule detection stage can discover features automatically.

In this paper, a two-stage method is proposed for lung nodule detection making the CAD system simpler and robust. The first stage of the system is to segment the lung parenchyma region, which was established with the help of morphological operations. The second stage is the nodule detection stage which will detect lung nodules using a novel stacked DCNN. The stacked DCNN is designed such a way that it will produce a small number of nodule candidates and hence eliminating the requirement of false-positive reduction stage.

2 Related Work

The detection of lung nodules using deep learning-based CAD systems utilizes DCNN in two stages. The first DCNN is used to detect lung nodules and the second DCNN is for false-positive reduction. In recent years, many DCNN is designed and utilized for this purpose. For example, A U-Net segmentation architecture based on residual dense structure was designed in [3] for lung nodule detection and 3D CNN with three network models were used for false-positive reduction. The outputs of three network models in false-positive reduction are combined to give final probabilities. A deep transfer convolutional neural network (DTCNN) and extreme learning machine (ELM) were used in [4]. In this method, a pre-trained deep convolutional neural network (DCNN) model like ResNet50, Xception, NASNetMobile, MobileNetV2 and EfficientNet-B5 is used to extract universal features for common image classification, and the target DCNN model is used for false-positive reduction to classify nodules. In [5], a faster R-CNN was proposed which composed of region proposal network (RPN) for generating region proposals and a detector for classifying the region proposals as nodules. A 3D faster R-CNN architecture was used for nodule detection in [6] with customized mixed link network (CMixNet) architecture is then used to learn the features of nodules and finally gradient boosting machine (GBM) was used for classification. A faster R-CNN with feature extraction network, region proposal network and region-of-interest classifier was designed in [7]. Feature extraction network was implemented using a VGG16 convolutional network with five group convolutions and a deconvolution layer to obtain a 148×148 feature map. The system uses two region proposal networks which are concatenated to the deconvolution layer and the middle convolution layer. Anchors of size 12 × 12, 18 × 18, 27 × 27, 36 × 36, 51 × 51, 75 × 75 and 120 × 120 are designed to predict multiple region-of-interest of different size termed as nodule candidates. In [8] pre-trained CNN like VGG16, VGG19, MobileNet, Xception, InceptionV3, ResNet50, Inception-ResNet-V2, DenseNet169, DenseNet201, NASNetMobile and NASNetLarge were used as features extraction networks and features are classified using Naive Bayes, multi-layer perceptron (MLP), support vector machine (SVM), near neighbors (KNN) and random forest (RF) classifiers. A deep residual 3D faster R-CNN architecture with transposed convolutional layers was proposed in [9] for lung nodule detection, and a 3D DCNN classifier as false-positive reduction network. A three-dimensional deep convolutional neural network (3D DCNN) with shortcut connections was used to extract features, and a 3D DCNN with dense connections was used for lung nodule classification in [10]. The shortcut connections and dense connections were used to alleviate the gradient vanishing problem by allowing the gradient to pass quickly and directly. A multi-level network based on the 2D U-net network, which is named as modified U-Net (MU-Net) was proposed in [11] for nodule detection. Three 2D MU-Net networks were trained for detecting nodules of different sizes. The non-detected nodules from previous MU-Net were used to fine-tune another MU-Net network to detect smaller nodules. A faster region-based convolutional neural network (faster R-CNN)-based DCNN was designed in [12] to extract nodule candidates regions from images. This DCNN consists of three sub-networks which are feature extraction network, region proposal network and final region-of-interest classifier. For feature extraction, a VGG16 with five group convolutions is used with a deconvolution layer to obtain a 148 × 148 feature map. The system used two region proposal networks, which are concatenated to the deconvolution layer and the middle convolution layer, respectively, to obtain different useful information about the nodules. Finally, in the region-of-interest classifier, seven anchors with different sizes 12 × 12, 18 × 18, 27 × 27, 36 × 36, 51 × 51, 75 × 75 and 120 × 120 were designed to predict multiple region-of-interest (nodule candidates) of different size. A three-dimensional (3D) DCNN based on 3D convolutional autoencoder is used in [13] for lung nodule detection. This 3D DCNN consists of three 3D convolutional layers with 3 × 3 × 3 as a convolutional kernel. It also had two 3D max-pooling layers with a kernel size of 2 × 2 × 2. Finally, a fully connected layer is used to change the size of the 13 × 13 × 5 × 64 feature map equal to 54,080 features to 1024. The ReLU was used as an activation function for all the inner layers, and softmax was used as an activation function for the output layer. The SVM was used as a classifier to classify the image as nodule and non-nodule from 1024 features extracted using 3D DCNN. The stacked DCNN model proposed in this paper is a stack of three DCNN models with different kernels for convolution.

3 Proposed Method

The system we proposed for lung nodule detection consists of two stages, namely lung parenchyma segmentation stage and lung nodule detection stage. At the lung parenchyma segmentation stage, we segment the lung parenchyma by applying different morphological operations. During the second stage, the lung parenchyma is put into stacked DCNN to detect lung nodules. The details of these stages are described in the following sections.

3.1 Lung Parenchyma Segmentation

The measurement units for CT scans are the Hounsfield unit (HU), which is a measure of radiodensity. The HU values in CT scan generally vary from −1000 to +2000. The complete process of lung parenchyma segmentation is shown in Fig. 1. To remove irrelevant objects such as bones, tissues or any external objects image slicing between 1000 and 400 HU is used. Then, the image is converted into a binary image using 600 HU as a threshold which is shown in Fig. 1b. The region proposal algorithm is applied to binary image to get different regions in an image, and only two regions with the largest area are extracted, which is shown in Fig. 1c. Next, the erosion operation is applied to remove any blob in the image as shown in Fig. 1d. To include nodules attached to the walls of lung, dilation and closing operation is applied as shown in Fig. 1e and f. Finally, the hole filling operation is performed to generate a complete mask, as shown in Fig. 1g. The segmented lung parenchyma after superimposing mask on the original image is shown in Fig. 1h.

3.2 Lung Nodule Detection

A window of 32×32 is run over segmented image to form the label. The label consists of 256 entries, each entry for each 32×32 window. The label entry is one if the nodule is present in that particular window else the entry is marked as zero. The segmented image with 32×32 windows and its label is shown in Fig. 2.

The segmented images with labels are used to train the stacked DCNN for nodule detection. Three DCNN architectures are designed and stacked together to detect lung nodules. Each DCNN architecture consists of eight convolutional layers. The first five convolutional layers are associated with max-pooling layers. The batch normalizing layers are implemented after every convolutional layer to minimize overfitting. Finally, the output of each architecture is flattened. All the three architectures are the same except the kernel size as $3 \times 3, 7 \times 7$ and 11×11. These kernels are selected to detect different size nodules. These three architectures are combined using averaging layer, and finally a dense layer activated by softmax is to give 256 probabilities for

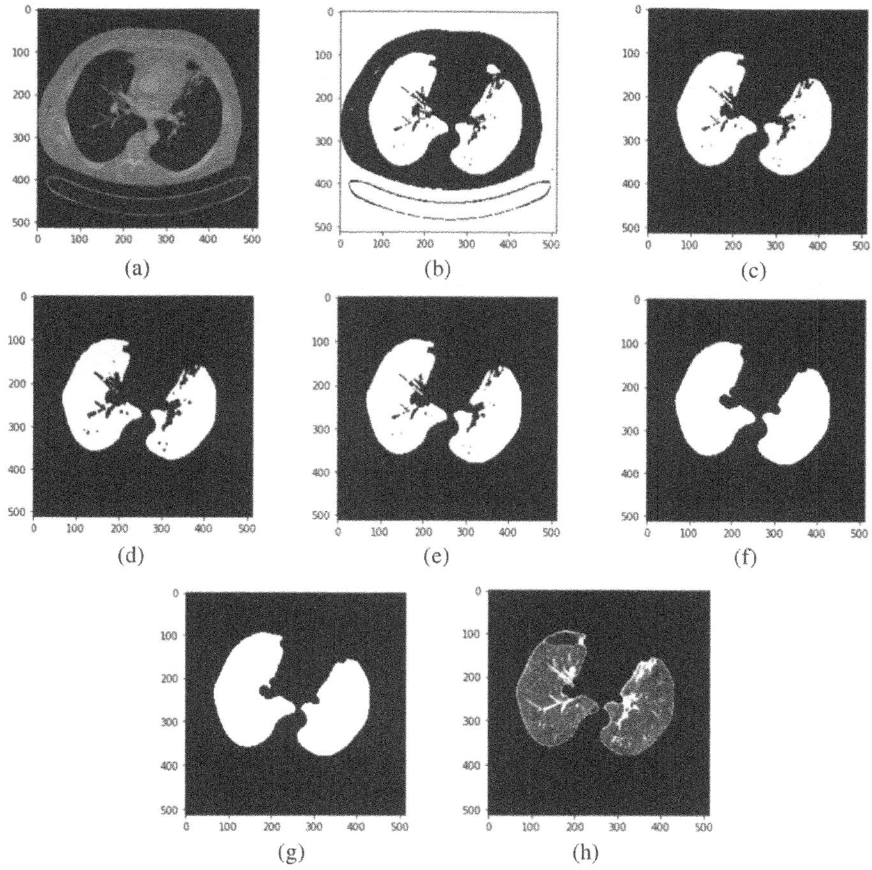

Fig. 1 Steps in lung parenchyma segmentation **a** original image, **b** binary image, **c** image with two largest region, **d** image after erosion operation, **e** image after dilation operation, **f** image after closer operation, **g** binary mask after applying hole filling operation, **h** lung parenchyma segmented image after superimposing binary mask on original image

lung nodule presence at each 32 × 32 window. The total trainable parameters in the network were nearly 3.4 million. The categorical cross-entropy was used to calculate the loss for the network. The architecture of the stacked DCNN is shown in Fig. 3.

4 Experimentation and Results

The dataset which we have used for experimentation is a subset of LIDC-IDRI [14] dataset containing 9107 images selected from the first 500 patients CT scans. Each of these scans was labeled by four experienced radiologists [15]. The dataset is

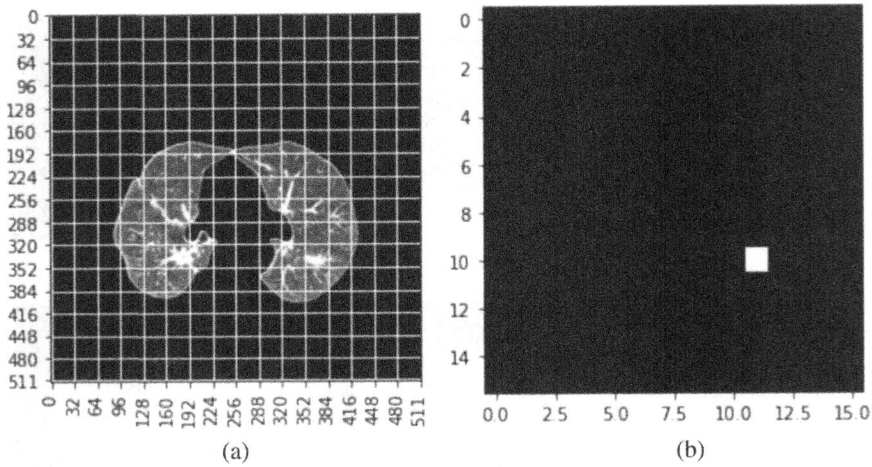

Fig. 2 Label creation process. **a** Segmented image with 32 × 32 windows, **b** label created for the image

divided into two parts, 30% dataset (2733 images) for testing and the 70% dataset (6374 images) for training purpose. The training dataset is further divided into 80% training dataset (5098 images) for training the stacked DCNN and the remaining 20% (1276 images) for validation. The stacked DCNN was trained with learning rate as 0.0001 for 40 epochs, and the Adams optimizer is used to update the model parameters. We minimized the categorical cross-entropy to maximize the probability of lung nodule presence of each training samples. The output of the proposed stacked DCNN on one of the test sample is shown in Fig. 4, along with ground truth.

The experimentation results on testing dataset show that the sensitivity of the proposed stacked DCNN reached more than 94% when the false-positive rate of 1 for each image and the accuracy reached more than 96% for 5 candidates per scan. Table 1 illustrates the sensitivity of the proposed model at different false positives per scan. The FROC curve of the proposed model is shown in Fig. 5.

We compare our proposed system with the lung nodule detection techniques discussed in related work to verify the effectiveness of our proposed system. Table 2 shows the comparison of the different DCNN-based lung nodule detection techniques. It can be seen from Table 2 that the performance of the proposed system is slightly better than the existing lung nodule detection systems.

5 Contributions

The contribution of the work proposed is to convert the problem of lung nodule detection as the presence of lung nodule in each window of the image. This enabled the proposed method to closely resemble the pipelining adopted in most of the

Fig. 3 Proposed stacked DCNN for lung nodule detection

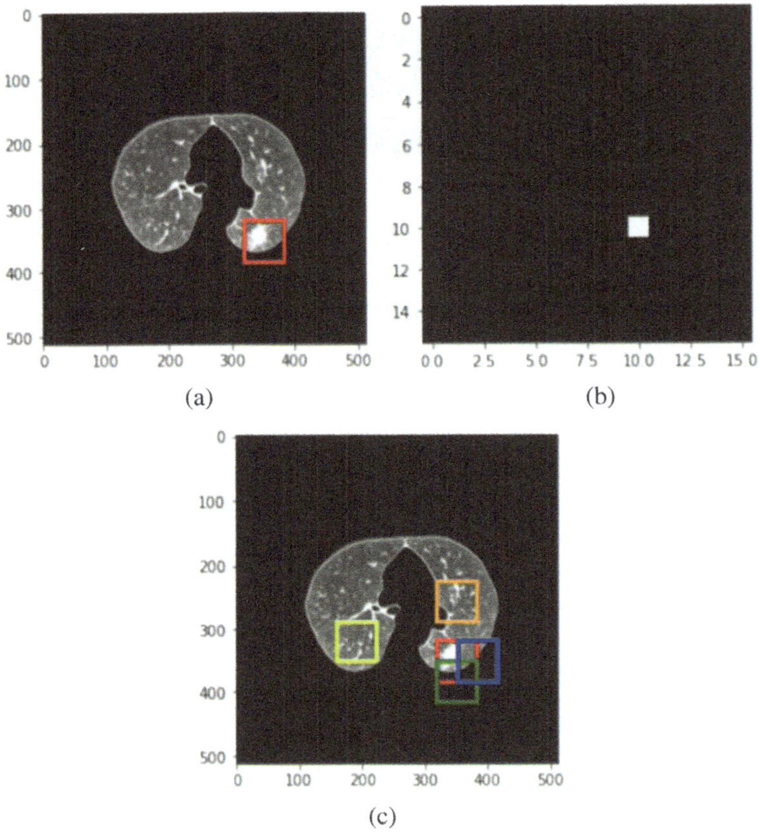

Fig. 4 Output of the proposed stacked DCNN **a** CT scans slices with ground truth, **b** prediction probabilities using the proposed stacked DCNN, **c** top 5 predictions with top 1 to top 5 shown as red, green, yellow, blue and orange bounding boxes

Table 1 Sensitivity of proposed stacked DCNN for lung nodule detection on the testing dataset

False positives/scan	Sensitivity
1	94.18
2	95.68
3	96.12
4	96.23
5	96.27
6	96.63
7	96.78
8	96.82

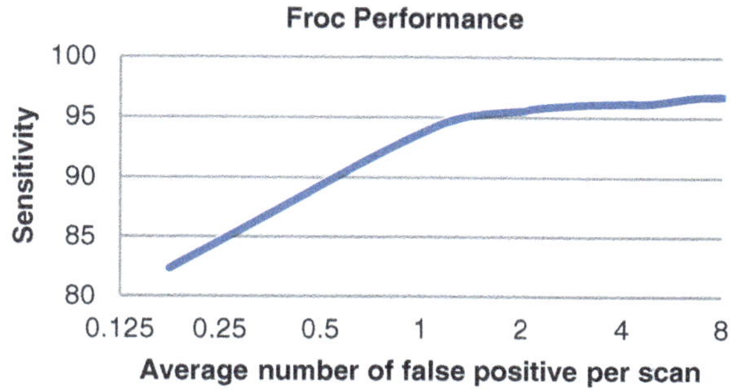

Fig. 5 FROC curve of the proposed stacked DCNN for lung nodule detection

Table 2 Comparison of the proposed system with the systems discussed in related work

Methods	Sensitivity (%)	False-positive/scan
TSCNN [3]	96.0	8
Faster R-CNN [5]	85.2	3.47
3D CMixNet [6]	94.21	8
Faster R-CNN + Boosting [7]	83.4	8
OURS	96.82	8

image detection methods. The proposed work also adopts a simple lung segmentation method which makes the system cost-effective.

6 Conclusion

In this paper, a novel and effective lung nodule detection method based on stacked DCNN is proposed. The proposed stacked DCNN-based lung nodule detection system is composed of two stages, the first stage is the lung parenchyma segmentation and the second stage is the lung nodule detection. In the first stage, the lung parenchyma segmentation is done using a combination of some basic morphological operations. In the second stage, we proposed a novel stacked DCNN for lung nodule detection which composed of three DCNN with 3×3, 7×7 and 11×11 kernel size to detect nodules of different size. The LIDC-IDRI dataset is used for the experimentation. The performance of the method demonstrates that the method can detect lung nodules accurately. The method was able to achieve a sensitivity of 96.82% at 8 false positives per scan without false-positive reduction stage. The proposed system does not classify the detected nodules as benign or malignant nodules. In future, a classification network can be designed and implemented to classify nodules as benign or malignant. The proposed system can be implemented to assist radiologist for the correct diagnosis of lung nodules. It would be interesting to see the effects of different kernel size of DCNN on the results of the system and also would be interesting to perform experiments on different datasets.

References

1. Lung Cancer, 2019 Fact Sheet. [Online]. Available: https://ginasthma.org/world-lung-cancer-day-2019/
2. Mahender G. Nakrani, Ganesh S. Sable, Ulhas B. Shinde, Detection of lung nodules in computed tomography image using deep machine learning: a review. in *Proceedings of International Conference on Communication and Information Processing* (ICCIP) (2019)
3. H. Cao et al., Two-stage convolutional neural network architecture for lung nodule detection. IEEE J. Biomed. Health Inf. (2020)
4. X. Huang, Q. Lei, T. Xie, Y. Zhang, Z. Hu, Q. Zhou, Deep Transfer Convolutional Neural Network and Extreme Learning Machine for Lung Nodule Diagnosis on CT images (2020), arXiv:2001.01279
5. Y. Li, L. Zhang, H. Chen, N. Yang, Lung nodule detection with deep learning in 3D thoracic MR images. IEEE Access **7**, 37822–37832 (2019)
6. N. Nasrullah, J. Sang, M.S. Alam, M. Mateen, B. Cai, H. Hu, Automated lung nodule detection and classification using deep learning combined with multiple strategies. Sensors **19**, 3722 (2019)
7. H. Xie, D. Yang, N. Sun, Z. Chen, Y. Zhang, Automated pulmonary nodule detection in CT images using deep convolutional neural networks. Pattern Recogn. **85**, 109–119 (2019)
8. R.V.M. Nóbrega, S.A. Peixoto, S.P.P. Silva, P.P.R. Filho, Lung nodule classification via deep transfer learning in CT lung images. in *IEEE 31st International Symposium on Computer-Based Medical Systems (CBMS)* (Karlstad, 2018), pp. 244–249
9. H. Tang, D.R. Kim, X. Xie, Automated pulmonary nodule detection using 3D deep convolutional neural networks. in *IEEE 15th International Symposium on Biomedical Imaging (ISBI 2018)* (Washington, DC, 2018), pp. 523–526
10. H. Jung, B. Kim, I. Lee, J. Lee, J. Kang, Classification of lung nodules in CT scans using three-dimensional deep convolutional neural networks with a checkpoint ensemble method. BMC Med Imaging (2018)
11. Z. Hu, A. Muhammad, M. Zhu, Pulmonary nodule detection in CT images via deep neural network: nodule candidate detection. in *ICGSP'18, Proceedings of the 2nd International Conference on Graphics and Signal Processing* (2018), pp. 79–83
12. N. Sun, D. Yang, S. Fang, H. Xie, Deep convolutional nets for pulmonary nodule detection and classification. in *Knowledge Science, Engineering and Management, KSEM 2018*, ed. by W. Liu, F. Giunchiglia, B. Yang. Lecture Notes in Computer Science, vol. 11062, (Springer, Cham, 2018)
13. L. Fan, Z. Xia, X. Zhang, X. Feng, Lung nodule detection based on 3D convolutional neural networks. in *International Conference on the Frontiers and Advances in Data Science (FADS)* (Xi'an, 2017), pp. 7–10
14. S.G. Armato III. et al., The lung image database consortium (LIDC) and image database resource initiative (IDRI): a completed reference database of lung nodules on CT scans. Med. Phys. **38**, 915–931 (2011)
15. K. Clark et al., The cancer imaging archive (TCIA): maintaining and operating a public information repository. J. Digit. Imaging **26**(6), 1045–1057 (2013)

Deep Learning Classification to Improve Diagnosis of Cervical Cancer Through Swarm Intelligence-Based Feature Selection Approach

S. Priya and N. K. Karthikeyan

Abstract Cervical cancer is one of the predominant cancers that cause death in women globally. This disease progresses slowly and is a curable cancer, if detected well in advance. Various studies on different data mining models used for diagnosing this disease have been carried out. Different approaches including support vector machine and recursive feature elimination (SVM-RFE) and support vector machine and principal component analysis (SVM-PCA) had been designed for cervical cancer diagnosis. However, these pose various challenges including low accuracy and high processing time for classification. The proposed system addresses these issues by designing a long short-term memory with artificial bee colony (LSTM-ABC) algorithm for cervical cancer detection. The study takes the cervical cancer dataset as an input and uses synthetic minority oversampling technique (SMOTE) for solving the class imbalance issue. From the preprocessed data, the feature selection is performed using artificial bee colony (ABC) algorithm. Long short-term memory (LSTM) scheme is then employed for classifying cervical cancer based on the selected features. Experimental outcomes demonstrate that the proposed system delivers superior results than the previous works with respect to sensitivity, accuracy, and specificity.

Keywords Cervical cancer · Artificial bee colony (ABC) · Synthetic minority oversampling technique (SMOTE) · Long short-term memory (LSTM)

S. Priya (✉)
Department of Computer Science and Engineering, Coimbatore Institute of Technology, Coimbatore, India
e-mail: priya.s@cit.edu.in

N. K. Karthikeyan
Department of Information Technology, Coimbatore Institute of Technology, Coimbatore, India

M. Paprzycki et al. (eds.), *Intelligent Systems, Technologies and Applications*, Advances in Intelligent Systems and Computing 1353,
https://doi.org/10.1007/978-981-16-0730-1_17

1 Introduction

World Health Organization (WHO) states that the foremost life killer cancer in women globally is cervical cancer [1]. Cervical cancer is detected by scanning the cervical region of women, which is lower part of the vagina. By NCCC, in 2010, it is identified as human papilloma virus (HPV) is one of the major cause of cervical cancer among the other causes. Some of the other causes include number of pregnancies, multiple sexual partners, smoking, etc. Initially, this cancer affects the cervical region, and then over a period of time, it affects the other parts of the body [2]. This is one of the deadliest diseases, because this disease cannot be detected at an earlier stage and the patient does not experience any symptom until the disease reaches its final stage. Therefore, by detecting this disease at an earlier stage, death ratio owing to it can be decreased.

The perfect kind of screening test have to be minimally invasive, simple, effective, acceptable to the subject, and affordable in detecting the ailment in its early invasive stage, when it can be easily treated. Four different types of screening techniques are currently available. They are cervical cytology, biopsy, Schiller, and Hinselmann techniques [3]. Cytology screening technique, also referred as Pap smear test, is a microscopic investigation of a sample of cells from the cervix for identifying cancerous or precancerous states of the cervix [4]. Biopsy technique is a surgical procedure, where a living tissue sample is taken for detecting the disease [5]. Hinselmann test is based on the application of a solution of iodine in the cervix to visually inspect the chances in tissues. In Schiller technique, Lugol's iodine is applied to the doubtful area of the cervix to facilitate visually inspecting the tissue changes [6].

Computing techniques like classification, clustering, etc., through machine and deep learning algorithms are used by various health-related organizations to seek improvement in medical diagnosis. In under developed countries and for the economically poor section of people around the world, it is not possible to go for the medically diagnostic system due to the lack of medical equipments. Such is the motivation toward computer-aided screening techniques that help in early diagnosis of the disease and enhance the longevity of women. Screening of cervical cancer requires multi-variate analysis on the values. The solution space is mostly limited in the available dataset which is unlikely for obtaining accurate information out of it. Challenges lie in methods to identify techniques to produce samples for applying screening techniques and produce better convergence to solution. Optimizations techniques alongside filter, wrapper methods can be applied for feature selection to enhance the accuracy of results.

Summary: The remaining part this paper is arranged as follows: The existing cervical cancer classification techniques are briefly discussed in Sect. 2, the model proposed for cervical cancer dataset in the current paper is discussed in Sect. 3, the experimental results in Sect. 4, and this study's conclusions are discussed in Sect. 5.

2 Literature Survey

Various researchers work on the niche of cervical cancer classification. All the below techniques use various feature selection and classification methods to nullify classification errors owing to false negative and false positive documentation, to improve accuracy, and to recognize the major risk factors of the disease. Alam et al. [7] designed a new cervical cancer detection via various screening techniques such as decision jungle, decision forest, and boosted decision tree algorithms. Initially, the preprocessing is done based on the median values to improve data efficiency. The second step contains building effective classification according to the boosted decision tree, decision forest, and decision jungle algorithms for prediction [7]. However, need to improve the true positive rate. Rustam et al. [8] designed a cervical cancer classification technique by employing Gauss–Newton representation-based algorithm (GNRBA). It has less complexity and is easier and more effective to execute. However, it only achieves 93% of accuracy [8].

Li et al. [9] designed a Bayesian possibilistic C-means (BPCM) clustering technique for detecting cervical cancer. A Bayesian version of possibilistic C-means clustering algorithm was designed to address the profound data insufficiency. It could vigorously detect worthy patterns for enhanced data imputation. Then, screening was done using a two-stage fuzzy ensemble learning procedure. During stage one, application of the bootstrap aggregation technique was done to sample all of the class-imbalanced dataset into many class-balanced subsets. During stage two, each subset is trained using many weak classifiers. And to evaluate the weak classifiers' classification results, and to get the final classification result, a fuzzy logic-based approach is designed. Experiments were carried out on a dataset of 858 participants. The result it is seen that the designed BPCM could detect the underlying patterns effectively and therefore is trustable for finding the missing attributes [9]. However, the designed system only achieves 76% of accuracy and 79% of sensitivity in cervical cancer screening.

Yang et al. [10] designed a machine learning model for cervical cancer detection. It was real data driven and determined the predictive accuracy of the major existing cervical cancer prediction techniques via multi-layer perceptron (MLP). Ultimately, random forest (RF) was used to analyze the important index of the disease's risk factors. A close connection between cervical cancer and several risk factors was revealed by the experiment outcomes. The risk factors that influenced the most included hormonal contraceptives, number of sexual partners, and age. This technique, hence, not only improved cervical cancer risk detection, but also inspired the designing of a random forest and multi-layer perceptron (MLP) approach-based pathological model [10]. It is hard to scale the results of multi-layer perceptron (MLP).

Benazir and Nagarajan [11] designed a system for cervical cancer detection by employing data mining techniques such as feature selection and classification techniques. The designed framework classified the disease by employing two stages including feature selection stage and classification stage. In the first stage, genetic algorithm was utilized to decrease the feature space's size. In the classification stage, artificial neural network (ANN) was trained to classify the dataset into two classes viz., abnormal and normal [11]. The ANN is computationally intensive to train. Tseng et al. [12] proposed three classification models including extreme machine learning, support vector machine (SVM), and C5.0 to forecast the recurrence of cervical cancer, and to detect the most related risk factors like cell type, radiation therapy, age, tumor size, and tumor grade from clinical dataset. The dataset had 168 patients' details and belonged to the Medical University Hospital, Chung Shan. The experiments detected two risk factors, namely radiation therapy and cell type. C5.0 was found to generate superior classification accuracy ratio than that of the other two classifiers [12]. However, a small change in the data could lead to a large change in the structure of the C5.0, making it relatively inaccurate often.

Wu and Zhou [13] designed a support vector machine-based approach for predicting cervical cancer. They applied support vector machine and recursive feature elimination (SVM-RFE) and support vector machine and principal component analysis (SVM-PCA), to investigate the presence of cervical cancer in a dataset from the UCI repository. The outcomes revealed that the classification of cervical cancer was successfully done by SVM technique. Besides, its combination with PCA algorithm and RFE algorithm enabled the extraction of highly influential risk factors and the reduction of the computation burden, respectively [13]. However, better mechanisms are required to improve the classification results.

Thus, the previous research works have introduced a decision jungle, decision forest, boosted decision tree, Gauss–Newton representation-based algorithm (GNRBA), Bayesian possibilistic C-means (BPCM) clustering, artificial neural network (ANN), multi-layer perceptron (MLP), extreme machine learning, support vector machine (SVM), and C5.0 approaches for cervical cancer prediction. These methods do not produce a satisfactory results in terms of accuracy and sensitivity. The existing techniques available in the literature are not sufficient to classify the cervical cancer with high accuracy; hence, it is necessary to develop an automated approach to reduce the false positive rate and to classify the cervical cancer with high accuracy.

3 Proposed Methodology

The proposed system focuses on diagnosis of cervical cancer based on long short-term memory with artificial bee colony (LSTM-ABC) approaches. The flow diagram of the proposed work is shown in Fig. 1.

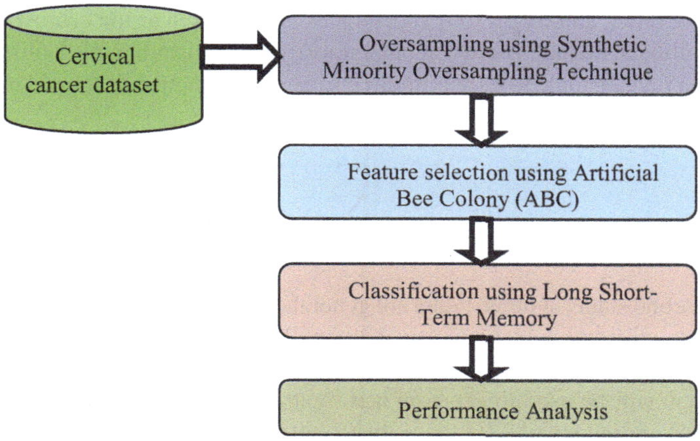

Fig. 1 Flow diagram of the proposed system

3.1 Cervical Cancer Dataset

The cervical cancer dataset used for the current study was obtained from a Caracas-based hospital called Hospital 'Universitario de Caracas' [14]. The data included 32 risk factors, encompassing details of historic medical records, patient's habits, and demographic information. The target labels or variables include cytology, Hinselmann, biopsy, and Schiller. While cytology, Schillers test, and biopsy use iodine for performing colposcopy, Hinselmanns test uses acetic acid. Further, since a few patients have not answered all the questions owing to privacy concerns, the dataset requires pre-treatment to handle the missing values. Oversampling procedure is used as the pre-treatment measure as the dataset belongs to imbalanced data.

3.2 Synthetic Minority Oversampling Technique

The input cervical cancer dataset belongs to imbalanced data. Hence, the pre-treatment process used is over sampling. Over sampling in data analysis are techniques applied to accommodate the class distribution of a dataset. It randomly copies the minority samples by increasing the number of minority class samples, thus balancing sample size of the majority class and the minority class. The most frequently used oversampling techniques for solving imbalance issues is synthetic minority oversampling technique (SMOTE). It interpolates the minority class to produce the synthetic training samples. The SMOTE algorithm consists of two stages [15].

In the first stage, k-nearest neighbors are found by using Euclidean distance calculation from each minority data with respect to all other minority data and then sorted in ascending order, and then k lowest distance data are taken as the nearest neighbors (kNN). Euclidean distance between one minority data (x) and another minority data (y) from the first attribute to n (maximum number of attributes) is defined in Formula (1)

$$d(x, y) = \sqrt{\sum_{a=1}^{n} (x_a - y_a)^2} \tag{1}$$

In the second stage, synthetic data are generated by using the interpolation method between two minority data. One of its kNN will be randomized to be candidates in synthetic data generation process. Thereafter, original minor data (x) and one chosen candidate (y) will be used to generate new synthetic data among x and y. Synthetic data formula among x and y for the a-th attribute is defined in formula (2)

$$\text{Synthetic Data}_a\,(x, y) = x_a + r\,.\,(x_a - y_a) \quad \text{for}\, 0 \leq r \leq 1 \tag{2}$$

where r-random number between 0 and 1.

The above formula is applied for n attributes. The process is performed again and again until the desired amount of synthetic data is obtained.

3.3 Synthetic Minority Oversampling Technique Feature Selection Using Artificial Bee Colony (ABC)

The proposed work uses artificial bee colony (ABC) algorithm for feature selection. It reduces the miss rate of the classifier. Artificial bee colony (ABC) algorithm is a population-based stochastic optimization. The ABC algorithm simulates honey bees' food-foraging behavior. In ABC, solutions are the term used to refer to the food source for bees [15–18]. ABC has three kinds of bees viz., the onlooker, the employed, and the scout. An ABC colony has equal number of onlooker and employed bees. Every employed bee gets to the food sources and gets back to the hive to provide this information to the onlooker bee through dancing on the dance area. Onlooker bee observes the dance moves and selects the food source. Once an employed bee abandons its food sources, it becomes a scout and begins looking for a new food source.

Initialization is the first step in the ABC algorithm. The position of features of cervical cancer dataset represents a possible solution to the issue of optimization. As each feature is made use of only by a single employed bee, the number of food sources and the number of employed bees (SN) are the same. Equation (3) is employed to generate a random initial population of SN features:

$$x_{i,j} = x_{\min,j} + \text{rand}(0, 1) \times \left(x_{\max,j} - x_{\min,j}\right) \tag{3}$$

where $i \in (i = 1, 2, 3, ..., SN)$ $j \in (j = 1, 2, ..., D)$ D-dimensional vector, where D is the number of parameters to be optimized.$x_{\max,j}$ and $x_{\min,j}$ are the upper and lower limits of $x_{i,j}$.

The population of the features positions is performed again and until the maximum cycle number (MCN), $C = 1, 2, \ldots$ MCN is obtained.

3.3.1 Employed Bee Phase

Each food source is allocated an employed bee, and a new solution is evaluated using the neighboring solution. Every employed bee alters the feature position using Eq. (4). A food source's nectar amount symbolizes the fitness of the considered solution. In this proposed research work, fitness is considered as accuracy. The accuracy of the feature subset is evaluated by employing the long short-term memory (LSTM) approach. An employed bee finds the new feature position by altering the feature position in her memory. Now compare the classification accuracy, if the new position's classification accuracy is higher than the earlier one, the bee retains the newly discovered feature position and erases the last discovered one. If not, she retains the former position in her memory.

$$v_{i,j} = x_{i,j} + \emptyset_{i,j}\left(x_{i,j} - x_{k,j}\right) \quad (4)$$

where $k \in \{1, 2, ..., SN\}$ and $j \in \{1, 2, ..., D\}$ are randomly chosen indexes; k is determined randomly and should differ from i, and $\emptyset_{ij}$ is a randomly generated number between [−1, 1].

3.3.2 Onlooker Bee Phase

Once the employed bees conclude their process of searching, they waggle dance to share the information related to cervical cancer with the onlooker bees. An onlooker bee analyzes the classification accuracy and chooses features with a probability, p_i, related to its classification accuracy following Eq. (5):

$$p_i = \frac{\text{fitness}_i}{\sum_{i=1}^{n} \text{fitness}_i} \quad (5)$$

where fitness_i is the fitness value of the solution i and n is the number of features in the cervical cancer data that is obtained by the following Eq. (6).

$$\text{fitness}_i = \begin{cases} 1 + |f_i|, \ f_i < 0 \\ \dfrac{1}{1 + f_i}, \ f_i \geq 0 \end{cases} \quad (6)$$

where f_i is the objective function value. In case, the new feature position has an equal or superior quality than the former feature position, the former one is swapped by the new one. Otherwise, the former one, which is the same as the employed bees stage, is retained.

3.3.3 Scout Bee Phase

Each bee (employee or onlooker) searches for detectable features for a limited and specific number of cycles. When there is no improvement in the accuracy, the employee bee turns into a scout bee. Then, the new features are selected by using Eq. (3). The parameter "limit" is the control parameter to ascertain the deserting of the features within the preset number of cycles (70 iterations). The complete flow diagram for the feature selection algorithm is shown in Fig. 2 (Table 1).

3.4 Classification Using LSTM

In this proposed research work, based on the selected features from ABC, long short-term memory (LSTM) is utilized for cervical cancer classification. LSTM is an extended version of RNN. It has three gates viz., an input gate, an output gate, and a forget gate. It is a two hidden layer neural network. Each layer contains 100 nodes. The activation signals are received by these units from varied sources. The designed multipliers control the activation of the cell. The altering of the memory cell contents for multiple times by the rest of the network is prevented by LSTM gates. In the current study, the features selected from ABC module are the input to LSTM.

The input gate layer is shown in Eq. (7). It ascertains the values that will be updated. The temporary cell state layer is shown in Eq. (8). It produces new candidate values in the form of a vector that could be included to the state.

$$i_t = \sigma(U_i . h_{t-1} + W_i x_t + b_i) \tag{7}$$

where i_t is the input layer, σ represents the input gate's sigmoid activation function, U_i represents weight connection to the former hidden states, W_i is weight connection to the input patterns, and b_i is the biases at the input layer.

$$\overline{C_t} = \tanh(U_c h_{t-1} + W_c x_t + b_c) \tag{8}$$

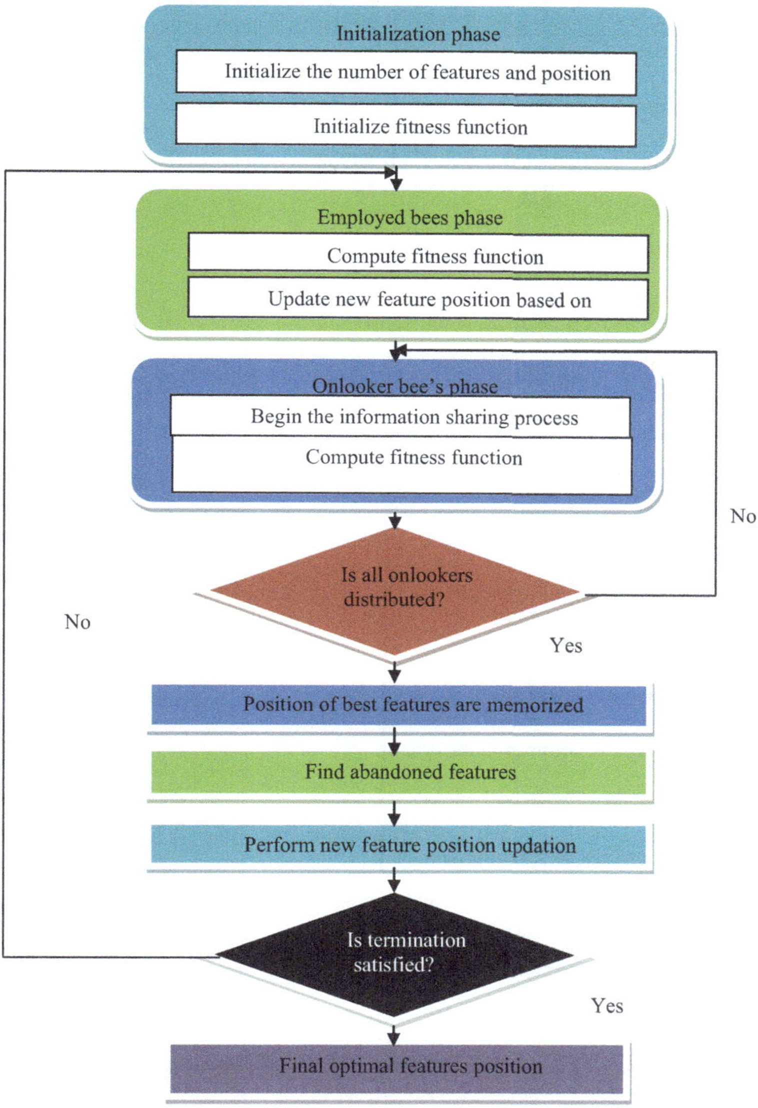

Fig. 2 Flow diagram of feature selection approach

where $\overline{C_t}$ is the temporary state layer, σ is the sigmoid function at the temporary state layer, U_c is the weight connection to the previous hidden states, W_c is weight connection to the input patterns, and b_c is the biases at the temporary state layer.

Table 1 Steps involved in ABC algorithm for implementation

Input: Number of features in the cervical cancer dataset
Output: Optimal features
1. Initialize the number of features , where $i = 1 \ldots SN$
2. Initialize feature position
3. Evaluate the $fitness_i$ value of the population
4. Set cycle =1
5. Repeat
FOR each employed bee
Produce new solution value v_i by using (4)
Calculate the $fitness_i$ function value
Apply greedy selection process
6. Calculate the probability p_i for the feature using (5)
7. FOR each onlooker bee
Choose a selected feature x_i depending on p_i
Produce new solution value v_i
Calculate the $fitness_i$ function using (6)
Apply greedy selection process
8. If the available solution is an abandoned solution for scout bees then replace it with new solution of features using (3)
9. Memorize the best selected features achieved so far
10. Update cycle = cycle + 1
11. Until cycle = MCN

Next, the new cell state is modified from the old cell state via the following equation:

$$C_t = f_t * C_{t-1} + i_t * \overline{C_t} \tag{9}$$

Thus, the output layer gate will be decided using Eq. (10):

$$o_t = \sigma(U_o.\, h_{t-1} + W_o x_t + b_o) \tag{10}$$

where o_t is the output layer, σ is the sigmoid function at the output layer gate, U_o is the weight connection to previous hidden states, W_o is weight connection to the input patterns, and b_o is the biases at the output layer.

At the end, the output of the LSTM can be determined using Eq. (11):

$$h_t = o_t * \tanh(C_t) \tag{11}$$

where tanh—hyperbolic tangent activation function.
x_t—number of samples at time t.
W—Weights (0–0.9).
b—Biases (0–0.9).

Further, the three gates in the memory cell of LSTM viz., input, output, and forget gate, expose, forget, and memorize the memory content in an adaptive manner. If the memory content is found to be crucial, the forget gate will be canceled, and the memory content will be carried to different time steps, that is when the long-term dependency will be captured. In this proposed work, consider $h_t = 10$, $C_t = 15$, $f = 5$. The output o denoted as number of classes. The output gate determines whether passing the output of memory is cervical cancer or not.

4 Experimental Results and Analysis

The experimental assessment of the existing and proposed methods is executed in the Java simulation environment. In this research work, cervical cancer dataset is collected from 'Hospital Universitario de Caracas' in Caracas, Venezuela. The performance of the proposed long short-term memory with artificial bee colony (LSTM-ABC) approach is compared with that of the existing support vector machine and recursive feature elimination (SVM-RFE) and support vector machine and principal component analysis (SVM-PCA) methods in terms of accuracy, sensitivity, and specificity.

4.1 Accuracy Comparison With and Without Feature Selection

The classification accuracy of the proposed approach before and after feature selection is shown in Fig. 3.

In this proposed research work, optimal feature selection is performed by using artificial bee colony (ABC) algorithm for improving the accuracy of the classifier. The experimental results show that the proposed system achieves higher accuracy while selecting optimal features through ABC methodology. It is clear that, without feature

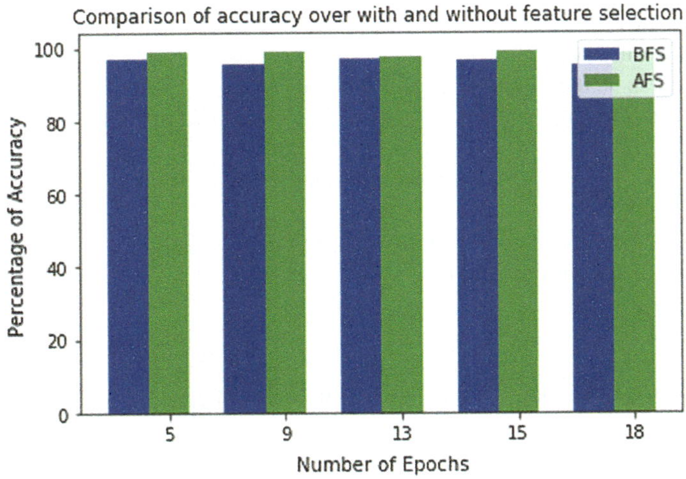

Fig. 3 Accuracy comparison with and without feature selection

Table 2 Accuracy comparison with and without feature selection

Number of features	Accuracy before feature selection (%)	Accuracy after feature selection (%)
5	96.91	98.90
9	95.59	98.81
13	96.74	97.43
15	96.53	98.80
18	95.02	98.49

selection approach, the accuracy of the classifier is not significant and potential. It clearly shows the importance of the feature selection approach. Table 2 gives the tabulated values of accuracy with and without feature selection.

4.2 Overall Classification Accuracy

Accuracy refers to a model's absolute correctness and is assessed as the total of the actual classification parameters (TP + TN) divided by the sum of the classification parameters (TP + TN + FP + FN) using the formula (12)

$$\text{Accuracy} = \frac{\text{TP} + \text{TN}}{\text{TP} + \text{TN} + \text{FP} + \text{FN}} \tag{12}$$

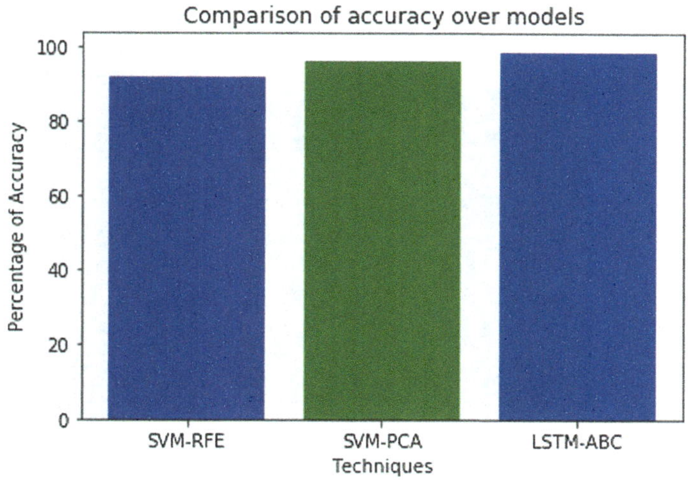

Fig. 4 Overall classification accuracy comparison

Table 3 Overall classification accuracy comparison

Techniques	Overall classification accuracy (%)
SVM-RFE	92.21
SVM-PCA	96.23
LSTM-ABC	98.82

Figure 4 shows the performance of SVM-RFE, SVM-PCA, and LSTM with ABC methods in terms of overall accuracy. In this research work, optimal features are selected by ABC algorithm. According to the selected features, the cervical cancer is classified using long short-term memory (LSTM) scheme to achieve higher accuracy. It can be inferred from the results that the proposed LSTM with ABC algorithm achieves 98.82% of accuracy, whereas other methods including SVM-RFE and SVM-PCA approaches attain 92.21% and 96.23%, respectively.

The experimental results show that the proposed LSTM with ABC algorithm gives greater accuracy than existing methods. This confirms the potential significance of the proposed LSTM-ABC approach. This significance is mainly due to the optimal features selected through ABC approach and also due to the potential classification capability of LSTM for the selected features. Table 3 shows the tabulated values accuracy comparison of proposed approach with the existing approaches.

4.3 Sensitivity

A test's potential to correctly recognize those with the disease (true positive rate) is referred to a sensitivity which is computed using the formula (13).

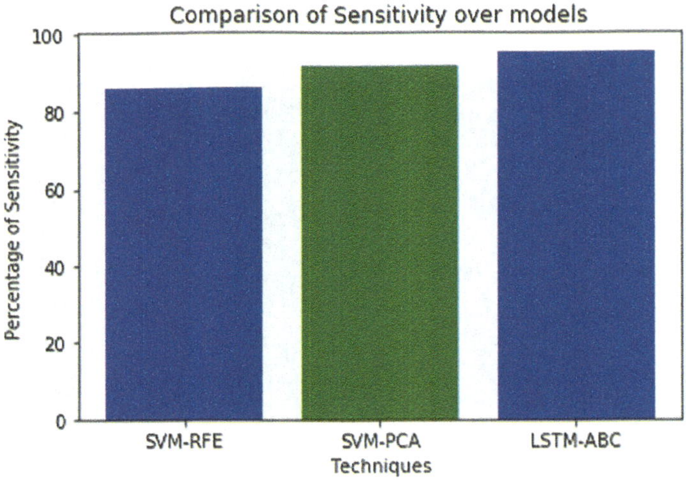

Fig. 5 Sensitivity comparison

Table 4 Sensitivity comparison

Techniques	Sensitivity (%)
SVM-RFE	85.56
SVM-PCA	91.45
LSTM-ABC	95.41

$$\text{Sensitivity} = \frac{\text{Number of true positives}}{\text{Number of true positives} + \text{Number of false negatives}} \tag{13}$$

The graph in Fig. 5 shows the sensitivity metric comparison for the methods. From the experimental outcomes, it can be concluded that the proposed LSTM with ABC algorithm attains 95.41% of sensitivity, whereas other methods including SVM-RFE and SVM-PCA achieve 85.56% and 91.45%, respectively. The proposed LSTM with ABC algorithm gives greater sensitivity than SVM-RFE and SVM-PCA approaches. Tabulated values of sensitivity comparison of different approaches are given in Table 4.

4.4 Specificity

A test's potential to correctly recognize those without the disease (true negative rate) is referred to as specificity.

$$\text{Specificity} = \frac{\text{Number of true negatives}}{\text{Number of true negatives} + \text{Number of false positives}} \tag{14}$$

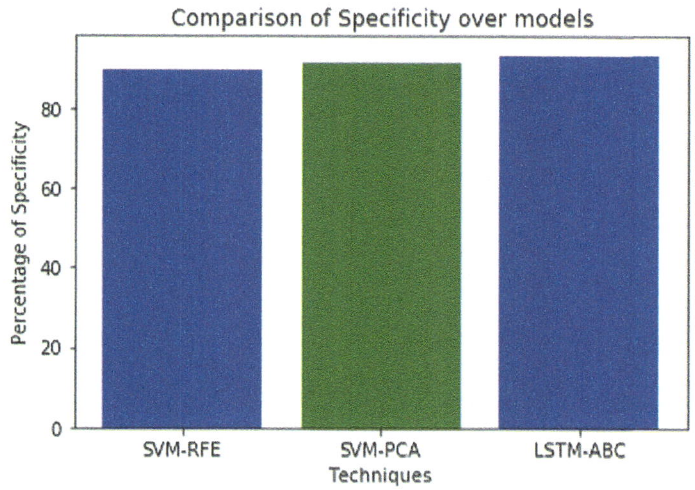

Fig. 6 Specificity comparison

Table 5 Specificity comparison

Techniques	Specificity (%)
SVM-RFE	89.57
SVM-PCA	91.67
LSTM-ABC	93.53

The specificity performances of the proposed LSTM with ABC and the existing SVM-RFE and SVM-PCA approaches are shown in Fig. 6. In this proposed research work, synthetic minority oversampling technique (SMOTE) is utilized for solving class imbalance problem in cervical cancer dataset. And also cervical cancer classification is performed by using LSTM network. It can effectively retain past information as well as accomplish the learning of long-term dependence information of text. Hence, it correctly estimates the likeliness of those patients without the disease. From Fig. 6, it can be stated that the proposed approach provides 93.53% of specificity, whereas other existing methods such as SVM-RFE and SVM-PCA attain 89.57% and 91.67%, respectively. The tabulation of the specificity values are given in Table 5.

4.5 ROC Curve

To better understand efficiency of the proposed LSTM-ABC approach, Fig. 7 presents the ROC curve of the proposed classifiers that better illustrate the precision of the proposed approach.

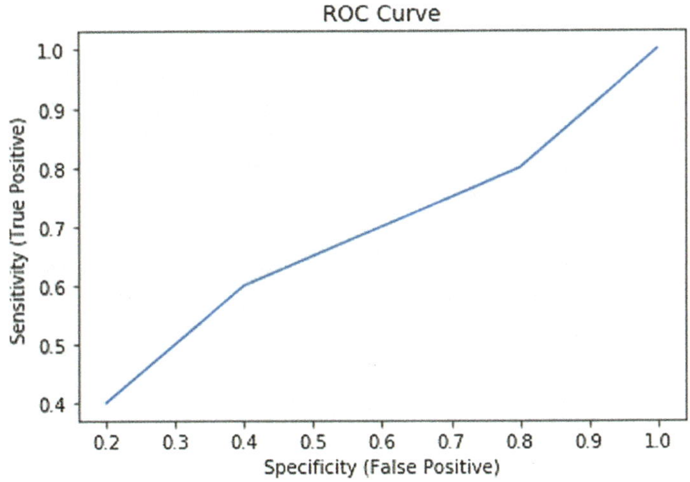

Fig. 7 ROC curve–LSTM-ABC approach

Table 6 LSTM-ABC ROC curve values

False positive rate	LSTM-ABC
0.2	0.45
0.4	0.58
0.6	0.69
0.8	0.85
1	0.96

The ROC curve gives a graphical graph that illustrates the performance of the proposed classifier. From the plot, we can easily evaluate the potential significance of the proposed approach. Table 6 gives the tabulated values of the ROC plot.

5 Conclusion

Cervical cancer is one of the deadliest diseases in women. Early and fast detection is vital for its cure via early treatment. However, its conventional screening involves a lot of time-taking clinical tests. In this era of extensive digitalization, one can look to machine learning to develop efficient techniques to quicken the diagnostic procedures. The proposed system has designed an algorithm named as long short-term memory with artificial bee colony (LSTM-ABC) algorithm for accurate cervical cancer detection. SMOTE is used for resolving problems arising due to imbalanced data set; and ABC algorithm is used for optimal feature selection to enhance the classifier's accuracy. The classification is performed using the long short-term memory

(LSTM) approach on the basis of the selected features. The experimental outcomes reveal that the proposed system generates better results when compared to existing models in terms of accuracy, sensitivity, and specificity. As a future enhancement, various classification methods such as ensemble classifiers and convolution neural network (CNN) are used to improve the classification accuracy.

References

1. J.C. Susan, H.K. Alex, Screening for cervical cancer US preventive services task force recommendation statement. JAMA **320**, 674–686 (2018)
2. C.S.C. Robbins, C. Pang, J. Leask, Australian newspaper coverage of human papillomavirus vaccination, October 2006–December 2009. J Health Commun. **17**, 149–159 (2011)
3. R.A. Kerkar, Y.V. Kulkarni, Screening for cervical cancer: an overview. J. Obstet. Gynecol. India **56**(2), 115–122 (2006)
4. G. Guvenc, A. Akyuz, C.H. Açikel, Health belief model scale for cervical cancer and Pap smear test: psychometric testing. J. Adv. Nurs. **67**(2), 428–437 (2011)
5. M.T. Galgano, P.E. Castle, K.A. Atkins, W.K. Brix, S.R. Nassau, M.H. Stoler, Using biomarkers as objective standards in the diagnosis of cervical biopsies. Am. J. Surg. Pathol. **34**(8), 1077 (2010)
6. H. Ramaraju, Y. Nagaveni, A. Khazi, Use of Schiller's test versus Pap smear to increase detection rate of cervical dysplasias. Int. J. Reprod. Contracept. Obstet. Gynecol. **5**, 1446–1450 (2017)
7. T.M. Alam, M.M.A. Khan, M.A. Iqbal, W. Abdul, M. Mushtaq, Cervical cancer prediction through different screening methods using data mining. IJACSA Int. J. Adv. Comput. Sci. Appl. **10**(2), (2019)
8. Z. Rustam, V.A.W. Hapsari, M.R. Solihin, Optimal cervical cancer classification using Gauss-Newton representation based algorithm. in AIP Conference Proceedings, Vol. 2168, No. 1, (AIP Publishing LLC, 2019 Nov), p. 020045
9. F.Q. Li, S.L. Wang, G.S. Liu, A Bayesian possibilistic C-means clustering approach for cervical cancer screening. Inf. Sci. **501**, 495–510 (2019)
10. W. Yang, X. Gou, T. Xu, X. Yi, M. Jiang, Cervical cancer risk prediction model and analysis of risk factors based on machine learning. in *Proceedings of the 2019 11th International Conference on Bioinformatics and Biomedical Technology* (2019), pp. 50–54
11. B. Benazir, A. Nagarajan, An expert system for predicting the cervical cancer using data mining techniques. Int. J. Pure Appl. Math. **118**(20), 1971–1987 (2018)
12. C.J. Tseng, C.J. Lu, C.C. Chang, G.D. Chen, Application of machine learning to predict the recurrence-proneness for cervical cancer. Neural Comput. Appl. **24**(6), 1311–1316 (2014)
13. W. Wu, H. Zhou, Data-driven diagnosis of cervical cancer with support vector machine-based approaches. IEEE Access **5**, 25189–25195 (2017)
14. K. Fernandes, J.S. Cardoso, J. Fernandes, Transfer learning with partial observability applied to cervical cancer screening, in *Proceedings of 2017 Iberian Conference on Pattern Recognition and Image Analysis*. Lecture Notes in Computer Science, vol. 10255, (2017), pp. 243–520
15. S. Barua, M.M. Islam, K. Murase, A novel synthetic minority oversampling technique for imbalanced data set learning. in *International Conference on Neural Information Processing* (Springer, Berlin, Heidelberg, 2011 Nov), pp. 735–744

16. M.S. Kıran, O. Fındık, A directed artificial bee colony algorithm. Appl. Soft Comput. **26**, 454–462 (2015)
17. G. Yu, A new multi-population-based artificial bee colony for numerical optimisation. Int. J. Comput. Sci. Math. **7**(6), 509–515 (2016)
18. S.S. Jadon, R. Tiwari, H. Sharma, J.C. Bansal, Hybrid artificial bee colony algorithm with differential evolution. Appl. Soft Comput. **58**, 11–24 (2017)

Overview of Deep Learning in Food Image Classification for Dietary Assessment System

Bhoomi Shah and Hetal Bhavsar

Abstract In modern days, people are very attentive about their health and diet. High-calorie intake food can be harmful and may lead to serious health condition. Food image identification plays a very important role in today's era. Food domain can be divided into two parts. First is to recognize food items, and second is to estimate the calorie. Accurate methods for food identification and calorie estimation can help people to fight against obesity which is the cause of overweight. So, recognition of food is first step for a successful healthy diet. Classification of food images is a very challenging task as the datasets of food images are not linear varying due to the large variations in food shape, volume, texture, color, and compositions. To recognize food items accurately, image processing can be used. Image processing techniques include image preprocessing, image segmentation, feature extraction, and image classification for recognition of food objects. Deep learning is an active research field nowadays in the field of object recognition, natural language processing, speech recognition, and many more. This paper defines the role of deep learning techniques based on a convolutional neural network for food object recognition. Many papers have been studied that have used deep learning as a tool for food object recognition and calorie estimation and achieved impressive results. This encourage us to do a detailed study in the field of food domain through deep learning. This paper analyzes, a deep learning framework, services, food items datasets, methods of segmentation and classification, and various food recognition techniques. Every method has its pros and cons as well. The main idea of this survey is to carry out a detailed study of current food item recognition techniques through deep learning. The result achieved by deep learning for recognition of food images will attract more researcher to put efforts in the field of food domain through deep learning in the future.

Keywords Computer vision · Convolutional neural network · Deep learning · Food image recognition · Object detection

---

B. Shah · H. Bhavsar (✉)
Department of Computer Science and Engineering, The Maharaja Sayajirao University of Baroda, Vadodara, Gujarat, India
e-mail: hetal.bhavsar-cse@msubaroda.ac.in

M. Paprzycki et al. (eds.), *Intelligent Systems, Technologies and Applications*, Advances in Intelligent Systems and Computing 1353,
https://doi.org/10.1007/978-981-16-0730-1_18

1 Introduction

Food is an essential element of everyone's life. Obesity is increasing at a higher rate day by day risking the lives of many people. It is necessary to control the amount of calorie intake to prevent obesity and many other diseases. Also, analyzing food images and calorie estimation can help people to follow healthy food diet. It can also be useful for normal people to maintain their day-to-day diet [1].

In today's world when people are more conscious about their diet, food recognition and calorie estimation has gained popularity. This can be achieved in two steps. First is to recognize food image successfully, and in second step, calorie can be estimated. This paper only focuses on various approaches developed using different technologies for food image recognition to get maximum possible accuracy over a wide range of food items and cuisines.

The contents of this paper are divided in following order. Section 2 discusses the challenges in food object recognition. Section 3 gives introduction to deep learning, review of deep learning networks and review of available frameworks, and techniques in deep learning for object identification. Section 4 discusses all the datasets for food image recognition. Section 5 compares different segmentation and classification techniques used for food image recognition. Finally, Sect. 6 concludes this article and gives future direction in this area.

2 Challenges in Food Image Recognition

Classification of food images is a very challenging task as the dataset of food images are not linear. There are so many varieties of food, so it is very difficult to make such a system that can identify each type of food. Some of the challenges faced while designing such a system has been shown in Fig. 1 and can be discussed as under [2–8].

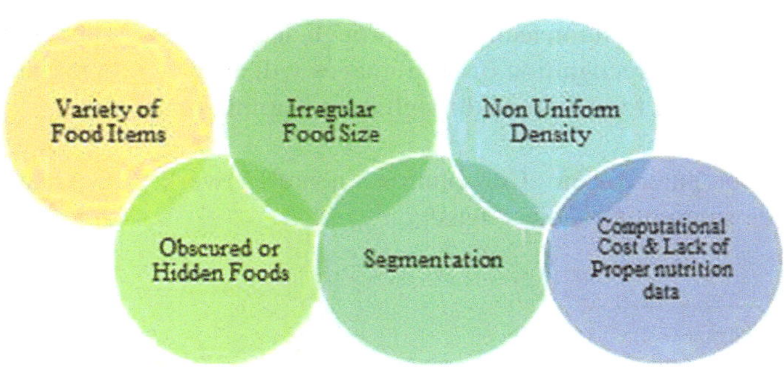

Fig. 1 Challenges for food image recognition

Fig. 2 Variety of food items

2.1 Variety of Food Items

There are so many varieties of the same food item that can be varied by different cooking method, camera quality, illumination and presentation. The varieties of rice have been shown in Fig. 2.

2.2 Obscured or Hidden Food Item

As shown in Fig. 3, here all dry fruits are not visible inside kheer and it is hard to identify the vegetables in rice.

Fig. 3 Obscured or hidden food item

Fig. 4 Irregular food size

2.3 Irregular Food Size

Food items do not have regular shapes in general. For example, the volume of a burger using a known geometric formula from a single image as shown in Fig. 4 cannot be estimated.

2.4 Segmentation

It is sometimes impossible to separate food items from the background due to the nonuniformity of the background. If the system uses a reference object for camera calibration, then segmentation is needed for that object too.

2.5 Liquid Food

Liquid food items do not have a fixed shape. To estimate liquid volume, one need to know the shape of the container holding the liquid and depth of liquid in that container.

2.6 *Non-uniform Density*

For image-based calorie estimation, it is assumed that calorie in a unit volume of food is constant. But, in real life, the density of food items can vary significantly. This variation in density can lead to some errors in calorie estimation.

2.7 *Computation Cost*

If automatic calorie estimation system to be deployed on a client-side device and produce output quickly, the computation cost should lie within a reasonable limit. For example, it is possible to improve the accuracy of volume estimation using 3D volume reconstruction from multiple images. In this case, computation cost can be very high.

2.8 *Lack of Proper Nutrition Data*

For calorie estimation from a portion of food, a large database covers every kind of food item. Since food items can be prepared in various ways, following different recipes are difficult to prepared such database. Even calorie content varies for the same food item with the different recipes. For example, calorie of a fried chicken differs from the calorie of a grilled chicken. So, this is a big challenge to accurately measure calories.

To deal with these challenges, deep learning is a recent advancement in the field of food domain and has achieved remarkable performance. The main two criteria needed for using deep learning are a large number of labeled data and a high processing machine, easily available nowadays. The main reason for the succession of deep learning is the ability to learn from extracted features directly from the images.

3 Overview of Deep Learning

Deep learning is an advanced technology for image processing, speech recognition, object detection, and food science and engineering [1, 9–12]. It works with artificial neural networks, which are designed to imitate how humans think and learn. Deep learning is a subset of machine learning. Unlike machine learning, in deep learning, basic details about the data need to be given that process through many layers and the computer trains to recognize the patterns by its own. Deep learning techniques are very successful, the reason is, availability of large number of datasets and the availability of high processing GPU [12]. We surveyed lots of recent articles in the food domain including food recognition and calories estimation. In this paper, we investigate the specific problems related to different types of food, the different food datasets used, and the preprocessing methods. Also, different networks available and its comparison, various deep learning frameworks, the segmentation and classification methods, the performance achieved in terms of Top-1 and Top-5 accuracy, and the comparison of all the CNN architecture with its advantages and disadvantages were studied. Firstly, it is very important to understand the types of learning techniques. Table 1 describes different categories of learning techniques with its definition, subcategories, methods and approach used, on which dataset it works, support which model and helpful in which situation has been defined [1, 13].

Machine learning techniques can be divided into mainly three categories.

1. Supervised learning
2. Unsupervised learning
3. Reinforcement learning

This research work is concentrated mainly on supervised learning.

Table 1 Supervised Vs. unsupervised learning Vs. reinforcement learning

Differ by	Supervised Learning	Unsupervised Learning	Reinforcement learning
Definition	The machine learns patterns by using labeled data	The machine is trained by Unlabeled data	No training dataset, so bound to learn from its experience
Data	Labeled	Unlabeled	No predefine data
Dataset	Use training dataset	Use just input dataset	Not predefine
Uses of	Prediction	Analysis	Consequence or behavior
Divided into	Classification and regression	Clustering, density estimation and dimensionality reduction	Reward-based
Approach	Maps the labeled input to the known output	Understand patterns and discovered output	Follows the trial and error methods
Models/networks	CNN, RNN, RvNN	DBN, DBM, GAN, VAE	Markov Decision process learning
Helpful in areas	Image Recognition, speech recognition, forecasting	Pre-process the data, pre-train supervised learning algorithms	Warehouses, Inventory management, aircraft control, robot motion control, chess game

3.1 Deep Learning Networks

Deep learning networks are the mathematical models that work like human brains. This mathematical model is created in form of neural network that consists of neurons. The neural network is divided into three major layers that are input layer (first layer of neural network), hidden layer (all the middle layer of neural network), and the output layer (last layer of the neural network.). The most popular deep learning networks for supervised learning can be described as follows.

3.1.1 CNN: Convolutional Neural Network

CNN is the main category to do the image recognition and image classification. CNN takes an input, process it, and classifies it under certain categories. The main advantages of CNN are parameter sharing, sparse interactions, and equivalent representations [9, 12–15].

A layer in the neural network is nothing but a collection of neurons that takes an input and provides an output. The input of each of these neurons is processed through the activation function assigned to the neurons.

3.1.2 RNN: Recurrent Neural Network

It is a very popular deep learning model that uses recursion techniques to build models. RNN saves the output of current layer which will be input to the next layer. It can memorize previous inputs due to its internal memory, because of this property it is especially used as language model. It is mostly used in natural language processing and speech processing [13, 18], text analysis and machine translation [16, 17].

3.1.3 RvNN: Recursive Neural Network

RvNN can handle the inputs of different modalities [13]. RvNN has been especially successful in NLP.

RvNN separate the image into different segment and form a syntactic tree [19].

For image classification, most of the research work has been used CNN and RNN [9–17]. Table 2 shows how convolutional neural network (CNN) differs from recursive neural network (RNN).

Many successful research works that have been done on food object recognition through CNN proved that CNN gives the best result in terms of accuracy and error rate for object recognition [9–17]. This research work is varied by the type of CNN, the dataset, segmentation, and classification techniques used.

Table 2 Comparison between CNN and RNN

CNN	RNN
Suitable for spatial data such as an image. Popularly used for image	Suitable for temporal data also called sequential data. Widely used for
Considered more powerful than RNN	Less powerful than CNN
Many research works have been done using CNN for image classification	It can be used for image classification, but theoretically, only a few researchers about RNN image classifier can be found
CNNs are special for video processing and image processing	RNN works primarily on time series information on the past influence of the customer
The interconnection consumes a finite set of input and generates a finite set of outputs according to inputs	RNN can allow arbitrary input length and output length

Table 3 briefly describes types of CNN with its main features, error rate, the number of parameters used, input size, number of convolution layers and stride based on recently published papers [20, 21].

Table 3 defines the most popular types of CNN. Many researchers have successfully experiment on GoogLeNet and ResNet model [27–29]. According to research, these two networks give best classification accuracy when used with food object recognition [27–29].

3.2 *Deep Learning Frameworks*

Deep learning frameworks is an interface or a tool which is easy to understand and combines deep learning algorithms, pre-built models and optimized components. Instead of writing hundreds of lines of code, deep learning framework builds a model quickly which provides good community support and parallelize the process to reduce computations [30–44].

Table 4 discusses and compares some of the famous deep learning frameworks like Torch, Theano, Tensorflow, Chainer, Keras, Apache Singa, MXNet, Caffe, Microsoft Cognitive Toolkit CNTK, Deep learning 4j, Neon.

The frameworks are studied from the date when the first and stable version was released, language supported, operating system supported, type of library support and the support for graphics processing unit (GPU), central processing unit (CPU) or tensor processing unit (TPU). The list of deep learning framework is very exhaustive. Every few months, new deep learning frameworks are introduced. Table 4 contains an analysis of the famous deep learning networks introduced till 2018. All the frameworks listed in Table4 support both the CNN and RNN models.

Table 3 Classification of CNN

Year	CNN	Main features	Top 5 error rate %	No. of parameters	Input size	No. of convolution layer	Stride
1998	LeNet [22]	First popular CNN architecture originally trained to classify handwritten digits	NA	0.060 M	28 × 28	2	1
2012	Alex Net [23]	Winner of ImageNet ILSVRC-12, it is deeper as compared to LeNet	15.3	60 M	227 × 227	5	1.4
2014	GoogLeNet/Inceprion n/w [24]	Winner of ILSVRC2014 competition. It has introduced block concept	6.67	4 M	224 × 224	21	1.2
2014	VGG Net [25]	Runner up of the ILSVRC-2014 competition. Homogeneous topology Uses small size kernels	7.3	138 M	224 × 224	16	1
2015	ResNet [26]	Relu is used identify mapping-based skip connections and implement heavy batch normalization	3.6	25.6 M, 1.7 M			

4 Dataset

A strong collection of images, Dataset, is the key element to achieve best classification accuracy [45]. A dataset is created considering the number of food classes and the type of food. Table 5 describes the number of available datasets with the food content including the number of food classes and number of food images. Since

Table 4 Deep learning frameworks

Name	Initial	Stable release	Language supported	Operating system supported	Type of library support	Support for GPU/CPU/TPU
Torch [30]	2002	2017	Lua, LuaJIT, C, CUDA, C++	Linux, Android, MacOS X iOS	Deep learning and machine learning	CPU and GPU
Theano [31, 32]	2007	2019	Python,CUDA	Linux, MacOS, Windows	Machine learning	CPU and GPU
Tensorflow [33, 34]	2015	2020	Python, C++, CUDA	Linux, MacOS, Windows, Android	Machine learning	CPU and GPU and optimize for TPU
Chainer [35]	2015	2019	Python	–	Deep learning	Best for GPU
Keras [36]	2015	2019	Python	Ios, Android	Deep learning and machine learning	GPU and TPU
Apache Singa [37]	2015	2020	C++, Python, Java	Linux, MacOS, Windows, Android	Machine Learning	CPU and GPU
MXNet [38, 39]	–	2020	C, Python, R, Java, Julia, Javascript Scala Go Perl	Linux, MacOS, Windows	Deep learning and machine learning	GPU
Caffee [40, 41]	–	2017	C++	Linux, MacOS, Windows	Deep learning	CPU and GPU
Microsoft cognitive ToolKit CNTK [42]	2016	2019	C++	Windows, Linux	Deep learning and machine learning	GPU
Deeplearning4j	–	2019	Java, CUDA, C, C++	Linux, MacOS, Windows, Android	Deep learning and machine learning	CPU and GPU
Neon [44]	–	2018	Python	–	Deep learning	–

Table 5 Summary of Food image databases

Name of dataset	Year	#images	#Food classes	Food content
Food85 [2]	2010	8500	85	Japanese food
Chen [3]	2012	5000	50	Chinese food
UNIMIB2016 [4]	2016	1027	73	Pictures captured by a digital camera in dining hall
Food524DB [5]	2017	247,636	524	Merging food classes from existing database vireo, food-101, food50 and modified version of UECFOOD256
FFOcat [6]	2018	58,962	156	Selected food images Downloaded from Web
Foodx-251[7]	2019	158,000	251	Selected food items like cakes, pasta, soups, etc
FooDD [8]	2015	3000	30	Single and mixed food images including fruits
PFID [46]	2009	1098	61	Fast food items from USA
TADA [47]	2009	256 + 50 replica	–	Common food in USA
Food50 [48]	2009	5000	50	Japanese Food
UEC-Food100 [49]	2012	9060	100	Popular Japanese Food
Food101 [50]	2014	101,000	101	Popular food in USA
UECFood256 [51]	2014	31,397	256	Famous foods in Japan & other Countries
UNICT-FD889 [52]	2014	3583	899	Different Nationality dishes like Italian, Thai., Japanese, etc
Diabetes2 [53]	2014	4868	11	Selected food
New Dataset [54]	2014	5000	11	Central European Food
Menu-match [55]	2015	646	41	Food from three restaurants (Asian, Italian, Soup of 10 types)
VIREOFood-172 [56]	2016	110,241	172	Popular Chinese dishes from "go cooking" and web
ChineseFoodNet [57]	2017	185,000	280	Food images either taken from real dishes or recipe pictures or selfies

deep learning is data hungry, a large collection of food images is required to train a food-classification model. Food image datasets vary in many aspects such as, a single food image, mixed food image, non-mixed food, several food groups, liquid food image, type of cuisine and total images per food class. Table 5 summarizes of different food image databases with their respective features.

Other than these, Anthimopoulos et al. [54] used one visual database created with 5000 food images and organized into 11 classes reflecting the nutritional habits in central Europe in 2014. Also, in 2017, Paritosh et al. [58] proposed work on the Indian food dataset containing 100 Indian food images of 50 different classes. It is the first Indian food dataset that is available to download.

5 Segmentation and Classification

5.1 Segmentation Techniques

In computer vision, image segmentation is the process of partitioning a digital image into multiple segments (sets of pixels, also known as super pixels). The goal of segmentation is to simplify and/or change the representation of an image into something more meaningful and easier to analyze. Image segmentation is the process of assigning a label to every pixel in an image such that pixels with the same label share certain characteristics. Segmentation means to divide parts of a scene [59, 60]. Segmentation is the last step before food classification where the identification of various food images takes place.

In segmentation, the image is partitioned in equal segments. After segmentation, the boundary detection of irregular food portions becomes easy, and it gives better detection of the food portion. Matsuda et al. [61] discussed to separate the food images from the background regardless of the lighting conditions or if the food is mixed or not. According to Yang et al. [62], it is very difficult to separate those food images which does not contain any specific attribute. Ramadevi et al. [63] showed a synergy between segmentation and object recognition is done using the EM algorithm, OSTU, and genetic algorithm. They have also discussed the difference between region-based and edge-based segmentation. There are different types of segmentation techniques, and it can be divided mainly into two categories: layer-based segmentation methods and block-based segmentation methods. Block-based segmentation can be partitioned into region-based segmentation and edge-based segmentation. Also, other types are segmentation based on weakly supervised learning in CNN, threshold segmentation methods, segmentation based on clustering, etc. Yan Hao [64] have evaluated and recap different image segmentation algorithms and compared them with pros and cons. A similar approach has been used by Shakuntala and Surendra [65], after studying different types of image segmentation, they have concluded that "all the works done in the field of image segmentation are needed to be monitored manually there is no such method which can detect the objects with precision and without any database, which obviously takes time to get build." Another similar approach has also been found by Vairaprakash and Subbu [66]. W. Liming et al. [67] used an object detection method which is a combination of top-down recognition with bottom-up image segmentation has been developed. On the other hand, Wataru and Keiji [68] have proposed neither a traditional proposal approach nor the fully convolutional approach, but a middle approach. The approach combined fully convolutional networks and back-propagation-based approach. However, Bryan et al. [69] computed multiple segmentations of images and then learned the object classes to choose the correct segmentation. A similar approach can also be found by Zhu et al. [70], They have combined two concepts: A set of segmented objects can be partitioned into classes based on global and local features; and perceptually, similar object classes can be used to assess the accuracy of image segmentation. They have shown improvement in the accuracy of segmenting food images using a segmentation compared to normalized cut without classifier feedback when there is no prior information about the scene. This idea has improved the overall accuracy of classification. Several methods for segmentation have been studied with the approach used, and performance of each has been summarized in Table 6.

Apart from this, Sandhya et al. [76] used different approaches like traditional methods and modern deep learning techniques for food image recognition which have been discussed, and the proposed work has been implemented on Indian cuisine, whereas Adnan et al. [77] discussed different calorie estimation techniques with pros and cons. Zhao et al. [78] proposed an algorithm on deep learning which can correctly identify food and analyze nutrition. Doyen et al. [79] developed a deep learning-based system known as "FoodAI" for recognition of food smartly, and the system has been tasted on commonly consumed food in Singapore. Ashutosh et al. [29] experiment classification of food or non-food item based on deep convolution neural network. They have used GoogLeNet model and achieved 99.2% classification and 83.6% in recognition of food. Simon and Barbara [80] proposed a novel approach, based on deep convolution neural network for food and drink image recognition, known as "NutriNet," and Achieved 86.72% classification accuracy. Md et al. [81] also used convolution neural network for classifying food images and achieved 92.86% accuracy.

Since there is no general solution to the image segmentation problem, these techniques often have to be combined with domain knowledge in order to effectively solve an image segmentation problem for a problem domain [11, 82].

5.2 Image Classification Techniques

Food image recognition can be achieved with machine learning-based approach or deep learning-based approach [10]. Recently, deep learning has become very famous as it gives impressive result due to the feature that it can learn features automatically just like a human does [83]. It is probably the best approach in the cases when we do not have enough pre-defined features. As this research work focuses on a deep learning approach, this paper discusses all of the work done till now for food image recognition through deep learning. Detailed studies have been shown in Table 7.

This section studied all the recent articles on food image classification. Many research works have been done on food image recognition that has used convolution neural network with different combined approaches and different datasets as shown

Table 6 Segmentation methods

Reference	Year	Approach	Performance
[15]	2015	Have proposed a DCNN-based food image segmentation which requires no pixel-wise annotation	It has been found that the proposed method gives better food region detection as compared to other RCNN
[61]	2012	Have used mixed techniques such as whole image, circle detector and JSEG segmentation	Top 1 accuracy is 21%, and top 5 is 45%
[71]	2017	Introduced a new efficient image segmentation method which consider various feature vectors such as color, texture, etc.	It has been found that the suggested method is an innovative segmentation method that needs No other information about images 70% of tested images gives better result as compared to ACT algorithm
[72]	2009	Have suggested a region-based approach which is a combination of object detection and image segmentation	Have proposed a modular energy function that is easy to evaluate and enhances computer vision technologies. The author has also proposed a number of future works
[73]	2016	Have collaborated image segmentation with machine learning techniques	The problem can be best solved with the ensemble model of random forest, ANOVA RBF SVM, and generalized linear model, giving the accuracy of 93.11111%
[74]	2013	Have proposeda mobile food recognition system which uses methods such as Grub Cut, SURF based, bag-of-features, SVM and fast χ^2 kernel	The suggested method gives 81.55% of top five accuracy
[75]	2005	Have suggested a segmentation algorithm that can be used for color food images having background	The assessed segmentation performance computed from the area under the receiver operation characteristic (ROC) curve was 0.9982

in Table 7. It has been observed from the studies that convolution neural network is a hot topic in food object recognition as it reduced the complex problem of food image recognition. Many researchers have tried to combined method based on convolution neural network with architectures such as Inception and GoogLeNet. It has also been proved that classification accuracy has been increased as the model gets deeper.

Table 7 Food classification

Reference	Year	Approach	Dataset	Top 1	Top 5
[27]	2016	Used deep learning approach for the classification and fine-tuned the image recognition architecture Inception	UEC Food-100	81.5	97.3
			UEC-Food-256	76.2	92.6
			Food-101	88.3	96.9
[28]	2016	Proposed a food recognition method based on CNN, and the architecture is based on Inception-ResNet and InceptionV3 model	Food-101	72.55%	91.31%
[54]	2014	Two ANN models were used. One is without hidden layer and another is with one hidden layer. Classification was done using three supervised methods SVM, ANN, and random forests (RF)	Diabetes	75.0%	–
[58]	2017	Constructed a multilayered deep convolutional neural network (CNN) architecture, also proposed new Indian Food dataset	Indian Food database	73.50	94.40
			ETH Food-101	72.12	91.61
[59]	2016	Proposed deep architecture, namely Arch-D, that defines relationship between food and ingredients label through multitask learning	UEC-Food-100	82.1	97.3
			VIREO	82.1	95.9
[84]	2014	Used the approach called random forest discriminant components (RFDC) and compared it with various other methods, also have introduced Food-101 dataset having 101,000 images	Food-101	56.4%	–
[85]	2015	Used deep convolutional neural network for food photograph recognition task in the ImageNet and have implemented this combination on Twitter photograph data. Achieved high level of accuracy proving DCNN gives best result on large-scale image data	UECFood-100	78.7%	–
			UEC-Food-256	67.57	89.0

(continued)

Table 7 (continued)

Reference	Year	Approach	Dataset	Top 1	Top 5
[86]	2016	Proposed new algorithm based on CNN and achieved impressive result on two datasets, namely Food-101 and UEC-FOOD-256	Food-101	77.4%	93.7%
			UEC-Food-256	54.7	81.5
[87]	2015	food size and labels ad apply this method to a dataset of images from 23 different restaurants	Food-101	79.0%	–
			Food201 segmented	76.0	–
			Menu-Match	81.4	–
[88]	2015	Used CNN-based classifier to estimate, used a deep convolution neural network with sic layers, used to classify food image patches. Experiments have achieved attractive result	Own database with 573 food items	84.90	–
[89]	2016	Used the graph cut method and uses deep convolution neural network to classify food images and have achieved remarkable accuracy for a single food image	Own database with 10,000 high-resolution images	99.0	–
[90]	2014	Used deep convolutional neural network with Fisher vector with HOG and color patches	UEC-Food-100	72.3	92.0
[91]	2016	Proposed a new dataset on which food recognition algorithms can be tasted	UNIMINB2016	78.3	–
[92]	2019	Developed a model with five-layer CNN, and first ever combining bag-of-features model with support vector machine to achieve high level of accuracy	ImageNet	74	–
[93]	2017	Gives overview on different approaches of food detection and model used for comparison is based on GoogLeNet architecture, principle component analysis and support vector machine	FCD	98.81%	–
			Ragusa DS	95.78%	

(continued)

Table 7 (continued)

Reference	Year	Approach	Dataset	Top 1	Top 5
[94]	2017	combination of multiple classifiers based on different convolutional models that complement each other hence improved performance	Food-101 Food-11	–	–
[95]	2018	Efficiently combined visual content, context and external knowledge and tested on different datasets	Different dataset	–	–
[96]	2018	Proposed a segmentation algorithm based on random forest and has used boundary detection and filling methods. Also, compared the proposed algorithm with three existing methods	Food-101	90.5	–

6 Conclusion and Future Work

This paper mainly focuses on how deep learning techniques can be used for food object recognition. It has been surveyed a wide range of articles designed for automated food image recognition. The paper covers popular networks for deep learning, mostly used frameworks, famous services available for object detection and Standard datasets. Different segmentation and classification techniques have also been discussed. The challenges in food object recognition have also been studied. According to the authors' knowledge, this is the first article that has fully focused on food object recognition through deep learning and has studied every phase of food object recognition. Also, it has been cleared from the challenges and studies that there is no such model exists that can recognize each and every type of food. Every model has its own limitation due to the nonlinear nature of food. It has been concluded that food image recognition is a hot topic in computer vision, and the use of convolution neural network has improved the result accuracy of food image recognition. It has been observed that there are variety of food domains that has not been touched yet as not much work done on recognition of liquid food items, and also, there is very less work done till now for European food and Indian cuisine. In fact, there are very limited number of dataset which exist for Indian food and European food. So, in the future, more work in this area needs to be focused. The result achieved by deep learning for recognition of food images will attract more researchers to put efforts in the field of food domain through deep learning in the future.

References

1. G. William, Y. Wei, A survey of deep learning: platforms, applications and emerging research trends. Special section on human-centered smart systems and technologies. IEEE Access **6** (2018)
2. H. Hajime, J. Taichi, Y. Keiji, Image recognition of 85 food categories by feature fusion, in *IEEE International Symposium on Multimedia* (2010), pp. 296–330
3. M. Chen, H. Yung, J. Chia, H. Shih, M. Shane, C. Eugene, H. Che, O. Ming, Automatic Chinese food identification and quantity estimation, in *SIGGRAPH 2012*, vol. 29 (Springer, Asia, 2012), pp. 1–4
4. C. Gianluigi, N. Paolo, S. Raimondo, Food recognition: a new dataset, experiments, and results. IEEE J. Biomed. Health Inform. **21**(3), 588–598 (2017)
5. C. Gianluigi, N. Paolo, S. Raimondo, Learning CNN-based features for retrieval of food images, in *ICIAP Workshops* (2017), pp. 426–434
6. D. Ivan, D. Mauro, Ontology-driven food category classification in images, in ICIAP (2019), pp. 607–617
7. K. Parneet, S. Karan, W. Weijun, B. Serge, D. Ajay, foodx-251: a dataset for fine-grained food classification, in *Computer Vision and Pattern Recognition* (CVPR19) (2019)
8. P. Parisa, Y. Abdulsalam, Foodd: food detection dataset for calorie measurement using food images, new trends in image analysis and processing, in *ICIAP Workshops* (2015), pp. 441–448
9. P. Ajeet, P. Manjusha, R. Siddharth, Application of deep learning for object detection, in *International Conference on Computational Intelligence and Data Science* (Elsevier, Amsterdam, 2018)
10. V. Bruno, S. Resende, C. Juan, A survey on automated food monitoring and dietary management systems. J. Health Med. Inform. (2017)
11. D. Li, A tutorial survey of architectures, algorithms,and applications for deep learning. APSIPA Trans. Signal Inform. Process. (2014)
12. Z. Lei, Z. Chu, L. Fei, Q. Zhengjun, H. Yong, Application of deep learning in food: a review. Comprehensive Rev. Food Sci. Food Safety (2019)
13. P. Samira, T. Haiman, A survey on deep learning: algorithms, techniques and applications. ACM Comput. Surv. **51**(5) (2018)
14. G. Ian, B. Yoshua, C. Aaron, *Deep Learning* (MIT Press, Cambridge, 2016)
15. S. Wataru, Y. Keiji, CNN-based food image segmentation without pixel-wise annotation, in *ICIAP* (Springer, Berlin, 2015), pp. 449–457
16. L. Pengfei, Q. Xipeng, H. Xuanjing, Recurrent neural network for text classification with multi-task learning, in *Proceedings of the Twenty-Fifth International Joint Conference on Artificial Intelligence* (2016)
17. S. Ilya, M. James, H. Geoffrey, Generating text with recurrent neural networks, in *Proceedings of the 28th Inter-national Conference on International Conference on Machine Learning (ICML'11)* (2011), pp. 1017–1024
18. C. Kyunghyun, M. Bart, G. Caglar, B. Dzmitry, B. Fethi, S. Holger, B. Yoshua, Learning phrase representations using RNN encoder-decoder for statistical machine translation, in *The Conference on Empirical Methods in Natural Language Processing* (2014), pp. 1724–1734
19. S. Richard, C. Cliff, C. Lin, Y. Andrew, Parsing natural scenes and natural language with recursive neural networks, in *International Conference on Machine Learning* (Omnipress, 2011), pp. 129–136
20. K. Asifullah, S. Anabia, Z. Umme, Q. Aqsa, A survey of the recent architectures of deep convolutional neural networks, in *Computer Vision and Pattern Recognition* (2019)
21. A. Zahangir, M. Tarek, Y. Christopher, W. Stefan, S. Paheding, N. Shamima, C. Brian, A. Abdul, K. Vijayan, The history began from alexnet: a comprehensive survey on deep learning approaches, in *Computer Vision and Pattern Recognition* (2018)
22. K. Fukushima, Neocognitron: A hierarchical neural network capable of visual pattern recognition. Neural Netw. **1**(2), 119–130 (1988)

23. A. Krizhevsky, I. Sutskever, G.E. Hinton, ImageNet classification with deep convolutional neural networks, in *NIPS* (2012), pp. 1106–1114
24. S. Cristian, L. Wei, J. Yangqing, S. Pierre, R. Scott, A. Dragomir, E. Dumitru, V. Vincent, R. Andrew, Going deeper with convolutions, in *Proceedings of the IEEE Conference on Computer Vision and Pattern Recognition* (2015), pp. 1–9
25. H. Tang, W. Wang, K. Gimpel, K. Livescu, Discriminative segmental cascades for feature-rich phone recognition, in *ASRU* (2015)
26. H. Kaiming, Z. Xiangyu, R. Shaoqing, S. Jian, Deep residual learning for image recognition, in *Proceedings of the IEEE Conference on Computer Vision and Pattern Recognition* (2016)
27. H. Hassannejad, G. Matrella, P. Ciampolini, I. Munari, M. Mordonini, Food image recognition using very deep convolu-tional networks. Intl Work Multime Ass Die Mana **1**, 41–49 (2016)
28. Y. Qian, M. Dongyuan, W. Jingfan, Deep learning based food recognition (2016)
29. S. Ashutosh, Y. Lin, E. Touradj, Food/non-food image classification and food categorization using pre-trained Google Net model, in *Conference, the 2nd International Workshop* (2016)
30. Torch, https://torch.ch/
31. Theano, https://deeplearning.net/software/theano/
32. The Theano Development Team: Theano: A Python framework for fast computation of mathematical expressions (2016)
33. Tensorflow, https://www.tensorflow.org/
34. A. Martin, A. Ashish, B. Paul, B. Eugene, C. Zhifeng, C. Craig, Tensorflow: large scale machine learning on heterogeneous distributed systems (2016)
35. Chaier, https://chainer.org/
36. Keras, https://keras.io/
37. Apache Singa, https://singa.incubator.apache.org/
38. C. Tianqi, L. Mu, L. Yutian, L. Min, W. Naiyan, W. Minjie, Mxnet: a flexible and efficient machine learning library for heterogeneous distributed systems (2015)
39. Mxnet, https://mxnet.io/
40. Caffe, https://caffe.berkeleyvision.org
41. J. Yangqing, S. Evan, D. Jeff, K. Sergey, L. Jonathan, G. Ross, G. Sergio, D. Trevor, Caffe: convolutional architecture for fast feature embedding, in *Proceedings of the 22nd ACM International Conference on Multimedia* (2014), pp. 675–678
42. Microsoft Cognitive Toolkit CNTK, https://www.microsoft.com/en-us/research/product/cognitive-toolkit/
43. Deeplearning4j, https://deeplearning4j.org
44. Neon, https://neon.nerva-nasys.com/docs/latest
45. A. Krizhevsky, I. Sutskever, H. Ge. ImageNet classification with deep convolutional neural networks (2012), pp. 1097–1105
46. C. Mei, D. Kapil, W. Wen, Y. Lei, S. Rahul, Y. Jie, PFID: Pittsburgh fast-food image dataset, in *ICIP* (2009), pp. 289–292
47. M. Anand, B. Marc, Z. Fengqing, J. Carol, A. Deborah, S. David, J. Edward, Personal dietary assessment using mobile devices. Comput. Imaging, **7246** (2009)
48. J. Taichi, Y. Keiji, A food image recognition system with Multiple Kernel Learning, in *ICIP* (2009), pp. 285–288
49. K. Yoshiyuki, Y. Keiji, Real-time mobile food recognition system, in *CVPR Workshops* (2013), pp. 1–7
50. B. Lukas, G. Matthieu, J. Luc, G. Van, Food-101 - Mining discriminative components with random forests. ECCV **6**, 446–461 (2014)
51. K. Yoshiyuki, Y. Keiji, Foodcam-256: a large-scale real-time mobile food recognition system employing high-dimensional features and compression of classifier weights. ACM Multimedia 761–762 (2014)
52. Giovanni, M., Farinella, D., Filippo, S.: A Benchmark Dataset to Study the Representation of Food Images. ECCV Workshops, pp.584–599, (2014).
53. D. Joachim, A. Marios, S. Sergey, G. Stavroula, Two-view 3D reconstruction for food volume estimation. IEEE Trans. Multimedia **19**(5), 1090–1099 (2017)

54. M. Anthimopoulos, L. Gianola, L. Scarnato, P. Diem, S. Mougiakakou, A food recognition system for diabetic patients based on an optimized bag-of-features model. J. Biomed. Health Inform. 1261–1271 (2014)
55. O. Beijbom, N. Joshi, D. Morris, S. Saponas, S. Khullar, Menu-match: restaurant-specific food logging from images in 2015. IEEE **25**, 844–851 (2015)
56. C. Jingjing, W. Chong, Deep-based ingredient recognition for cooking recipe retrieval. ACM Multimedia 32–41 (2016)
57. C. Xin, Z. Yu, Z. Hua, D. Liang, W. Dongyan, ChineseFoodNet: a large-scale image dataset for chinese food recognition, in *Computer Vision and Pattern Recognition* (2017)
58. P. Paritosh, D. Akella, T. Bappadi, N. Puhan, FoodNet: recognizing foods using ensemble of deep networks. IEEE Signal Process. Lett. **24**(12) (2017)
59. J. Chen, C. Ngo, *Deep-Based Ingredient Recognition for Cooking Recipe Retrieval* (ACM, 2016), pp. 32–41
60. G. Farinella, M. Moltisanti, S. Battiato, Classifying Food Images Represented as Bag of Textons (IEEE, 2014), pp. 5212–5216
61. Y. Matsuda, H. Hoashi, K. Yanai, Recognition of multiple-food images by detecting candidate regions, in *IEEE International Conference on Multimedia and Expo* (2012), pp. 25–30
62. S. Yang, M. Chen, D. Pomerleau, R. Sukthankar, Food recognition using statistics of pairwise local features. IEEE **1**, 2249–2256 (2010)
63. Y. Ramadevi, B. Kalyani, T. Sridevi, Synergy between object recognition and image segmentation. Int. J. Comput. Sci. Eng. **2**(8), 2767–2772 (2010)
64. Y. Song, H. Yan, Image segmentation algorithms overview, in *Computer Vision and Pattern Recognition* (2017)
65. S. Shakuntala, K. Surendra, A review paper on image segmentation and object recognition procedures. Int. J. Sci. Eng. Technol. **4**(1) (2016)
66. G. Vairaprakash, K. Subbu, Review on image segmentation techniques. Int. J. Sci. Mod. Eng. **1**(8), 2319–6386 (2014)
67. W. Liming, S. Jianbo, S. Gang, S. Fan, Object detection combining recognition and segmentation, in *Asian Conference on Computer Vision* (Springer, Berlin, 2007), pp. 189–199
68. S. Wataru, Y. Keiji, Foodness proposal for multiple food detection by training of single food images, in *Proceedings of the 2nd International Workshop on Multimedia Assisted Dietary Management* (2016), pp. 13–21
69. C. Bryan, A. Russell, S. Josef, T. William, A. Freeman, Using multiple segmentations to discover objects and their extent in image collections, in *Proceedings of the 2006 IEEE Computer Society Conference on Computer Vision and Pattern Recognition* (2006)
70. Z. Fengqing, B. Marc, K. Nitin, B. Carol, J. Edward, Multiple hypotheses image segmentation and classification with application to dietary assessment. IEEE J. Biomed. Health Inform. **19**(1)
71. A. Roohollah, Q. Kazem, R. Mohammad, An efficient evolutionary based method for image segmentation, in *Computer Vision and Pattern Recognition* (2017)
72. G. Stephen, G. Tianshi, K. Daphne, Region-based segmentation and object detection, in *Proceedings of the 22nd International Conference on Neural Information Processing Systems* (2009), pp. 655–663
73. S. Ashima, S. Sahil, Image segmentation and object recognition using machine learning, in *International Conference on Smart Trends for Information Technology and Computer Communications* (Springer, Berlin, 2016), pp. 204–210
74. K. Yoshiyuki, Y. Keiji, Real-time mobile food recognition system, in *Proceedings of the 2013 IEEE Conference on Com-puter Vision and Pattern Recognition* (2013), pp. 1–7
75. M. Domingo, P. Franco, Segmentation of colour food images using a robust algorithm. J. Food Eng. 353–360 (2005)
76. A. Sandhya, C. Gauri, C. Devangi, D. Eesha, J. Shevi, Survey of different approaches used for food recognition, in *Information and Communication Technology for Competitive Strategies* (Springer, Berlin, 2018), pp. 551–560
77. S. Adnan, Z. Nauman, H. Mir, Food item recognition and intake measurement techniques, in *11th International Conference on Machine Learning and Computing* (2019)

78. Y. Zhao, J. Ming, Y. Chen, C. Cao, C. Forde, N. Wah, S. Tat, Food photo recognition for dietary tracking: system and experiment, in *International Conference on Multimedia Modeling* (Springer, Berlin, 2018), pp. 129–141
79. S. Doyen, H. Wang, K. Shu, X. Wu, L. Hung, A. Palakorn, P. Ee, C. Steven, FoodAI: food image recognition via deep learning for smart food logging, in *Conference on Knowledge Discovery and Data Mining* (2019)
80. M. Simon, K. Barbara, NutriNet: a deep learning food and drink image recognition system for dietary assessment. Nutrient (2017)
81. T. Md, B. Islam, K. Nafiz, S. Siddique, T. Rahman, Image recognition with deep learning, in *International Conference on Intelligent Informatics and Biomedical Sciences* (IEEE, 2018)
82. M. Nida, M. Zaitouna, Survey on image segmentation techniques, in *International Conference on Communication, Management and Information Technology* (Elsevier, Amsterdam, 2015), pp. 797–806
83. C. Vishwanath, C. Priyadarshini, Food image segmentation using edge adaptive based deep-CNNs. Int. J. Intell. Unmanned Syst. 2049–6427 (2019)
84. L. Bossard, L., M. Guillaumin, Lv. Gool, Food-101-mining discriminative components with random forests in computer vision, in *ECCV* (2014), pp. 446–461
85. K. Yanai, Y. Kawano, *Food Image Recognition Using Deep Convolutional Network with Pre-training and Fine-Tuning* (IEEE, 2015), pp. 1–6
86. Yan, L., Guanling, C., Vinod, V., Yunsheng, M., Chang, L.,Yu, C.: Deep food: deep learning-based food image recognition for computer-aided dietary assessment., pp.37–48, (2016).
87. A. Meyers, Im2calories: towards an automated mobile vision food diary, in *IEEE International Conference Computer Vision* (2015), pp. 1233–1241
88. S. Christodoulidis, M. Anthimopoulos, S. Mougiakakou, Food recognition for dietary assessment using deep convolutional neural networks, in *ICIAP* (2015), pp. 458–465
89. P. Pouladzadeh, P. Kuhad, S. Peddi, A. Yassine, S. Shirmohammadi, *Food Calorie Measurement Using Deep Learning Neural Network* (IEEE, 2016), pp. 1–6
90. Y. Kawano, K. Yanai, *Food Image Recognition with Deep Convolutional Features* (ACM, 2014), pp. 589–593
91. G. Ciocca, P. Napoletano, R. Schettini, Food recognition: a new dataset, experiments and results., IEEE J. Biomed. Health Inform. **99** (2016)
92. L. Yuzhen, Food image recognition by using convolutional neural networks, in *Vision and Pattern Recognition* (2019)
93. A. Eduardo, A., B. Marc, R. Petia, exploring food detection using CNNs, in *Computer Aided Systems Theory (EUROCAST 2017)* (Springer, Berlin, 2017), pp. 339–347
94. A. Eduardo, B. Marc, R. Petia, Food recognition using fusion of classifiers based on CNNs, in *Image Analysis and Processing (ICIAP 2017)* (Springer, Berlin, 2017), pp.213–224
95. H. Luis, M. Weiqing, J. Shuqiang, Food recognition and recipe analysis: integrating visual content, context and external knowledge, in *Computer Vision and Pattern Recognition* (2018)
96. I. Sanasam, S. Ayan, K. Pritee, Classification of food images through interactive image segmentation, in *Asian Conference on Intelligent Information and Database Systems (ACIIDS)* (Springer, Berlin, 2018), pp. 519–528

Comparison of Face Embedding Approach Versus CNN-Based Image Classification Approach for Human Race Detection from Face

Rupesh Wadibhasme, Amit Nandi, Bhavesh Wadibhasme, and Sandip Sawarkar

Abstract Human race detection using face with deep learning technique is an active research area. It helps expand growing areas like human–computer interface, understanding user demographics. It provides great insight into a better understanding of demographics and diversity among the population. Understanding ethnic diversification among user base can help many commercial applications to improve and optimize their products and services better suited for the community needs. Development around race detection is already an active area of research and improving its performance with speed is one of them. In this study, we compared FaceNet architecture-based features extraction technique to detect the race and compared with plain CNN-based classification techniques. Comparative results support the claim that race detection problem is better handled by embedding-based approach than a plain image classification approach. Embedding-based techniques also provide a competitive edge over other methods used in this comparative study.

1 Introduction

An ethnic group, or the ethnicity is the class of people who identify with each other from similarities, e.g., general origin, language, history, country, culture, or nation [7]. Ethnicity is normally the inherited condition of the world in which one lives. Membership of the ethnic group tends to be determined by the shared ethnic heritage,

---

github: https://github.com/rupesh-wadibhasme/race-detection.

R. Wadibhasme (✉) · A. Nandi · B. Wadibhasme
CMS, Pune University, Pune, India
e-mail: amit.g.nandi@pwc.com

B. Wadibhasme
e-mail: wadibhasme.bhavesh.x@principal.com

S. Sawarkar
IIT Bombay, Mumbai, India
e-mail: 174020014@iitb.ac.in

M. Paprzycki et al. (eds.), *Intelligent Systems, Technologies and Applications*,
Advances in Intelligent Systems and Computing 1353,
https://doi.org/10.1007/978-981-16-0730-1_19

lineage, origin story, history, country, language or accent, symbolic organization, e.g., ritual, cuisine, dressing style, art, and physical appearance. Human face is rich source of key demographic information such as race/ethnicity. The term race is used to group humans. We evolved with different races due to various geographical reasons which reflects in various physical traits such as face, skin color, and hair texture. Out of which skin color and hair texture could get modified easily as we grow in modern world with sophisticated techniques, facial features still remain the most significant differentiator among the different human races. Race detection has great implication in research and business. It provides great insight into better understanding of demographics and diversity among population. Many retail brands gain from the wide understanding of customers and their consumer needs.

1.1 Related Work

Previously many different approaches have been proposed to identify the ethnicity. One such approach is to identify the special facial features such as LDA (Fisher Linear discriminant analysis) proposed by Lu et al. [17]. Attempts have made to combine LDA with principal component analysis (PCA) scheme. LDA features are extracted on images scaled at different scales, and an ensemble is constructed by integrating the classification results to arrive at final decision. However, the dataset used in this work had only two classes, namely Asian and non-Asian.

Hosoi [15] provided an approach to identify the ethnicity by extracting the key facial features by combining Gabor wavelets transformation and retina sampling. These features are then used to train the support vector machine (SVM) classifier. This classification is done for three categories such as Asian, African, and European. However, this approach has a limitation in generalizing it to the other ethnic groups. Also, the requirement of handcrafted features can be a drawback and makes this approach dataset specific.

Recently, ethnicity detection from face using deep convolution neural network has shown great improvement. With the latest advancement in CNN-based frameworks and availability of more compute power researchers are preferring such approaches, and it is found to outperform the previous state-of-the-art algorithms [15, 17]. Using one of such approach, Vo et al. [16] proposed race recognition framework using CNN. This race recognition framework uses deep network which is inspired from VGG network [9]. Along with an image, augmentation method is to generate the synthetic data to train the network. This framework uses three modules: an information collector, face detection and preprocessing, and RR.

2 Proposed Approach

From the latest research [16], it is found that CNN-based approaches work better or even outperforms the state-of-the-art methods. CNN-based approach has succeeded in extracting large number of features which helps for better classification. Till now researchers have approached the ethnicity detection problem as standard image classification problem. But this approach is likely to run into issue of overfitting. This is because convolutional neural network (CNN) is known for its capability to extract the low-level and high-level features, and in case of features from face, this will mostly end up in extracting the features like nose, eyes, iris, lips, and forehead. This means the similar features will be extracted from all the classes in training sets as all faces despite any ethnic group it may belong has the same features. Hence, the framework/approach which uses the CNN as feature extractor and classifier (fully connected network with SoftMax) on top, it will not generalize well in real-time scenario.

We propose that embedding extraction-based technique inspired from FaceNet is better suited for race detection problem. In this study, we compared the CNN-based image classification approach using popular networks such as VGG16, RESNet50 [14], and MobileNet with embedding extraction approach. Results are compared on common training and test dataset (validation set is chosen randomly from training set by each model during training). Later its shown that the results from embedding extraction-based technique are better than other conventional CNN-based image classification approach.

2.1 Embedding Extraction for Race Detection

While the race detection from face sounds like classical multi-class image classification problem, race detection problem has some unique characteristics, i.e., finding similarity between two comparable images/things. Similar to face recognition system, race detection from face also has to deal with the extraction of key facial features/embedding. Here we used Inception_v1 network to extract the facial embedding per class and is trained to maximize squared L2 distance between classes. Similar to FaceNet [8] approach, training input is selected as batch of triplets using online triplet selection [8, 10, 12] method. The triplet loss function is used to allow the network to learn better race embedding. Once the embedding has been created, the final classification is performed using nonlinear support vector machine (SVM).

The study has also proved that skin color is not a distinguishing parameter in different races [3]; hence, we choose to preprocess all the images to black and white so that color would not be used as one of the feature for classification. This helps with reducing the computational overheads although this is not meant to handle different illuminations and is left for algorithm to come up with efficient feature embedding to handle the same. The skin color may be a useful parameter, but faces detected in

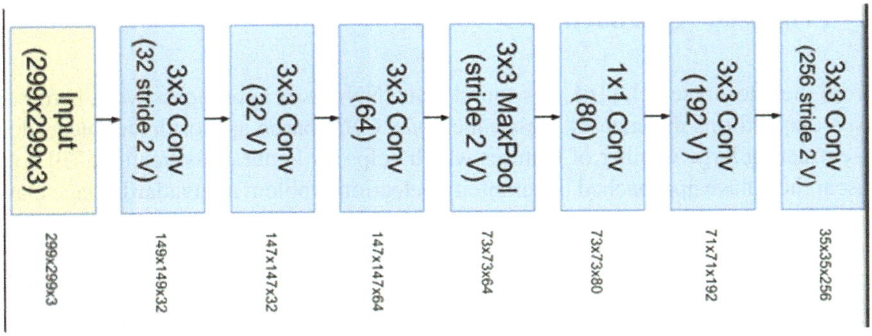

Fig. 1 Shown is stem of inception_resnet_v1. This kind of network employs the idea of using filters of different sizes in one layer. The details of the model parameters are available in paper [8] and same are used in this study as well

unnatural color or lightening are difficult to estimate the skin color [4], and it needs special low-level image processing. Furthermore, this feature could be included in future research, whereas we prefer to continue with just the facial structure.

The experimental framework is developed using Keras [5] (for Vgg , ResNet50, MobileNet) and TensorFlow [1] (for race embeddings) library. The training was done using GeForce GTX 1060 GPU (Fig. 1).

3 CNN-Based Networks

3.1 Why These Networks?

This comparative study is aimed to validate that race detection problem is better handled by face embedding-based approach than plain CNN classification approach. Following networks are chosen because these are representative of their respective deep network design approach. VGG16 network is deep and wide network which tends to extract large number of low- and high-level features. MobileNet architecture employs depthwise separable convolutions, and ResNet uses identity shortcut connection. Comparison against these networks would set good baseline against large number of available CNN networks and hence are good candidates for comparison against embedding-based approach.

3.2 VGG-16 Network

As name suggests, this network has 16 convolutional layers arranged in sequence and has 144 million parameters. Convolution filters of size 3*3 were along with five maxpooling layers of size 2*2 size. This is then followed by a fully connected layer. SoftMax is used at final layer to get the classification probability. It uses rectified linear units (ReLU) as activation function and dropout in between fully connected layers to keep overfitting in check. VGG -16 network [9] was introduced by visual geometry group in 2014, and due to its very deep architecture, it is also used as base network for extracting useful image features.

3.3 MobileNet Network

MobileNet is an architecture built with depthwise separable convolutions [11] which helps in reducing the number of trainable parameters when compared to general convolutional networks of same depth. The convolutional features are essential for computer vision-related tasks but expensive to compute. In MobileNet architecture, those are replaced by depthwise separable convolutions. In traditional convolutional operation, a kernel/filter is applied across all channels of input image. This means convolution operation combines output of all the input channels of image into one; thus, if input image has three input channels, the single convolution operation results the output image with single channel. The MobileNet architecture also does the standard convolution but it does it with the combination of depthwise convolution and pointwise convolution. First the depthwise convolution takes effect by performing convolution on each channel separately. This means the output image has same channels as that of input image, and each channel of image gets its own set of weights learned. The depthwise convolution is then followed by pointwise convolution operation. This is a convolution operation with 1*1 kernel. This operation could be termed as simple weighted sum and hence combines the output of depthwise convolution over each channel to create new feature. This split operation is shown in Fig. 2.

Another good thing about rending the regular convolution operation into two steps is that currently we will apply a ReLU [2] two times rather than just the once. The activation employed by MobileNet is ReLU6. This can be just like the well-known ReLU; however, it prevents activation's from changing into too massive.

$$y = \min(\max(0, x), 6) \tag{1}$$

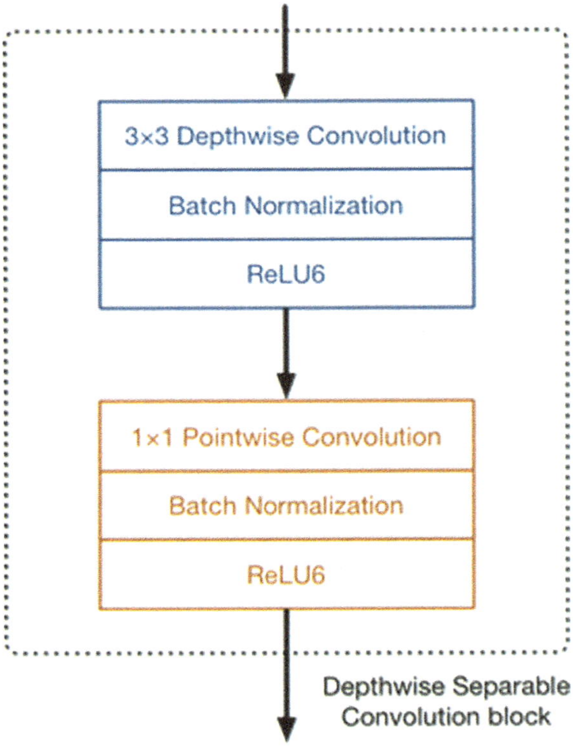

Fig. 2 Split operation in MobileNet architecture

3.4 *Transfer Learning*

Most recognized neural networks such as VGG16, RESNet50, MobileNet are typically considered as very deep neural networks. These networks require substantial computation power to train. Due to very deep architecture, they also have huge number of training parameters typically in millions, and this means they require large dataset to train. Training such networks on comparatively small dataset may lead to overfitting of model. Fortunately, transfer learning [6] method comes handy in such situation. Transfer learning uses the basis that for most of image classification tasks, the features learned by initial layers of networks also called as low-level features will roughly be the same regardless of dataset. This allows us to initialize such networks with parameter values learned from different dataset such as ImageNet [13]. These parameters are also called as pretrained weights, and such network would then be called as pretrained networks.

For our purpose, we used the pretrained version of VGG16, RESNet50, MobileNet networks provided in keras library. We choose to initialize our models with pretrained

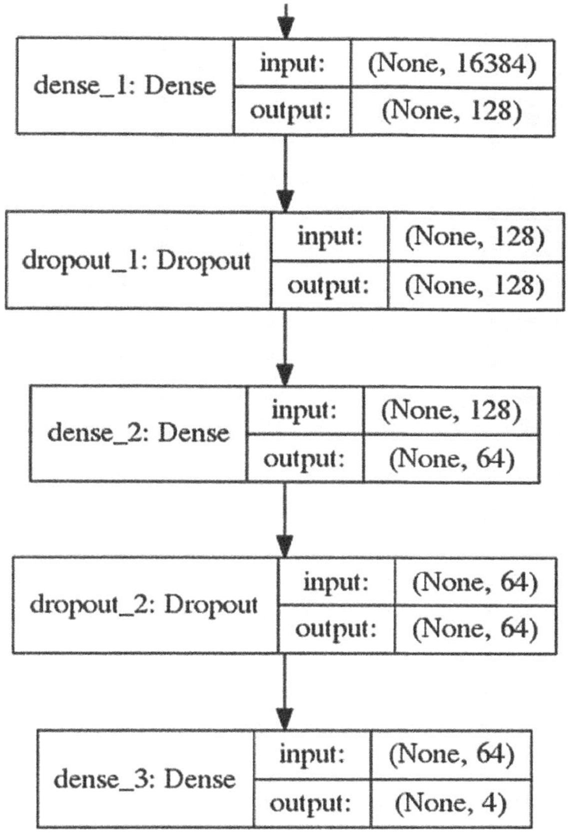

Fig. 3 Details of stack of fully-connected layers followed by a SoftMax activated layer. This stack is same for all CNN-based classification models used in this comparative study

weights and added fully connected network at the end and allowed it to train for our dataset used for race detection. To make the comparison fair, we kept the fully connected network same for all the networks used for comparison (Fig. 3).

4 Dataset Details and Data Preprocessing

To train the neural networks, it is a necessary requirement to pass input of the same size; hence, the resizing of training images is required. We chose to resize all the training images to 256 * 256 size. Though this could be one of the hyperparameters to choose, the dataset used is UTK faces [18] dataset which was perfect for this study. This dataset comes with aligned and cropped faces with classes as "White, Black, Asian, Indian". The dataset is mix of images of people from different age group from 0 to 116 years which gives excellent variance for model to learn rich facial features and good hard negative pair samples for online triplet selection.

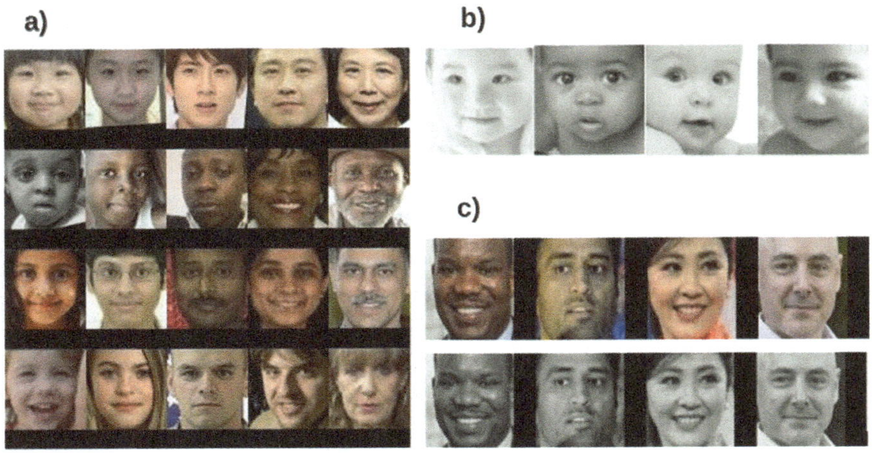

Fig. 4 **a** Snippet of UTK faces dataset. **b** Newborn faces for age upto 2 years. **c** Actual versus greyscale images

Figure 4a shows the snippet of UTK faces dataset. Each row shows face images of a particular race with individual age ranging from 5 to 80 years (left to right). Row 1–4 represents different races such as Asian, Black, Indian, and White, respectively.

Figure 4b shows sample face images of newborn babies between age upto 2 years of different race. As in newborn babies, the distinctive facial features do not get developed upto certain age, and this will prove good sample for choosing the anchor images for triplet selection. Selection method for anchor images used is online triplet generation mentioned in [8] .

As stated in introduction part, skin color is not considered as parameter for race detection; hence, we converted each image into black-and-white color scale forcing neural network not to use face/skin color as input feature (Fig. 5).

5 Results and Discussion

For training purpose, we used the UTK-faces dataset and some part (20%) is held out randomly (validation set) for the evaluation of those trained networks. We also created test dataset which was not included in training. Due to unavailability of other suitable dataset except UTK faces which has annotation for race, we manually scrapped the images from web by keeping the distribution of samples similar across all four classes. [Dataset will be made available upon request.] Various models were compared on basis of performance parameter called accuracy.

$$\text{Accuracy} = \frac{\text{TP} + \text{TN}}{\text{TP} + \text{TN} + \text{FP} + \text{FN}} \tag{2}$$

where TP = true positive, TN = true negative, FP = false positive, FN = false negative

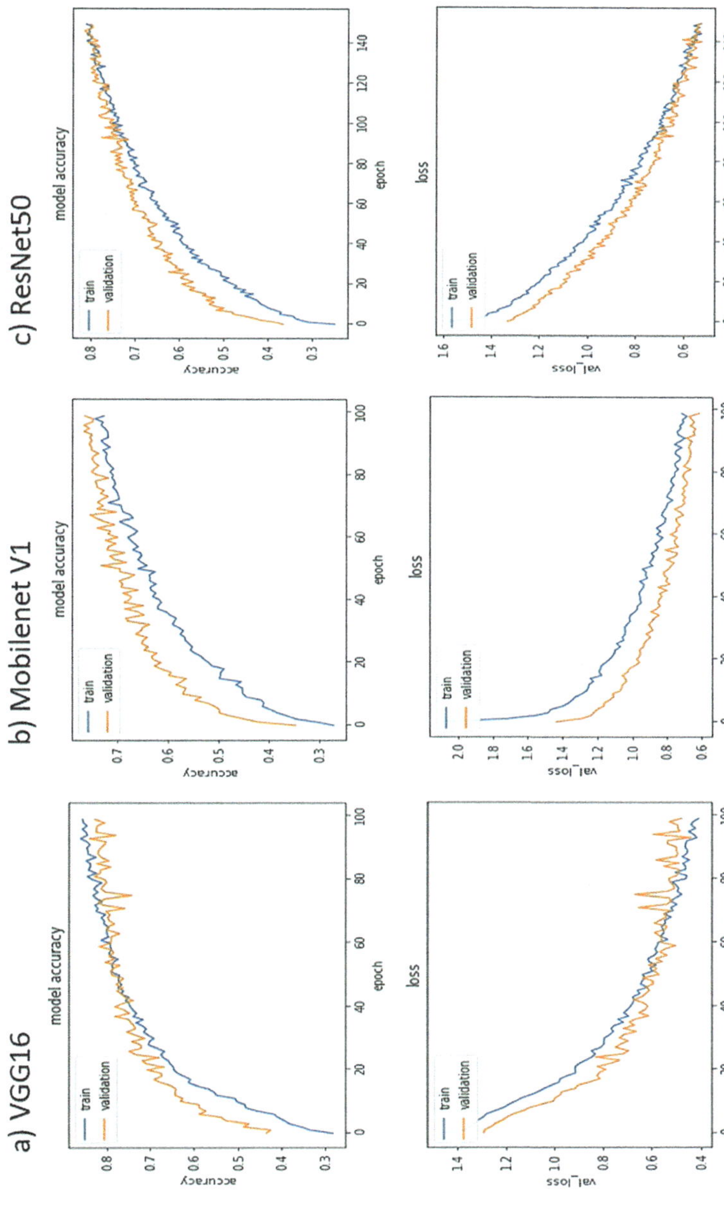

Fig. 5 Plot for accuracy and loss per epoch during training for CNN classification-based models. The graphs on top show accuracy Vs epoch and the one below shows loss versus epoch. Training accuracy and loss per epoch for model **a** VGG16, **b** MobileNet-V1, **c** ResNet50

For better comparison, we have also reported class-wise accuracy for all the models used. The classical CNN models are trained with transfer learning with categorical cross-entropy as loss function.

$$L(y, \hat{y}) = \sum_{j=0}^{M} \sum_{i=0}^{N} (y_{ij} \times \log(\hat{y}_{ij})) \quad (3)$$

In this comparative study, race embeddings-based approach has performed better compared to the other CNN-based models. This helps to validate that the race embeddings-based approach is better suited for race detection than treating general image classification approach. The online triplet selection method inherited from FaceNet also found to be better suited for our approach. While all the models used in this comparative study use CNN for the feature extraction, the triplet selection and triplet loss method found to be key differentiator for improved accuracy across classes. Although CNN-based models appear to perform better on training and validation set (both subset of training dataset), those do not generalize well on unknown test dataset. On contrary, the proposed approach performed much better (Tables 1 and 2; Fig. 6).

Table 1 This table compares class-wise accuracy of race embeddings approach with other CNN classification models over the test dataset also shown overall accuracy of each model

Model name	Accuracy (%)				
	Overall	White	Black	Indian	Asian
VGG16	59.25	68	56	58	55
MobileNetV1	60	64	54	60	62
ResNet50	59.50	55	61	59	63
Race Embeddings	**76.25**	**82**	**74**	**73**	**76**

Table 2 This table shows training time required for each model used in this comparative study. Training time is based on common training set used for all the models

Model	Training time (min)
VGG16	41
MobilenetV1	32
ResNet50	35
Race Embeddings	10

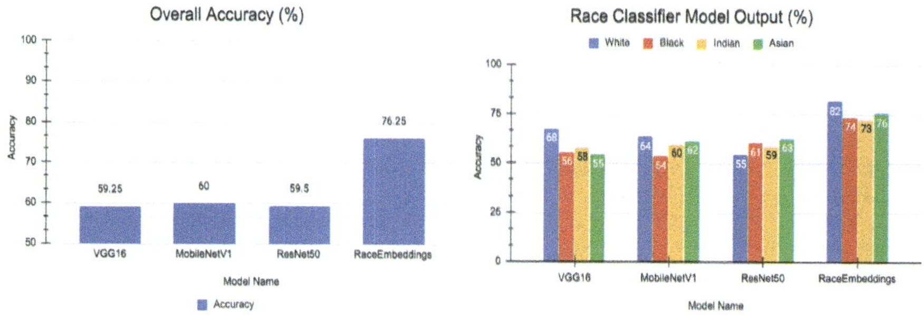

Fig. 6 Figure on left shows overall accuracy comparison among the models used in this study. The figure on right shows class-wise accuracy achieved by models used in this study

6 Conclusion

In this paper, the comparative study proves that embedding extraction method generalize better than CNN-based classification approach for race detection problem. It also found to be better in terms of training time required. We also ensured the effect of skin color not to be taken into account, and thus, the results are purely based on facial structure. We proved that race embeddings-based approach can learn efficient features compared to other plain CNN-based models and are better suited for real-life application.

References

1. M. Abadi, P. Barham, J. Chen, Z. Chen, A. Davis, J. Dean, M. Devin, S. Ghemawat, G. Irving, M. Isard, et al., Tensorflow: a system for large-scale machine learning, in *12th {USENIX} Symposium on Operating Systems Design and Implementation ({OSDI} 16)* (2016), pp 265–283
2. A. F. Agarap. Deep learning using rectified linear units (relu) (2018). *arXiv preprint* arXiv:1803.08375
3. Y.G. Bar-Haim, T. Saidel, G. Yovel, The role of skin colour in face recognition. Perception **38**(1), 145–148 (2009). https://doi.org/10.1068/p6307
4. M. Bindemann, A. Mike Burton, The role of color in human face detection. Cogn. Sci. **33**(6), 1144–1156 (2013). https://doi.org/10.1111/j.1551-6709.2009.01035.x
5. F. Chollet et al., Keras, in *GitHub* (2015)
6. C. Tan, F. Sun, T. Kong, W. Zhang, C. Yang, C. Liu, A survey on deep transfer learning, in *The 27th International Conference on Artificial Neural Networks* (2018). arXiv:1512.03385
7. Oxford Dictionaries, in *Ethnicity: Definition of Ethnicity* (Oxford University Press, Oxford, 2013)
8. D. Kalenichenko, F. Schroff, J. Philbin, in facenet: a unified embedding for face recognition and clustering, in *IEEE Conference on Computer Vision and Pattern Recognition (CVPR)*, Boston, MA (2015), pp 815–823
9. K. Simonyan, A. Zisserman, Very deep convolutional networks for large-scale image recognition, in *International Conference on Learning Representations* (2015)

10. E. Hoffer, N. Ailon, Deep metric learning using triplet network, in *International Workshop on Similarity-Based Pattern Recognition* (2015), p. 2
11. A.G. Howard et al., mobilenets: Efficient convolutional neural networks for mobile vision applications. ArXiv abs/1704.04861 (2009), pp. 145–148
12. T. Leung, C. Rosenberg, J. Wang, J. Philbin, B. Chen, J. Wang, Y. Song, Y. Wu, Learning fine-grained image similarity with deep ranking, in *Conference on Computer Vision and Pattern Recognition (CVPR)* (2014)
13. J. Yosinski, J. Clune, Y. Bengio, H. Lipson, How transferable are features in deep neural networks?, in *Advances in Neural Information Processing Systems* (2014), pp. 3320–3328
14. K. He, X. Zang, S. Ren, J. Sun, Deep residual learning for image recognition, in *IEEE Conference on Computer Vision and Pattern Recognition (CVPR)* (2016)
15. E. Takikawa, S. Hosoi, M. Kawade, Ethnicity estimation with facial images, in *Sixth IEEE International Conference on Automatic Face and Gesture Recognition*, Seoul, South Korea (2004), pp. 195–200
16. T. Vo, T. Nguyen, C.T. Le, Race recognition using deep convolutional neural networks. Symmetry (2018). 10.564.10.3390/sym10110564
17. A.K. Jain, X. Lu, Ethnicity identification from face images, in *Proceedings of SPIE. Biometric Technology for Human Identification* (2004). https://doi.org/10.1117/12.542847
18. Z. Zhang, Y. Song, H. Qi, Age progression/regression by conditional adversarial autoencoder. *IEEE Conference on Computer Vision and Pattern Recognition (CVPR)* (2017), pp. 5810–5818

An Effective Real-Time Approach to Automatic Number Plate Recognition (ANPR) Using YOLOv3 and OCR

Harsh Sinha, G. V. Soumya, Sashank Undavalli, and R. Jeyanthi

Abstract This paper aims to develop an efficient real-time automatic number plate recognition (ANPR) system to mitigate theft of services at various self-serviced locations such as gas stations. A thorough assessment of the system existing at such self-serviced gas stations has been done to find out the loopholes. An extensive literature review is completed to come up with possible solutions to the problem of non-payment by customers. After analysing limitations in the existing solutions, an end-to-end ANPR system based on convolutional neural network, using custom trained You Only Look Once version 3 (YOLOv3) and Tesseract Optical Character Recognition (OCR), has been developed. This system is capable of performing object detection, object localisation, image processing and optical character recognition. It can be interfaced with a physical camera and deployed as such in a real-world scenario. The system has been trained regressively on an exhaustive dataset of vehicle images to achieve accurate results. Testing revealed that the system performs much better in terms of speed as compared to other solutions, while still maintaining good accuracy.

Keywords YOLO · YOLOv3 · OCR · ANPR · Object detection · Character recognition · CNN · Deep learning

1 Introduction

In countries around the world, self-serviced petrol bunks or gas refuelling stations are a common sighting. This system has been effective in reducing the cost of operation for fuel companies and has also put an end to the customer complaints of gas station employees filling less gas than what is paid for by the customer. Although this seems to be a win-win situation, over the years this system has been time and again

H. Sinha · G. V. Soumya · S. Undavalli · R. Jeyanthi (✉)
Department of Electronics and Communication Engineering, Amrita School of Engineering, Amrita Vishwa Vidyapeetham, Bengaluru, India
e-mail: r_jeyanthi@blr.amrita.edu

M. Paprzycki et al. (eds.), *Intelligent Systems, Technologies and Applications*, Advances in Intelligent Systems and Computing 1353,
https://doi.org/10.1007/978-981-16-0730-1_20

misused and fuel theft has become a major concern. Now with the increasing number of vehicles, it has become more important than ever to establish control and maintain a record of the vehicles coming over for a refill.

To deal with such issues of theft at service locations, security issues, better traffic management, car speed monitoring, prevention of car theft and misuse of present systems, various approaches to automatic number plate recognition(ANPR) system have been developed and adopted widely. The ANPR system has been broadly classified as object detection and text recognition problem.

The conventional object detection algorithm of ANPR systems had been built on various neural networks such as artificial neural network (ANN) [1], convolutional neural network(CNN) [2–5] and regional convolutional neural network (R-CNN) [6]. Simpler, less accurate ANPR systems have also been developed solely on image processing techniques [7–12]. Although these techniques give a fairly accurate result, they are much slower than the expected milestone. Due to this, none of these ANPR systems are real time and hence cannot be effectively deployed in real-world situations.

This paper proposes an ANPR system that uses YOLOv3 [13] technique for number plate detection and OCR [11] on the open-source engine Tesseract [14] for character recognition, both seamlessly interfaced through Python scripts to produce real-time output. YOLOv3 has been used to optimise detection algorithm as it can look at object detection as a single regression problem and not a classification-based problem. It is implemented on a CNN with the initial network layers extracting image features, while the fully connected layers predict the output probabilities and coordinates. After YOLOv3 identifies the number plate in the image, the number plate is localised and used for text recognition.

The proposed ANPR system in this paper can produce real-time outputs with great accuracy and can be deployed for various applications such as to prevent theft at service locations like fuel station, for better traffic management, for vehicle speed monitoring, prevention of vehicle theft, etc.

2 Proposed Methodology

The proposed ANPR system aims to provide an end-to-end solution for various applications of ANPR. This proposed system is divided into four modules: number plate detection, interfacing of YOLOv3 and OCR, image processing followed by character recognition and database operation. This entire process is depicted in Fig. 1.

2.1 Number Plate Detection

Object detection was conventionally characterised as a classification problem, but this yielded slow object detectors. YOLO treats object detection as a regression

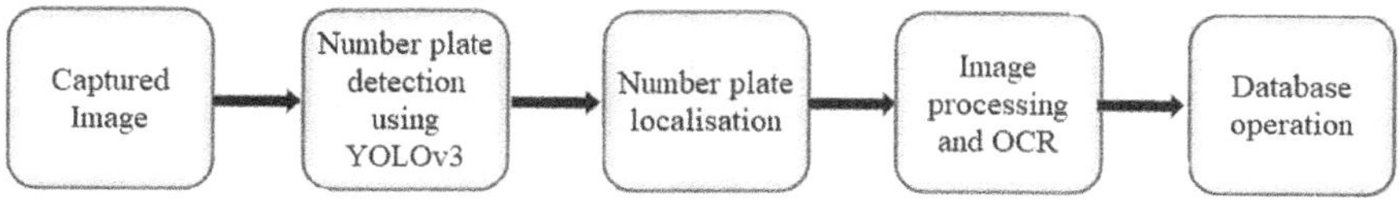

Fig. 1 Proposed ANPR system

problem [15, 16]. It gained in speed because it was built as a single convolutional network. It performed convolutions on the entire image as one, rather than sections of the image. This feature of YOLO facilitated contextual information about the classes and appearance to be encoded at once. The proposed ANPR system uses YOLOv3, a variant of YOLO that is originally based on Darknet-53. Darknet-53 is the backbone CNN of YOLOv3. It uses 53 convolutional layers for training on images and 53 more layers for detection [17].

Features of YOLOv3 The input image is divided into a $n \times n$ grid cell known as a classifier which is responsible for generating 'k' number of bounding boxes around potential objects whose ground truth falls within the grid classifying it as an object. YOLOv3 makes three predictions, and the box containing the object is called the bounding box. Each grid cell of the image predicts three bounding boxes by using the anchor boxes that have a predefined height and width. As detection happens in three scales of 32, 16 and 8, a total of nine predictions are made using nine anchor boxes. The cell consisting of the centre of the object is chosen as the correct bounding box [13].

Figure 2 demonstrates the prediction of three bounding boxes from a grid cell, in each stride. After obtaining the three scales, bounding boxes are predicted by [18]

$$b_x = \sigma(t_x) + c_x \tag{1}$$

$$b_y = \sigma(t_y) + c_y \tag{2}$$

$$b_w = p_w e^{t_w} \tag{3}$$

$$b_h = p_h e^{t_h} \tag{4}$$

Here, coordinates of the centre of the predicted box are given by b_x and b_y, whereas the height and width of the box correspond to b_h and b_w. The output of the network gives the centre coordinates as t_x, t_y and the dimensions as to t_w, t_h. c_x and c_y give the amount by which the cell in consideration might be offset from the first or the top left grid cell. The height and width of the box prior to bounding box or the anchor box are given by p_h and p_w.

These outputs at different scales are concatenated, and the most appropriate detection is made using non-maximal suppression (NMS) [19], removing other non-vital detections. Refer to [15] for a more detailed analysis of the bounding box prediction.

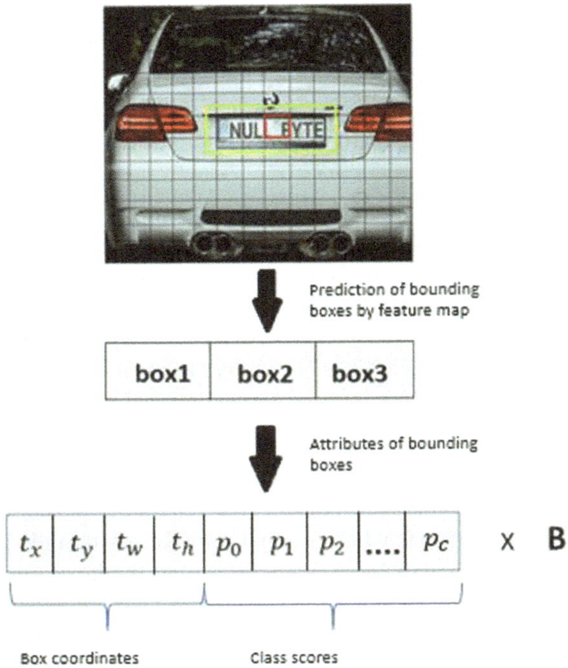

Fig. 2 Prediction of 3 bounding boxes from the ground truth cell. Each bounding box has 5 associated attributes. These are the x and y coordinates of the centre of the cell, height of the cell, width of the cell and detection confidence. Detection confidence is the probability that the chosen cell contains the centre of a particular object

2.2 Interfacing

The default output format of YOLOv3 object detection algorithm is an image with a bounding box detection around the region of interest. This output is not ideal for accurate results of further ANPR steps. This is due to the simple fact that the lesser the number of objects in the image fed to OCR, lesser will be the false-positive recognitions. Hence, the challenge is to localise the number plate detection and crop it out from the entire image. This was approached with a two-step strategy. In the first step, algorithmic modification to the YOLOv3 algorithm was made to output a JavaScript Object Notation (JSON) file containing the coordinates of the number plate. In the second step, a Python script is developed in the simulation environment which accepts the JSON file as input, extracts the coordinates of the number plate from the test image, localises the number plate and then crops the number plate out of the image. Since this entire process is used to interface YOLOv3 and OCR in a way that increases the accuracy of OCR, this step of the proposed methodology is referred to as the 'interfacing algorithm'.

2.3 *Optical Character Recognition (OCR)*

The advancement of computer vision over the past decade has led to several applications, with OCR being one of the very first. Traditionally, OCR has been used to convert a scanned image or a photograph into digital text that can be stored in the computer and thereby used for all kinds of applications. One of the major areas which have been developing as an application of OCR for quite some time now is recognition of characters from a number plate.

In this paper, Tesseract is proposed as the OCR engine to be used for character recognition, accompanied by YOLOv3 for object detection. Tesseract is an open-source OCR engine that is regularly upgraded and improved to give precise results with minimal computing requirements.

Text extraction from the number plate is a two-step process:

1. Image processing
2. Character recognition.

Image Processing This part of the OCR process is not mandatory, but is highly useful in providing accurate results. This is due to the simple fact that image processing can greatly reduce the noise in an image and thereby reduce the possibility of false positives.

In this phase, the cropped out image of the number plate is taken as input and it undergoes resizing, greyscaling and binarisation.

Figure 3 illustrates stepwise implementation of image processing. Firstly, the image is resized to an appropriate dimension. This dimension decision is taken after multiple trial and error to see which dimension yields the best results in a majority of the cases. The dimensions in the proposed system are set to resize the image to 90% of its length, while keeping the width intact. Resizing is a prerequisite for working with Tesseract OCR as Tesseract is incapable of reading characters from larger images. Then, the image is converted from RGB to greyscale. This reduces the complexity of the image while still retaining the varying intensity levels. The image is then binarised to de-noise it and make the recognition more robust by removing the spatial information that is not vital for the system. These three operations conclude the image processing phase of OCR.

Character Recognition The processed image obtained after the image processing phase is fit to undergo character recognition. Character recognition in itself is a five-step process [14].

The first step for character recognition is adaptive thresholding. Adaptive thresholding is an image processing operation which allows the image to be split into two layers of foreground and background objects. This separation is done based on the

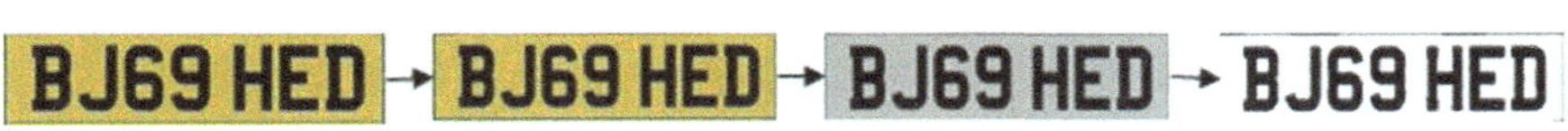

Fig. 3 Image processing pipeline

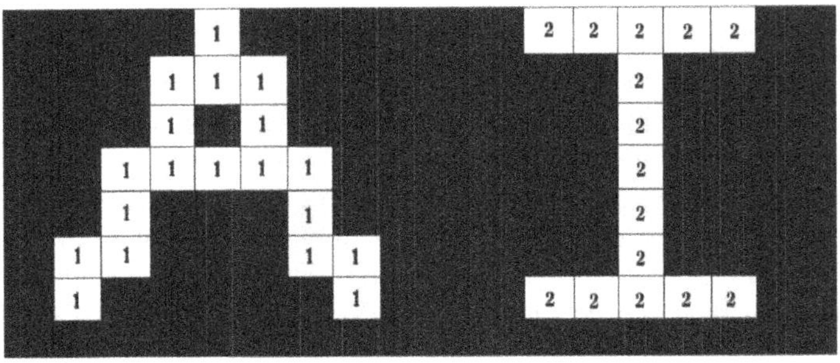

Fig. 4 The two tilted lines and one horizontal line in the alphabet 'A' are all connected and therefore will be labelled by the same unique number, which will be different than the labelling of a separately connected component 'I'

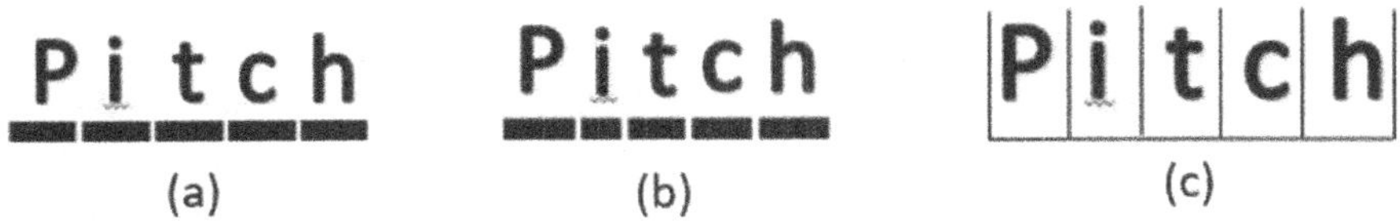

Fig. 5 **a–c** Depict fixed pitch, proportional pitch and chopping operation, respectively

pixel intensities. After this, the connected components are identified by connected component labelling. This operation labels or assigns a unique number to all the connected components of an image. Figure 4 depicts the connected component labelling.

After determining all the connected items by labelling, the outlines of all the connected components are stored and nested to form text lines. The text lines are further analysed for fixed pitch and proportional pitch. Fixed pitch text refers to the text written in a font that allocates equal space to every character, irrespective of the character width. Proportional pitch text refers to text written in a font that allocates each character with a different spacing, depending on the width of the character. Fixed pitch and proportional pitch are illustrated in Fig. 5a, b, respectively.

Based on whether the text was fixed pitch or proportional pitch, the text lines then undergo chopping. Chopping is the process of segmenting characters one by one in a word. Figure 5c shows the chopping operation for segmentation. The segmented characters then undergo recognition. Recognition happens in two passes. In the first pass, an attempt is made to read the word. Each recognised word is stored and used as training data to read the words better in future. Then, the second pass occurs during which any missed out character or blanks are re-read according to the training data newly added. This is known as adaptive classifier recognition technique [14].

2.4 Database Operation

The character recognition algorithm outputs the value of the number plate and stores it as a text string into a variable. The string format makes it convenient to perform further operations on the output. To do that, a SQL database is created. This database is instantiated in the ANPR pipeline, and the detected string is added to the database. Also, a time stamp that holds the date and time of the particular detection is added to the database. The database has a search feature embedded into it, and this opens up a lot of possibilities. The database can be used to maintain a count of all the times a particular vehicle undergoes number plate recognition. The database can also be cross-referenced to the state maintained database of vehicular criminal activities to generate red flags whenever a number that is present in the criminal database is detected.

2.5 Advantages of Proposed Methodology

The proposed system improves accuracy of object detection at different aspect ratios owing to the use of Darknet-53. There is also a significant decrease in localisation errors which in turn has led to an increase in the average precision[13]. Tesseract for OCR is a good choice because of its lightweight library and easier training dataset preparation. Also, Tesseract performs better at reading and recognising texts with thin font[11, 14].

3 Performance Measures

Any proposed system can only be considered fit for usage after a thorough analysis of its result. Traditionally, object detection algorithms such as YOLOv3 have been judged on two parameters, the average loss [15] and the mean average precision (mAP). However, the ANPR system proposed in this paper combines YOLOv3 for object detection along with Tesseract OCR for character recognition. This interfacing is done using an interfacing algorithm, and finally the output is stored on a server database. As a result, measuring the performance of only YOLOv3 trained on our custom dataset is not enough. It is vital to measure the performance of OCR as well, along with a performance analysis of the entire ANPR process. To quantify the performance of OCR and also to give a simpler performance parameter for a single-class YOLOv3, accuracy as a performance measure has been used in this paper.

3.1 Average Loss

The average loss is calculated as the sum-squared error between the prediction and the ground truth. This loss is the sum of three different loss components, namely classification, localisation and confidence loss. The average loss function for YOLOv3 is given by

$$\text{Average Loss (AL)} = \lambda \text{ const} \sum_{i=0}^{n^2} \sum_{j=0}^{B} \mathbb{I}_{ij}^{\text{object}} \left[(x_{pi} - x_{gi})^2 + (y_{pi} - y_{gi})^2 \right]$$
$$+ \lambda \text{ const} \sum_{i=0}^{n^2} \sum_{j=0}^{B} \mathbb{I}_{ij}^{\text{object}} \left[(\sqrt{w_{pi}} - \sqrt{w_{gi}})^2 + (\sqrt{h_{pi}} - \sqrt{h_{gi}})^2 \right]$$
$$+ \sum_{i=0}^{n^2} \sum_{j=0}^{B} \mathbb{I}_{ij}^{\text{object}} (C_{pi} - C_{gi})^2 + \lambda \text{ noobject} \sum_{i=0}^{n^2} \sum_{j=0}^{B} \mathbb{I}_{ij}^{\text{noobject}} (C_{pi} - C_{gi})^2$$
$$+ \sum_{i=0}^{n^2} \mathbb{I}_{ij}^{\text{object}} \sum_{c \in \text{classes}} (p_{pi}(c) - p_{gi}(c))^2 \tag{5}$$

Refer to [15] for a more detailed analysis of the average loss function.

3.2 Mean Average Precision(mAP)

Object detection algorithms have traditionally used mAP as a metric of choice to quantify detection performance. mAP is calculated as the area under the precision–recall curve. Precision and recall of any object detector are given by Eqs. 6 and 7, respectively.

$$\text{Precision } (P) = \frac{\text{TP}}{\text{TP} + \text{FP}} \tag{6}$$

$$\text{Recall } (R) = \frac{\text{TP}}{\text{TP} + \text{FN}} \tag{7}$$

where

TP is true positive
FP is false positive
FN is false negative.

The parameter used for classifying detections as TP, FP and FN is intersection over union (IoU). IoU is the ratio of overlapping area between ground truth and prediction to the total area of ground truth and prediction. Researchers choose IoU threshold of choice to quantify the performance of object detectors, but the most

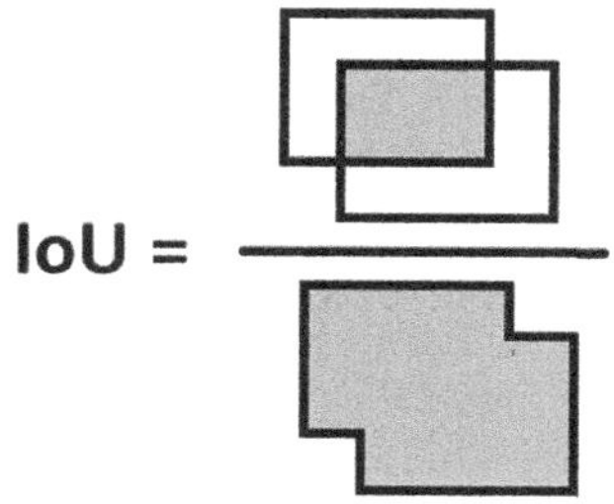

Fig. 6 Illustration of IoU. The square boxes represent the ground truth box and prediction box each. The numerator represents the common area between the two boxes, whereas the denominator represents the total area occupied by the two boxes

common thresholds are 0.50 and 0.75. An IoU threshold of 0.50 means that any prediction which has an IoU score of 0.50 or higher will be considered as a true positive. Figure 6 depicts IoU operation.

After the TPs are determined, the precision–recall curve is plotted and the area under the curve is calculated. In other words, mAP is given by Eq. 8.

$$mAP = \int_{0}^{1} p(r)\mathrm{d}r \tag{8}$$

Here, it should be noted that no distinction is being made between average precision (AP) and mAP. mAP is usually the average of AP over different classes of object detection. However, in the proposed system YOLOv3 is custom trained to detect only one class of objects, that is, the number plate. Hence for the proposed system, AP and mAP are treated as the same.

3.3 Accuracy

The performance measures discussed in Sects. 3.1 and 3.2 can only quantify the performance of YOLOv3. To measure the performance of OCR and also the performance of the entire ANPR system, accuracy as a performance measure is introduced. In terms of YOLOv3, we define accuracy to be the ratio of the number of correct number plate detections to the total number of image input to YOLOv3. In terms of OCR, we define accuracy as the ratio of the number of correct character recognitions to the number of correct number plate detections by YOLOv3. This is because the output of the custom-trained YOLOv3 is fed as the input to the OCR. Finally, the accuracy of the proposed ANPR system is defined as the ratio of the number of correct recognitions by OCR to the total number of image input to YOLOv3. Refer to Fig. 1 for the illustration of the proposed ANPR system as a process, for a better understanding of this step. No performance measures are needed for the interfacing algorithm and the database operation since they are not based on machine learning. They are implemented as stand-alone Python scripts and perform their task accurately 100% of the time.

4 Implementation

The proposed ANPR pipeline was executed on the simulation environment of our choice, Google Colab—a cloud platform built specifically for deep learning applications. The system specifications used on Google Colab were as follows:

- 12 GB RAM
- 70 GB solid-state drive (SSD)
- 12 GB NVIDIA Tesla K80 graphics processing unit (GPU).

4.1 Implementation of YOLOv3

For the sake of accurate number plate detection, we trained YOLOv3 on a custom dataset of 3000 vehicle images. The trained YOLOv3 algorithm is then used for number plate detection.

Custom Training To detect any specific class of objects, training is done on that particular object. The larger the training dataset and higher the iterations, the better the system performs. A dataset of 3000 car images of varying resolutions was collected [20]. It was divided into training and testing dataset in the ratio of 90:10, respectively. The training dataset had 2700 images, while the testing dataset had 300 images. The training dataset was then labelled and annotated by drawing bounding box around the number plate and specifying its label as 'number plate'. Labelling and annotation were done using the LabelImg tool which generates a text file as the output. The text file contains the four coordinates of the ground truth number plate. Figure 7 illustrates the process of labelling and annotation using LabelImg.

The algorithm was trained on this dataset regressively for 25,000 iterations. The batch size for custom training was set to 64, and subdivision was set to 16. As a result, the YOLOv3 detector was trained over 64 images, in groups of 16, 25,000 times. At the end of the training, the algorithm had trained on 1,600,000 instances of 3000 images, as can be seen from Eq. 9.

$$\begin{aligned} \text{Number of instances} &= \text{Batch size} \times \text{Number of iterations} \\ \text{Number of instances} &= (64 \times 25{,}000) = 1{,}600{,}000 \end{aligned} \tag{9}$$

Execution After training is completed, a weights file is generated. As the name suggests, a weights file gives a weightage to each neuron in the neural network.

Fig. 7 Labelling and annotation using LabelImg

Fig. 8 Output of object detection

Based on the learning gathered from the weights file, YOLOv3 makes a detection on the test image and outputs a JSON file with plate coordinates. Figure 8 shows a sample output of the object detection phase.

4.2 *Implementation of Interfacing Algorithm*

The JSON file is fed as input to the interfacing algorithm, which is a Python script. Figure 9 shows a sample JSON file.

Using the coordinates from the JSON file, it crops out the number plate from the test image. Figure 10 shows a sample output of the interfacing algorithm.

```
"frame_id":1,
"filename":"/content/test_image/test_2.jpg",
"objects": [{"class_id":0, "name":"Number
plate", "relative_coordinates":
{"center_x":0.715234, "center_y":0.547700,
"width":0.110277, "height":0.068866},
"confidence":0.980144}]
```


Fig. 9 A sample JSON file. The interfacing algorithm extracts the coordinates of the number plate from this JSON file

Fig. 10 Output of interfacing algorithm

4.3 Implementation of OCR

OCR on the open-source engine Tesseract is first fine-tuned to read character from small objects like number plates. The number plate is then fed to it for recognition.

Fine-Tuning The Tesseract OCR algorithm is a well-trained algorithm that does not require training from scratch, as in the case of YOLOv3. However, in some cases it may fail to detect characters due to certain issues such as unknown font styles and broken characters. To deal with this, the algorithm is fine-tuned on the 'Charles Wright' font dataset. 'Charles Wright' is the standard font used in UK number plates. A training data of 25 randomly skewed images was collected and trained for 400 iterations to give the desired result.

Execution After fine-tuning Tesseract OCR, the cropped out number plate is taken as the input image. The image undergoes image processing steps which include conversion to greyscale, binarisation and resizing of the image. After this, it is fed to OCR as an input. The characters are recognised by OCR and stored as text. The extracted text from the number plate is shown in Fig. 11.

4.4 Implementation of Database Operation

The output of OCR is stored in a string variable. The database is instantiated at this stage, and the value of the number plate is stored in the database. The time stamp of the detection of number plate is also stored in the database.

Fig. 11 Output of OCR

AHM3DS

5 Result and Analysis

The proposed ANPR system was custom trained on 3000 images for 25000 iterations. Any machine learning algorithm starts at a very high average error value, but a good training must be able to reduce the error progressively as the number of iterations increases. The increase in the average precision and corresponding decrease in average loss beyond a certain number of iterations take longer to happen. A stabilised average loss of 0.06 or below is deemed as an excellent outcome. The training on our custom dataset performs better than expected and goes further down the accepted levels of average loss, to 0.0318. Figure 12 shows the average loss curve over 25,000 iterations.

The proposed ANPR system was then evaluated on a test dataset of 300 images. YOLOv3 could successfully detect 295 images out of 300. Out of the 295 successful plate detections, OCR was able to predict accurate characters in 266 images, partially accurate prediction in 9 images and incorrect prediction in the remaining 20 images. Table 1 compiles the result of the proposed ANPR system over the three performance measures discussed in Sect. 3. Figures 13, 14 and 15 capture the output at the different stages of the proposed ANPR system.

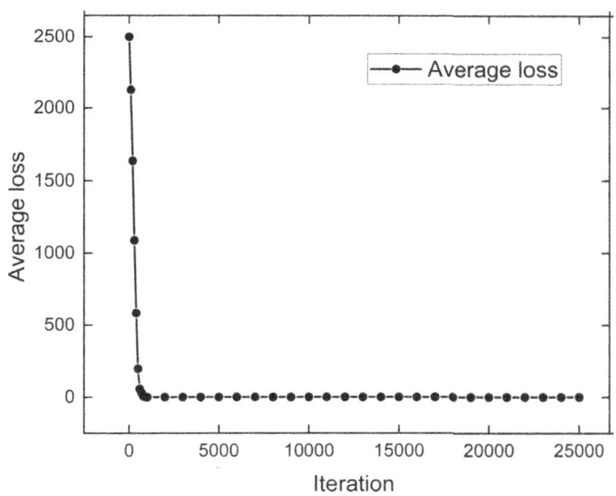

Fig. 12 The average loss decreases gradually over 25000 iterations. At the 1st iteration, the average loss is around 2500, but at the 25,000th iteration the average loss reduces drastically to 0.0318. Beyond first 1000 iterations, the rate of decrease in average loss reduces. This is because after the first 1000 iterations, the object detector algorithm has already trained on all the images in the dataset at least once

Table 1 Performance analysis of the proposed ANPR system

Technique	Performance measures		
	mAP@0.50 (%)	mAP@0.75 (%)	Accuracy (%)
YOLOv3	99.51	85.43	98.13
OCR	NA	NA	90.16
ANPR	NA	NA	88.67

NA Not applicable

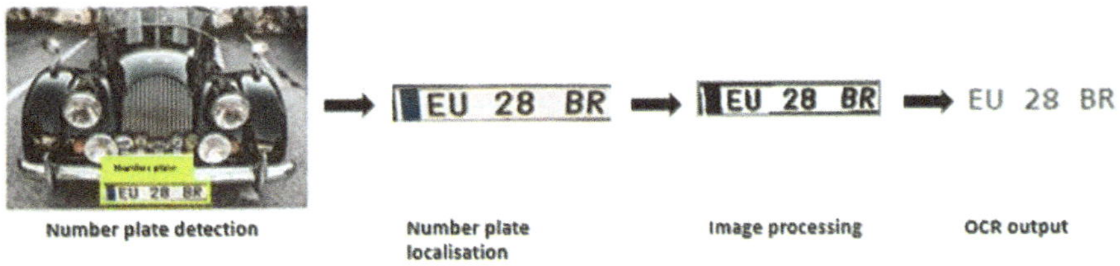

Fig. 13 Result of proposed ANPR system

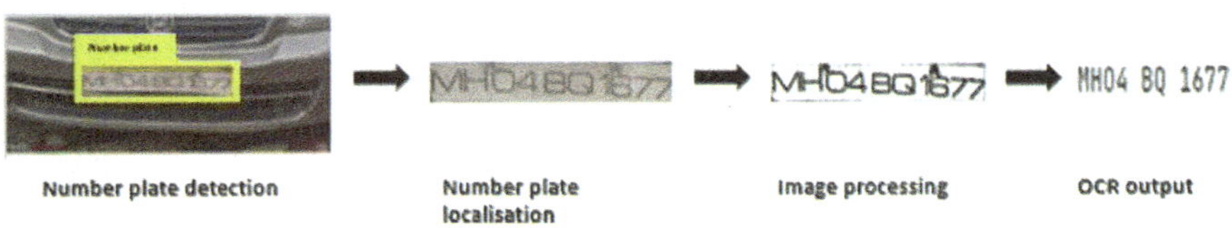

Fig. 14 Result of proposed ANPR system for non-standard (thin font) number plate

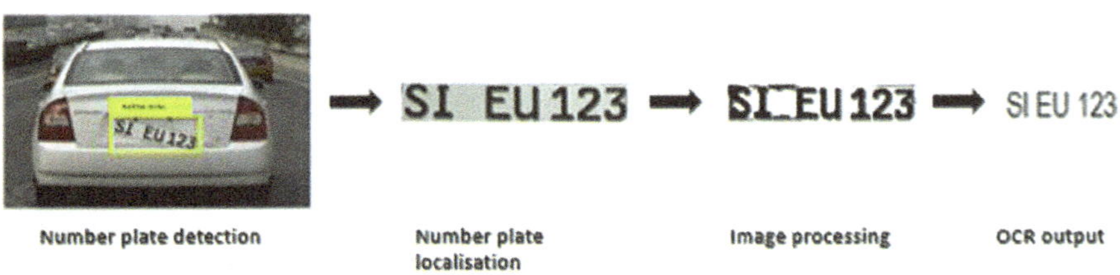

Fig. 15 Result of proposed ANPR system for tilted number plate

6 Conclusion

The proposed ANPR system does much of what we wanted while maintaining top-notch speed and accuracy. YOLOv3 for object detection turns out to be a very rewarding choice in certain scenarios, and this is one of them. It might be a few points short of some object detectors such as faster R-CNN when it comes to mAP, but it gains heavily in terms of speed and is able to optimise speed and accuracy. The exhaustive custom training that YOLOv3 went through makes sure that it does not fall behind in terms of accuracy when compared to any other object detector that could have been used to detect number plates. YOLOv3 being a single shot detector is also very light on computation and therefore turns out to be an ideal choice for deployment in real-time situations. The interfacing of YOLOv3 for object detection and Tesseract for OCR was done seamlessly. The interfacing algorithm written in Python took care of all the aspects that needed to be considered to pass the output of YOLOv3 as a valid input to OCR. OCR for character recognition from number plates proved to be a much more complex task. OCR, inherently, is built to read or recognise characters from documents such as pages and books. Fine-tuning OCR to read text from number plates was a challenge that was overcome by making certain parameter changes to the OCR algorithm. Also, after multiple trial and error, an image processing algorithm was created to feed images to OCR. This image pre-processing made things much easier for OCR leading to good enough accuracy to start with. Despite all this, OCR gave partially incorrect character recognition in certain situations such as number plates with graphical designs and multiple coloured backgrounds. An important point here would be to understand the fact that such number plates are deemed non-standard by state authorities and therefore exist in very small numbers and the observation to make here would be that further research in computer vision is needed.

References

1. H.E. Kocer, K.K. Cevik, Artificial neural networks based vehicle license plate recognition. Procedia Computer Sci. **3**, 1033–1037 (2011)
2. A. Menon, B. Omman, Detection and recognition of multiple license plate from still images, in *International Conference on Circuits and Systems in Digital Enterprise Technology (ICCSDET)* (2018). https://doi.org/10.1109/ICCSDET.2018.8821138
3. M. Mondal, P. Mondal, N. Saha, P. Chattopadhyay, Automatic number plate recognition using CNN based self synthesized feature learning, in *IEEE Calcutta Conference (CALCON)* (2017). https://doi.org/10.1109/CALCON.2017.8280759
4. S. He, C. Yang, J.-S. Pan, The research of Chinese license plates recognition based on CNN and length_featurein *Trends in Applied Knowledge-Based Systems and Data Science. IEA/AIE 2016*. Lecture Notes in Computer Science, vol. 9799 (Springer, Cham, 2016). https://doi.org/10.1007/978-3-319-42007-3_33
5. Wang et al.: CNN-based object detection solutions for embedded heterogeneous multicore SoCs, in *22nd Asia and South Pacific Design Automation Conference (ASP-DAC)* Chiba (2017), pp. 105–110. https://doi.org/10.1109/ASPDAC.2017.7858304
6. R. Girshick, J. Donahue, T. Darrell, J. Malik, Rich feature hierarchies for accurate object detection and semantic segmentation, in *2014 IEEE Conference on Computer Vision and Pattern Recognition*, Columbus, OH (2014)
7. A. Lazrus, S. Choubey, G.R. Sinha.: An efficient method of vehicle number plate detection and recognition. J. Mach. Intell. (2011). https://doi.org/10.1109/CICT.2015.13
8. B. Pechiammal, J. Arokia Renjith, An efficient approach for automatic license plate recognition system, in *Third International Conference on Science Technology Engineering and Management (ICONSTEM)* (2017). https://doi.org/10.1109/ICONSTEM.2017.8261267
9. A. Goyal, R. Bhatia, Automated car number plate detection system to detect far number plates. IOSR J. computer. Eng. **18**(4), 34–40 (2016)
10. A. Kashyap, B. Suresh, A. Patil, S. Sharma, A. Jaiswal, Automatic number plate recognition, in *2018 International Conference on Advances in Computing, Communication Control and*

Networking (ICACCCN), Greater Noida (UP), India (2018), pp. 838–843. https://doi.org/10.1109/ICACCCN.2018.8748287
11. M.T. Qadri, M. Asif, Automatic number plate recognition system for vehicle identification using optical character recognition, in *International Conference on Education Technology and Computer* (2009). https://doi.org/10.1109/ICETC.2009.54
12. S. Du, M. Shehata, W. Badawy, Automatic license plate recognition (ALPR): a state-of -the-art review. IEEE **23**(2) (2003). https://doi.org/10.1109/TCSVT.2012.2203741
13. J. Redmon, A. Farhadi, YOLOv3: an incremental improvement. arXiv (2018). https://arxiv.org/abs/1804.02767
14. R. Smith, Google Inc.: An overview of the tesseract OCR engine, in *Ninth International Conference on Document Analysis and Recognition (ICDAR 2007)*. https://doi.org/10.1109/ICDAR.2007.4376991
15. J. Redmon, S. Divvala, R. Girshick, A. Farhadi, You only look once: unified, real-time object detection, in *IEEE Conference on Computer Vision and Pattern Recognition (CVPR)*, Las Vegas, NV (2016), pp. 779–788. https://doi.org/10.1109/CVPR.2016.91
16. J. Redmon, A. Farhadi, YOLO9000: better, faster, stronger, in *Conference on Computer Vision and Pattern Recognition (CVPR)* (2017)
17. J. Redmon, Darknet: open source neural networks in c (2013–2016). http://pjreddie.com/darknet/
18. C. Liu, Y. Tao, J. Liang, K. Li, Y. Chen, Object detection based on YOLO network, in *Information Technology and Mechtronics Conference* (2018)
19. J. Wei, J. He, Y. Zhou, K. Chen, Z. Tang, Z. Xiong.: enhanced object detection with deep convolutional neural networks for advanced driving assistance. IEEE Trans. Intell. Transp. Syst. **21**(4), 1572–1583 (2020). https://doi.org/10.1109/TITS.2019.2910643
20. J. Krause, M. Stark, J. Deng, L. Fei-Fei, in *4th IEEE Workshop on 3D Representation and Recognition, at ICCV 2013 (3dRR-13)*, Sydney, Australia (2013)

Generating Audio from Lip Movements Visual Input: A Survey

Krishna Suresh, G. Gopakumar, and Subhasri Duttagupta

Abstract Generating audio from visual scene is an extremely challenging yet useful task as it finds application in remote surveillance, comprehending speech for hearing impaired people, or in silent speech interface (SSI). Due to the recent advancements of deep neural network techniques, there have been considerable research effort toward speech reconstruction from silent videos or visual speech. In this survey paper, we review several recent papers in this area and make a comparative study in terms of their architectural models and accuracy achieved.

Keywords Audio generation · Visual speech · Lipreading · Deep learning · CNN · LSTM

1 Introduction

Visual and audio are two important modalities of human perception and often one complements the other in human communication. Individually these modalities have progressed significantly in the research communities such as face/facial expression recognition [5, 25] and speech recognition [17]. Sometimes, we need to work on one modality to extract the second one to complement the communication process. For example, generating visual scenes with lip movements that synchronize with a given audio [7, 16] can be very helpful for comprehending speech, especially for those having hearing impairments. These methods often take still images of the target face and an audio speech segment as inputs, and try to generate audio-visual output as if it is an original recording by ensuring lip-audio synchronization.

Generating audio from visual scene is also equally important task and can find many applications. For example, in a noisy environment, we may not be able to

K. Suresh (✉) · G. Gopakumar · S. Duttagupta
Department of Computer Science and Engineering,
Amrita Viswa Vidyapeetham, Amritapuri, India
e-mail: gopakumarg@am.amrita.edu

S. Duttagupta
e-mail: subhasrid@am.amrita.edu

M. Paprzycki et al. (eds.), *Intelligent Systems, Technologies and Applications*,
Advances in Intelligent Systems and Computing 1353,
https://doi.org/10.1007/978-981-16-0730-1_21

capture clear audio but it can be reconstructed with the help of lip movements and expressions. Though this is a natural choice in such noisy environments, it is a challenging (even for humans) but very useful task. Another application could be in surveillance systems where the visual information of the speaker is only available due to the distance from the speaker or because of the non-availability of the audio recording device. A good application could be in silent speech interface (SSI) as part of sensor-based communication systems, in which a speaker articulates normally but does not activate their vocal chords. SSI systems are primarily based on ultrasound tongue imaging, electromyography, electromagnetic articulography, etc.

A standard speech production model accepts two features as input excitation and vocal tract (*spectral envelope*). Reconstructing audio speech from visual speech presents the challenge that the visual articulators provide information from only part of the vocal tract. In order to support real-time conversations, silent speech interfaces must ensure that the total processing delay is no more than 150 ms as specified by the upper bound for mouth-to-ear delay in conversational speech. Working on either of these modalities to generate the other is an extremely difficult task for humans since multiple phonemes often correspond to a single viseme (visual unit of speech) and multiple viseme corresponds to single phoneme.

Thanks to the recent advancements in computing facilities [1, 2], the availability of large datasets [12, 20], and the development of sophisticated deep neural network architectures [30], there has been considerable research to achieve the state-of-the-art results in almost all computer vision domain including the speech reconstruction from silent videos. There are increasingly large number of research activities in this domain but there is a lack of good review papers that compile these important research activities for a quick reference. In this work, we would like to review important articles that have focused on audio extraction from visual information. Specifically, we are considering only those that take the face and lip movement of the speaker though there are other means of extracting audio from silent videos even when the speaker is not present [11].

This review article is organized as follows. Section 3 discusses audio speech reconstruction from the visual speech, Sect. 4 deals with papers that produce audio from multiview visual feeds, and Sect. 5 presents papers related to lipreading where speech recognition is the focus. We also discuss the metric used to represent the generated audio quality and show a comparison table of important papers in Sect. 7.

2 Audio Model and Dataset

One of the important aspects of speech reconstruction is the audio speech model used in a system. It helps in identifying all the parameters useful for producing high-quality audio. Any technique for reconstructing audio speech from visual speech attempts to derive these parameters. A conventional speech coding application also relies on a number of acoustic features extracted from audio signal as given below:

1. a spectral envelope surface in time and frequency domain, $S(f, i)$;
2. a contour of fundamental frequency, f_0 ;
3. a measure of aperiodicity in time–frequency domain $A(f, i)$.

where f and i represent the frequency bin and frame index, respectively. But, many of the papers use only visual speech signal where certain acoustic features such as fundamental frequency and aperiodicity are unavailable. In that case, these values are derived from other parameters.

Next, we discuss three datasets that are widely used by many of the research papers we cover in this survey. Each of the datasets have certain unique properties.

GRID Dataset: Many of the papers dealing with extracting audio from the visual speech make use of GRID dataset [9]. GRID dataset constitutes of video and audio recordings of 34 different types of speakers (16 females and 18 males). For each speaker, there are about 1000 articulations and each of them is made up of a combination of 6 words formed from a vocabulary of 51 words. Video has a fixed duration of 3 sec with a resolution of 720 × 576 pixels.

OULU Dataset: OULU VS2 is an audio-visual database [4] which includes video recordings from more than 52 speakers with three different expressions. This database includes continuous digit strings, short phrases, and TIMIT sentences. This database constructs talking motion produced in a three-dimensional space which is done by using six cameras placed at different angles around the speaker to record from five different views taken at an angle 0°, 30°, 45°, 60°, 90°, respectively. This results in more than 20 thousand different recordings.

LRW Dataset: Many of the papers in Sect. 5 use LRW dataset [8]. LRW contains the video sequences extracted from BBC TV broadcasts. It contains 1000 articulations of 500 different words spoken by different speakers. Each video has 29 frames (of 1.16 sec duration) and the words appear in the middle of the video.

3 Generating Audio from Visual Speech

Lipreading or *Speechreading* refers to the process of obtaining phonetic information, thereby the speech uttered by a person simply by observing the speaker's lip movements or from a silent video. Speech reading is an enormously difficult task for humans to perform. Several neural network-based techniques are employed for generating audio of reasonable quality.

Cornu et al. [10] in their initial work use Gaussian mixture models, followed by deep neural networks (DNN) within a regression framework. But in their later work, a classification framework is used in which visual vectors are utilized for predicting a codebook corresponding to audio vectors. Codebooks help in restricting the derived estimates to be within a certain range. Their model consists of two major components: (1) estimating the static spectral envelope followed by (2) including temporal information. The first component requires a two-stage training—it first creates a vector quantization (VQ) codebook followed by passing it through a deep neural network (DNN)-based classifier to obtain a class label. Then, the class label and a vector quantization (VQ) codebook, C, are used as input for a simple lookup to obtain an estimated audio vector.

The second component in their architecture [10] incorporates *feature level* temporal information obtained using DNN. This stage groups audio and visual matrices and uses three sliding window-based techniques in order to convert estimated audio matrices to a sequence of audio vectors. Temporal information at the *Model level* uses recurrent neural networks (RNNs) as RNNs with long short-term memory (LSTM) architecture are very effective in predicting output sequences from input sequences.

Epharat et al. [13, 14] present an end-to-end model based on CNN to reconstruct speech for each input frame based on neighboring video frames. Their work also uses GRID dataset. In this work, authors represent speech using linear predictive coding (LPC) which represents the spectral envelope of a digital signal of speech in compressed format. Their model learns visual features from full frontal face as well as from cropped mouth region. This study reflects that the full frontal face gives more relevant information than the cropped mouth region. Since the authors consider overlapping audios, for each video frame, there are two feature vectors which are combined to form a sound vector. The authors also use this model to predict out-of-vocabulary (OOV) words which demonstrates good accuracy.

Another work on speech reconstruction from silent videos [3] uses a CNN, LSTM, auto-encoder, and fully connected layers. The auto-encoder is trained to extract the bottleneck features of the audio spectrogram and the decoder tries to reconstruct it. These bottleneck features are the target for the speech reconstruction network. In the model, CNN extracts the spatio-temporal features from the preprocessed video frames and using an enabled LSTM, the model obtains the temporal dimensions and order. These features are reshaped and flattened and fed to a trained decoder of the auto-encoder. Then, the reconstructed audio shows 98% correlation with the real audio.

The recently developed model on speech reconstruction [21] uses vocoders for speech reconstruction. Their work uses the vocoder features like spectral envelope, fundamental frequency, and aperiodic parameter. They also apply a voiced–unvoiced state (VUV). This model takes both cropped bottom portion of the face and the full frontal face for analysis.

The architecture model used by Michelsanti et al. [21] consists of a video encoder, gated recurrent unit (GRU) and five decoders: SP decoder, AP decoder, F0 decoder, VUV decoder, and a VSR decoder. There are two outputs for the model: (1) reconstructed speech and (2) text output corresponding to the speech. The video encoder extracts visual information whereas the GRU gives the sequential nature of video data. The output from the GRU is fed to all the five decoders. All the decoders except the VSR decoder are required to get a high-quality intelligible speech. The model compares its result with [29] which uses *Generative Adversarial Network* and this model outperforms in both quality and intelligibility for both speaker-dependent and speaker-independent cases.

4 Speech Reconstruction from Multiview Visual Feed

Before the advancement of machine learning, only the frontal view of the speaker was taken into account for the speech reconstruction model. But this type of system was not applicable at a massive range. Since most of the speakers in the real world cannot speak without a break or any movement continuously in front of a camera, the need for speech reconstruction from multiview images comes into play. This section deals with the various models where multiview visual inputs are considered for speech reconstruction system.

Initial work on multiview speech reconstruction model uses multiview silent video feed [32] as illustrated in Fig. 1. It requires as input—multiple views of the same speaker and the corresponding audio. Then, preprocessing of images is done to obtain better features. This model first uses a CNN to extract visual features from the images captured. Then, the long short-term memory (LSTM) layers are used for extracting the temporal dependency of both video and audio. The output of LSTM layers is passed through dense layers to produce audio corresponding to the image sequences. The complete architecture diagram is shown in Fig. 2.

The aim of this system is to map a single video frame to a single audio vector. However, rapid lip and mouth movements result in inefficiency of the system. To address this issue, the system maps a sequence of images to a certain portion of the corresponding audio vector such that the audio provides the temporal neighborhood for the image. As an additional parameter, the sequence length of the images is also considered and optimal value is taken that results in the best quality of sound. Further, the temporal dependency of the frame is decided based on five previous frames while constructing the speech. Thus, LSTM uses the timestep parameter as five images.

The work by Kumar et al. [18] exhibits a personalized speech reconstruction model. Like the previous paper, the architecture of this paper includes a combination of CNN, RNN (spatio-temporal convolution neural network (STCNN)) with

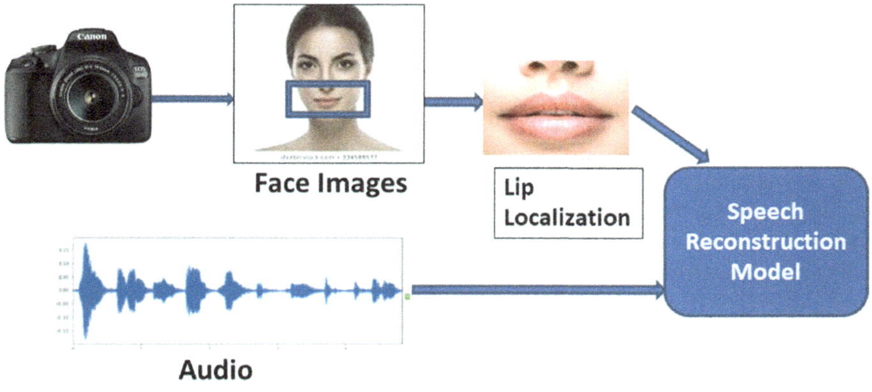

Fig. 1 Training of the speech reading with input as silent video feed and original audio

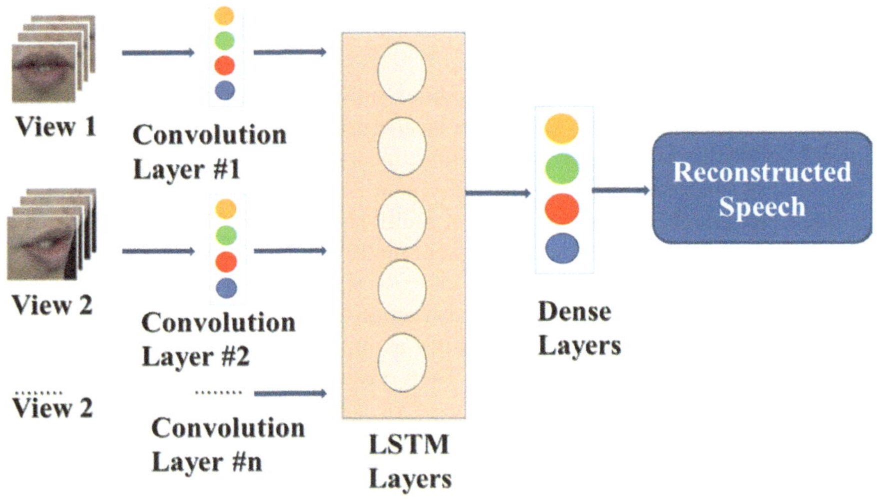

Fig. 2 CNN-LSTM architecture used for reconstruction of audio by speech reading [32]

bidirectional gated recurrent unit (BiGRU). Seven-layer STCNN brings out visual information related to lips, mouth, and teeth to reconstruct the speech. This step is followed by two layers of BiGRU which help to determine the temporal dependencies in speech. It is more convenient to use BiGRU as context is preserved better in this and the data present within the system (both audio and visual) depends on the context time. It is also shown to be more reliable than LSTM.

The output of STCNN-BiGRU is the corresponding audio of a given input which is represented by linear predicting code (LPC). LPC is a technique for representing the spectral envelope of speech in a compressed form using a predictive model that results in a high quality speech at a low bit rate. Line spectrum pairs (LSP) represent the LPC coefficients in a quantization robust manner. Thus, LSP proves to be more useful representation for greater reliability. The current audio pipeline offers the benefit of modeling of human emotion which is usually not taken into account in other speech reconstruction techniques.

The next paper [19] improves upon the above-mentioned technique. Here, the model provides two outputs corresponding to an input: (1) the audio and (2) the text. This system also follows the same STCNN-BiGRU architecture but unlike the above one, in this model the input is not given directly to STCNN. The system consists of a view classifier and a deciding logic prior to the STCNN. The view classifier classifies different types of views whereas the deciding logic chooses the best combination. The output of STCN-BiGRU is given to the audio decoder and the audio is represented in terms of linear predictive code (LPC). The generated audio output gets converted to text output using a text prediction model.

The paper [28] by Uttam et al. deals with the latest content on multiview speech reconstruction. This system comprises of four components:

1. *Pose classifier*: It is used to classify lip poses.
2. *Decision network*: It is to choose the correct encoder for the input video sequence.
3. *Video encoder*: It is used to extract spatio-temporal features of input which in turn gives the corresponding encoded representation.
4. *Audio auto-encoder*: This system represents the audio in the form of mel-frequency spectrogram.

The output of the video encoder is sent to the decoder of the audio auto-encoder to generate a spectrogram representation of the speech. In earlier papers, the audio is represented by linear predictive coding (LPC) whereas in this paper, the audio is represented by mel-frequency spectrogram, as this kind of depiction helps to represent speech without much dependency on speaker's characteristics.

Best View Combination:

The first paper on multiview speech reconstruction is limited only to two views on the speaker and the one which gives higher score on PESQ is ($0° + 30°$) combination whereas the remaining papers consider all the five different combinations of views. The papers also demonstrate that the combination of ($0° + 45° + 60°$) provides the best quality. However, the best view combination is ($0° + 30° + 60°$) in the work by S. Uttam et al.

5 Visual Speech Recognition

This section discusses papers related to lipreading where the output is represented using text. These models take speech as input and attempt to recognize the spoken words and output these words as text. This can be viewed primarily as a classification problem. A related useful application of security system deals with spoofing [15] in which the acoustic characteristics of speech signal is maintained but the data content in the speech utterances is modified and corrupted.

The paper [24] proposes the first model which takes the visual features (from different poses of face) as input for speech classification. The model uses bidirectional LSTM (BLSTM) followed by a softmax function. The encoder part of the model extracts the visual information from video of multiple views and the BLSTM combines information from all the encoders and gives a label to each input. The model uses Oulu VS2 dataset and considers five different poses of the speaker from 0° to 90°.

The model is first trained with single views independently and then it is used for multiviews. On the top of all these views, BLSTM is added to combine all the information of the single views. The best view combination of this paper is ($0° + 45° + 90°$).

The authors of [26] use a combination of spatio-temporal CNN (STCNN), residual network, and bidirectional LSTM (BLSTM). The model is trained and tested on LRW dataset. STCNN is used to extract spatial features from the video frames while the residual network progressively drops the dimensions of the extracted spatial features until it gets to one-dimensional vector and BLSTM models the sequential nature of the video. Each component in the proposed model is evaluated separately, then one by one other components are added to the model.

Almost every model for lipreading uses BGRU while B. Martinez et al. use a temporal convolution network (TCN) for speech recognition on LRW dataset. The latter model is able to generalize variable length video sequence using variable length augmentation. The model is a multiscale TCN, where the information from each TCN is mixed to get an output. The model is compared with BGRU models, and it outperforms all the state-of-the-art models with an accuracy of 85.3%.

The paper [31] uses two modalities for the speech recognition—both the video frames and audio. This is a two-stage algorithm wherein the first stage (audio enhancement), the sound is isolated from the noise using visual information from the face and in the second stage, both the enhanced audio and visual information are combined for the classification (multimodality speech recognition (MSR)). This model brings a pseudo-3D residual convolution (P3D) to extract selective visual features and for temporal tasks the temporal convolution blocks are used. Multimodality speech recognition model uses an encoder–decoder mechanism that is built using elementwise attention GRU (EleAtt-GRU) as MSR pays attention to each input. In the aforementioned paper, the audio is represented using short-term Fourier transform (STFT) and the model is evaluated on LRW dataset. Since the paper discusses a classification problem, the word error rate (WER) is used as the evaluation metric. The model achieves 84.8% accuracy for the classification of words.

6 Audio Reconstruction from Video

This section deals with a paper [29] that uses one of the latest deep learning technique called *generative adversarial network (GAN)* for speech reconstruction. Applications of GAN can be found in various domains such as hyperspectral image classifications [22] or as a conditional generative adversarial network for self-driving vehicles [23]. All the papers discussed earlier use spectrogram or LSPs to represent the audio signal, which do not hold complete information about the speech signal. This paper attempts to overcome this issue by performing a direct estimation of the time domain signal to reconstruct the speech. In this technique, the visual features from the input video are directly transformed to the corresponding audio waveform. The model comprises of three different components as shown in Fig. 3:

1. *Generator*—to generate audio waveform from the extracted visual information.
2. *Critic*—to check how real is the generated audio is.
3. *Speech encoder*—to conserve the speech content of the generated and the original audio.

The generator part of the network comprises a content encoder and an audio frame decoder. Content encoder contains a visual feature encoder and a RNN. Visual feature encoder is a five-layer 3D CNN to extract the visual information from the

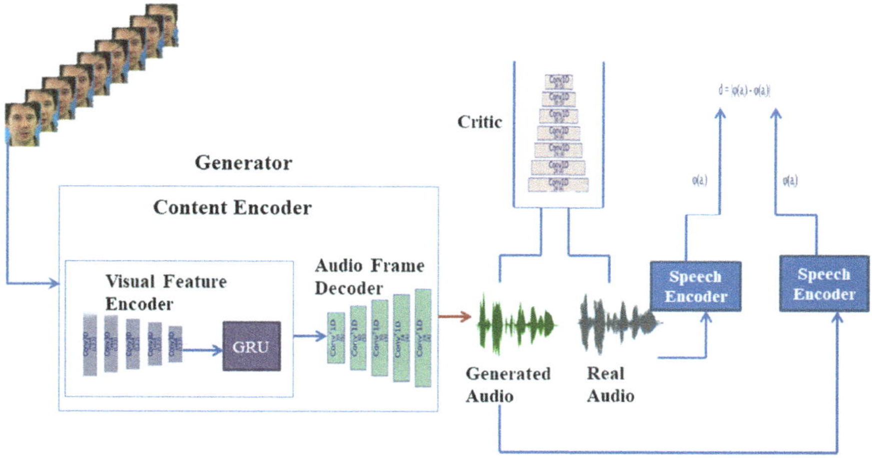

Fig. 3 GAN architecture used for speech reconstruction [29]

preprocessed video frames, and at each time step, these extracted visual features are fed to the recurrent neural network (RNN). The RNN outputs a series of features containing all the information about the video. All the features are given to the audio frame decoder which is a five-layer CNN and produces a waveform corresponding to the audio of the speaker.

The critic part of the network is a 1D CNN which takes both the real and reconstructed audio of a particular length. It is used to know the proximity of the generated audio to the real audio. Here, the Wasserstein distance between the generated and the original waveform is calculated. The third component which is the speech encoder is used while calculating the loss function. Speech encoder extracts the speech features from both the audios and calculates its perceptual loss. Speech encoder is a pretrained neural network. Perceptual loss is calculated by taking the L_1 distance between the speech features extracted from the real and generated audios.

7 Evaluation Metrics and Comparison Table

In this section, we mention important metrics used in the referred papers. The metric that is most commonly adopted in almost all the papers is the *perceptual evaluation of speech quality* (PESQ). This metric is used in all the work as it indicates the proper caliber and intellect of the speech. There exists a wide range of alternatives for speech quality apart from PESQ such as perceptual speech quality measure (PSQM) [6]) and perceptual evaluation of audio quality [27] (PEAQ) as proposed by ITU-T. The reason for wide acceptance for PESQ is that it not only reflects the perceived audio

Table 1 Comparison of papers using Ovulu dataset on multiview input

Reference	Architecture	PESQ score	Best combination
[32]	CNN, LSTM, dense layers	2.57	$0^\circ + 30^\circ$
[18]	STCNN, BiGRU	2.95	$0^\circ + 45^\circ + 60^\circ$
[19]	STCNN,BiGRU	2.315	$0^\circ + 45^\circ + 60^\circ$
[28]	CNN, LSTM, dense Layers	1.96	$0^\circ + 30^\circ + 60^\circ$

but also the system loss function. To achieve PESQ score, a comparison is done between the audio generated by the system and the original audio.

The process followed for obtaining PESQ consists of the following four steps: (1) aligning the level of two audios, (2) passing the signals through filters for temporal alignment (3) further, passing them through additive transformation, and (4) finally extracting two parameters reflecting the difference between the transformation of two signals.

At the final step, the signals are aggregated in frequency and time domain, and a mean value for the score is obtained. PESQ score can vary between 0.5 and 4.5 where the score of 4.5 reflects that the initial audio is reconstructed with high accuracy. The intelligibility of the reconstructed speech is done by human evaluation as stated by all the papers.

Next, we compare all the surveyed papers in two tables in Table 1 the papers on multiview are listed whereas in Table 2, the important papers discussed in other sections are listed. We analyze the papers based on the aspects such as video region used in the model, audio representation, architecture components, evaluation metric, and any specific feature of the proposed model.

8 Conclusion

Extracting audio from a visual input using neural networks and deep learning techniques is an active research area. In this survey paper, we have reviewed a number of research papers that deal with audio speech extraction from video. There exist multiple techniques which extract different visual features, pass through a number of learning layers before the audio reconstruction. This survey paper mainly analyzes three such approaches—generating audio from visual speech, reconstructing speech using visual feeds obtained from multiple views, speech recognition using lipreading. We discussed the architectures adopted in all these approaches. Finally, we also describe the audio metric used commonly to represent the speech quality.

Table 2 Comparison of research papers using GRID dataset (except the last row)

Reference	Video portion	Audio architecture	Model	PESQ score	Specific features
Cornu et al.	Mouth region	Spectral envelope (LPC)	DNN	1.9 (Regression)	Codebook finds a cluster that minimizes distance with an audio vector
			RNN, LSTM	2.05 (Classification)	
H. Akbari et al.	Mouth region	Auditory spectrogram	CNN, LSTM,	1.88	An auto-encoder extracts bottleneck features from spectrogram
A. Ephrat et al.	Full frontal face	Linear predictive coding (LPC)	CNN-based encoder–decoder	1.9	Use optical flow of the video as input
D. Michelsanti et al.	Full frontal face, Mouth region		CNN, GRU	1.90	Vocoder to reconstruct speech
K. Vougioukas et al.	Mouth region	time-domain signal	CNN, GRU	1.71	GAN for speech reconstruction
B. Martinez et al.	Lip portion	Mel-frequency chepstrum	Multiscale TCN	NA	85.3% accuracy on LRW dataset

References

1. Amazon web services for deep learning applications. https://aws.amazon.com/deep-learning/, accessed: 2020-08-01
2. Nvidia GPU cloud computing. https://www.nvidia.com/en-in/data-center/gpu-cloud-computing/. Accessed: 1 Aug 2020
3. H. Akbari, H. Arora, L. Cao, N. Mesgarani, Lip2audspec: speech reconstruction from silent lip movements video, in *IEEE International Conference on Acoustics, Speech and Signal Processing (ICASSP)* (2018), pp. 2516–2520
4. I. Anina, Z. Zhou, G. Zhao, M. Pietikäinen, Ouluvs2: a multi-view audiovisual database for non-rigid mouth motion analysis, in *2015 11th IEEE International Conference and Workshops on Automatic Face and Gesture Recognition (FG)*, vol. 1 (2015), pp. 1–5
5. S. Arya, N. Pratap, K. Bhatia, Future of face recognition: a review. Procedia Computer Sci. **58**, 578–585 (2015) [(Second International Symposium on Computer Vision and the Internet (VisionNet'15)]
6. J.G. Beerends, J.A. Stemerdink, A perceptual speech-quality measure based on a psychoacoustic sound representation. J. Audio Eng. Soc. **115–123** (1994)
7. L. Chen, Z. Li, R.K. Maddox, Z. Duan, C. Xu, Lip movements generation at a glance, in *Proceedings of the European Conference on Computer Vision (ECCV)* (2018), pp. 520–535
8. J.S. Chung, A. Zisserman, A.: Lip reading in the wild, in *Asian Conference on Computer Vision* (2016), pp. 87–103
9. M. Cooke, J. Barker, S. Cunningham, X. Shao, The grid audio-visual speech corpus (2006). https://doi.org/10.5281/zenodo.3625687
10. T.L. Cornu, B. Milner, Generating intelligible audio speech from visual speech. IEEE/ACM Trans. Audio Speech Lang. Process. **1751–1761** (2017)

11. A. Davis, M. Rubinstein, M., Wadhwa, N., G.J. Mysore, F. Durand, W.T. Freeman, A review on automatic facial expression recognition systems assisted by multimodal sensor data. ACM Trans. Graph. **33**(4) (2014)
12. J. Deng, W. Dong, R. Socher, L.J. Li, K. Li, L. Fei-Fei, ImageNet: a large-scale hierarchical image database, in *CVPR09* (2009)
13. A. Ephrat, T. Halperin, S. Peleg, Improved speech reconstruction from silent video, in *2017 IEEE International Conference on Computer Vision Workshops (ICCVW)* (2017), pp. 455–462. https://doi.org/10.1109/ICCVW.2017.61.
14. A. Ephrat, S. Peleg, Vid2speech: speech reconstruction from silent video, in *2017 IEEE International Conference on Acoustics, Speech and Signal Processing (ICASSP)* (2017), pp. 5095–5099
15. K.K. George, K.C. Santhosh, K.I. Ramachandran, A. Panda, Improving robustness of speaker verification against mimicked speech, in *Odyssey* (2016)
16. A. Jamaludin, J.S. Chung, A. Zisserman, You said that? Synthesising talking faces from audio. International Journal of Computer Vision **127**(11–12), 1767–1779 (2019)
17. B. Jolad, R. Khanai, An art of speech recognition: a review, in *2019 2nd International Conference on Signal Processing and Communication (ICSPC)* (2019), pp. 31–35
18. Y. Kumar, R. Jain, M. Salik, R. Shah, R. Zimmermann, Y. Yin, Mylipper: a personalized system for speech reconstruction using multi-view visual feeds, in *2018 IEEE International Symposium on Multimedia (ISM)* (2018), pp. 159–166
19. Y. Kumar, R. Jain, K.M. Salik, R.R. Shah, Y. Yin, R. Zimmermann, Lipper: synthesizing thy speech using multi-view lipreading, in *AAAI* (2019)
20. I. Matthews, T.F. Cootes, J.A. Bangham, S. Cox, R. Harvey, Extraction of visual features for lipreading. IEEE Trans. Pattern Anal. Mach. Intell. **24**(2), 198–213 (2002)
21. D. Michelsanti, O. Slizovskaia, G. Haro, E. Gómez, Z.H. Tan, J. Jensen, Vocoder-based speech synthesis from silent videos. ArXiv **abs/2004.02541** (2020)
22. S. Nirmal, V. Sowmya, K.P. Soman, Open set domain adaptation for hyperspectral image classification using generative adversarial network, in *Lecture Notes in Networks and Systems* (2020), pp. 819–827
23. S.O. Patil, V.V.S. Variyar, K.P. Soman, Speed bump segmentation an application of conditional generative adversarial network for self-driving vehicles, in *Fourth International Conference on Computing Methodologies Communication (ICCMC)* (2020)
24. S. Petridis, Y. Wang, Z. Li, M. Pantic, End-to-end multi-view lipreading, in *British Machine Vision Conference (BMVC)* (2017)
25. N. Samadiani, G. Huang, B. Cai, W. Luo, C.H. Chi, Y. Xiang, J. He, A review on automatic facial expression recognition systems assisted by multimodal sensor data. Sensors **19**(8), 1863 (2019)
26. T. Stafylakis, G. Tzimiropoulos, Combining residual networks with lSTMS for lipreading, in *International Speech Communication Association (ISCA)* (2017)
27. T. Thiede, W.C. Treurniet, R. Bitto, C. Schmidmer, T. Sporer, J.G. Beerends, C. Colomes, Peaq-the itu standard for objective measurement of perceived audio quality. J. Audio Eng. Soc. **3–29** (2000)
28. S. Uttam, Y. Kumar, D. Sahrawat, M. Aggarwal, R.R. Shah, D. Mahata, A. Stent, Hush-hush speak: speech reconstruction using silent videos, in *Proceedings Interspeech* (2019), pp. 136–140. https://doi.org/10.21437/Interspeech.2019-3269
29. K. Vougioukas, P. Ma, S. Petridis, M. Pantic, Video-driven speech reconstruction using generative adversarial networks, in *Proceedings of Interspeech* (2019)
30. X. Wang, Y. Zhao, F. Pourpanah, Recent advances in deep learning. Int. J. Mach. Learn. Cybern. **11**, 747–750 (2020)
31. B. Xu, C. Lu, Y. Guo, J. Wang, Discriminative multi-modality speech recognition, in Proceedings of the IEEE/CVF Conference on Computer Vision and Pattern Recognition (CVPR) (June 2020)
32. K. Yaman, M. Aggarwa, P. Nawal, S. Satoh, R.R. Shah, R. Zimmermann, Harnessing AI for speech reconstruction using multi-view silent video feed, in *ACM Multimedia Conference (MM '18)* (2018)

Improving the Performance of Imbalanced Learning and Classification of a Juvenile Delinquency Data

Takorn Prexawanprasut and Thepparit Banditwattanawong

Abstract The social environment has changed over recent years. Most parents have working lives and, as a result, they might not have enough time to pay sufficient attention to their children. In addition, the tendency of juvenile delinquency has increased over time. Although the policy of the Department of Juvenile Observation and Protection is to help children develop a more positive attitude about themselves and their world, the recidivism rates of children are still high. In order to solve this issue, the possibility of recidivism must be understood. The research aims to predict number of delinquents who would commit crime in the future. The experimental dataset, the Juvenile Delinquency Dataset, was considered as imbalanced data. The research employed a proposed algorithm, namely Cluster and Over (CaO), which is a notable resampling technique for imbalanced data. The results showed that the conventional prediction models without resampling techniques yielded both low accuracy and low recall. When CaO was applied with a decision tree, the values of Gmean increased by 5.3 and 3.9 percent compared with SMOTE and KmeansSMOTE, respectively. When CaO was conducted with KNN, the values of Gmean increased by 3.8 and 3.1 percent compared with SMOTE and KmeanSMOTE, respectively. Therefore, CaO yielded more efficient results than SMOTE and KmeansSMOTE.

Keywords Juvenile delinquency · Juvenile detention · Recidive · Machine learning · Imbalanced data · Resampling technique · Oversampling · Clustering

1 Introduction

The pattern of juvenile delinquency in Thailand has changed over time because of many reasons. The main reason is the revolution in the social environment that might

T. Prexawanprasut (✉) · T. Banditwattanawong
Kasetsart University, Bangkok, Thailand
e-mail: takorn.p@ku.th

T. Banditwattanawong
e-mail: thepparit.b@ku.th

M. Paprzycki et al. (eds.), *Intelligent Systems, Technologies and Applications*,
Advances in Intelligent Systems and Computing 1353,
https://doi.org/10.1007/978-981-16-0730-1_22

affect the way of living and also the behavior of children. There are many problems that can affect teenagers, such as drug abuse and addiction, lack of awareness, and the need for attention.

Since teenagers are influenced by what they see and learn in their society, there is likely to be a connection between recidivism and the nature of the society in which they live. When teenagers grow up in a "bad" environment, juvenile delinquency is more likely to happen [1]. Parents play an important role in children's lives and need to provide their children with a suitable environment and instruction. In order to solve recidivism, the environment of children is of primarily concern. Developing the environment of children or taking children out of the bad environment is a possible solution to prevent recidivism. In addition, encouraging teenagers to adopt a positive attitude is also important to reduce recidivism.

However, helping juvenile delinquents to think positively is still a challenging task. Not all delinquents understand and comply with social rules. Some delinquents who have been prosecuted continue to commit crimes and are often rearrested. In order to solve the recidivism problem, delinquents should be encouraged to do the right thing. Moreover, adopting a positive attitude is key to developing and maintaining good habits [2].

The data for this research was obtained from the Department of Juvenile Observation and Protection of Thailand. The data were collected from 2011 to 2015. The number of arrest cases did no significantly change and the number of cases of recidivism increased over the period. It was found that the percentage of recidivists increased from 15.86 to 19.03 in these five years. Table 1 shows the number of arrest cases, recidivism cases, and the percentage of recidivism.

According to the data in Table 1, not all delinquents recidivate. Only approximately, 20% of all first-time offenders were concerned. An interesting question arises from these data: What are the main factors that cause delinquency and recidivism? If these factors are better understood, better care could be taken of young people to prevent future recidivism.

The research focused on determining the characteristics of delinquents who were involved in recidivism cases. If delinquents were classified into two types, recidivist and non-recidivist, it would be easier to manage the more critical cases. The research constructed a model to predict the possibility of recidivism regarding the characteristics of delinquents. Machine learning techniques were employed in the proposed model.

Table 1 Number of cases

Case	Year				
	2011	2012	2013	2014	2015
Arrests	35,049	34,276	36,763	36,537	33,121
Recidivism	5559	4125	6849	7490	6302
Percentage of Recidivism	15.86	12.03	18.63	20.50	19.03

The data collected from the Department of Juvenile Observation and Protection of Thailand were considered to be imbalanced data. Most juveniles, roughly 80%, offended only once, while only 20% were recidivists. However, the existing classification techniques may not yield reliable results. Therefore, the research proposes a prediction model which attempts to address the issue of imbalanced data to yield more accurate results.

2 Literature Review

When the data are imbalanced, the classification model often yields unacceptable results. The true misclassification costs may be unknown at learning time [3]. For example, the dataset is composed of 1000 samples with 900 samples of class A and 100 samples of class B. The imbalanced ratio is 9:1. When applying a classification method to the specified dataset, the algorithm would learn more from class A than class B. As a result, the model would give a more reliable prediction for class A than class B.

Much research employed statistical techniques to deal with imbalanced data. In order to overcome the imbalance problem, there are two main methods, which are increasing the number of samples in the minority class and reducing the number of samples in the majority class [4, 5]. The oversampling technique means increasing the number of samples randomly, while the undersampling technique [6] means randomly decreasing the number of samples. Moreover, there are other techniques that are more sophisticated and yield better results.

The synthetic minority oversampling technique [7], SMOTE, is based on the assumption that the minority class specimens are grouped in terms of Euclidean distance, not mingled with the samples in another class. It is an oversampling technique using the methodology of k-nearest neighbor. Firstly, the algorithm randomly selects one sample from the minority class and its neighbors, which are also in the minority class. The algorithm then randomly generates a new sample in the position between the selected sample and its neighbors. Next, the first step is repeated where one sample and its neighbors are randomly selected and generate a new sample somewhere amongst them. The algorithm repeats these steps until it reaches the goal. The main idea of SMOTE is to increase the density of the members in the minority class. However, some groups of members in the minority class are ignored if the number of neighbors around the selected sample does not meet the criterion.

Another oversampling technique is called "Borderline SMOTE" [8]. Firstly, it classifies all instances into three types, which are safe instance, danger instance, and noise instance. A safe instance is an instance in the minority class that is surrounded by instances in the same class. A danger instance is an instance in the minority class that is surrounded by instances belonging to both the majority and minority classes. When the number of surrounding instances in the majority class is greater than the number of surrounding instances in the minority class, such an instance is called a dangerous instance. A noise instance is an instance in the minority class that is surrounded by instances in the majority class. Borderline SMOTE manages these three types of instance differently. Every safe instance is not processed because they are already in the proper condition. In the case of dangerous instances, Borderline SMOTE would perform an oversampling technique to increase the density of the minority class in the custom area. Finally, Borderline SMOTE would not perform anything with noise instances, and it would not consider such instances for the next iteration of the algorithm. Another method of identifying danger instances is the SVM implementation which can be found in synthetic informative minority oversampling (SIMO) [9, 10]

The random oversampling technique may lead to bias in the sampling process. In other words, a group of samples does not imply the real knowledge of the specified class. In order to solve such a problem, Resampling-Based Ensemble Methods [11] and ECO-Ensemble [12] are introduced. It supports multiple random rounds to reduce possible bias. Each round retains the same instances from the minority class but randomly selects the instances from the majority class. As a result, the number of instances from the majority class and minority.class is roughly equal in the sample.

Sampling with the data cleaning technique is conducted to solve two main issues, which are noise and overlapping areas of the majority class and the minority class. Wherever noise or overlapping areas occur, they are transformed or eliminated. Therefore, the learning efficiency of sampling with data cleaning is better than other standard sampling techniques. Examples of sampling with data cleaning are condensed nearest neighbor (CNN) [13], one side selection (CNN + Tomek) [14], edited nearest neighbor (ENN) [13], and SMOTE + ENN [13].

The cluster-based sampling technique is developed from the standard clustering algorithm (unsupervised learning). It performs well when the data are imbalanced. The first step of cluster-based sampling is to randomize a certain number of samples and determine the centroid of these samples. The first step is repeated until all samples have their own centroids. The next step is to calculate the distance between an instance and every centroid. The instance would be assigned to the new centroid if the distance is shorter than the current one. The algorithm would continually randomize a certain number of instances and calculate the distance to determine the centroid. All processes are terminated when nothing has changed. When the above algorithm finishes, well-defined clusters are presented. Finally, the oversampling technique is applied to every cluster in order to maintain a class distribution. Examples of cluster-based sampling are adaptive weighted oversampling [15], DBSM [16], and the heuristic oversampling method based on k-mean and SMOTE [17].

Recent studies have found that imbalanced problems may arise at specific times of incremental learning. Changing data concept over time requires a unique learning technique called concept drift [18]. Solving imbalanced problems in concept drift depends on proper design of algorithm. Ensemble techniques are used to reduce the variation of learning concept drift in forest-based classifier [19]. In addition, imbalanced problems that occur in multiclass classifiers have also been studied and has been observed that class imbalance may cause a significant deterioration in the performance attainable by standard supervised methods [20].

Table 2 List of features

No.	Meaning	No.	Meaning
1	Recidivism	14	Careless
2	IQ level	15	Resist law enforcement officers
3	Department code	16	No expression
4	Anti-social record	17	Seek power and control over others
5	Anti-social behavior	18	Underestimate the situation
6	Relationship with anti-social friends	19	Ignorance
7	Education or career	20	Unable to handle the problems
8	Addictive substance	21	Take advantage of others
9	Married life	22	Lawless
10	Hobby	23	Excitement seeking
11	Thoughts of perpetrator	24	Ask to be excused
12	High self-esteem	25	State of preparedness
13	Relationship with friends	26	Level of violence risk

3 Juvenile Delinquency Dataset

The Juvenile Delinquency Dataset was employed in the research to determine which characteristics relate to the risk of recidivism. The data were collected from delinquents when they first entered the juvenile justice system. The data were separated into two parts which were general information and attitude score. The examples of general information were gender, age, type of crime, and IQ level. The attitude score was evaluated by psychologists in seven dimensions, which were social personality, characteristics of friends, offender concepts, family characteristics, study background, hobby style, and drug-use behavior.

It was found that the data were not complete and included some errors. For example, the age of one offender was over 100 years old. The number of times of recidivism for some delinquents was more than 100 times. Therefore, these errors needed to be eliminated from the data. Some errors should not be deleted but should be corrected to retain the information. The list of features employed in the research is shown in Table 2.

4 Evidence of Imbalanced Effect

The early experiment was conducted using machine learning techniques to construct a classification model. The data used in the research came from the Juvenile Delinquency Dataset. The model attempted to define the possibility of recidivism based on general information and the attitude score of offenders. There are two types of target classes: class 0 and class 1. Class 0 means that it was the first time to commit

Table 3 Results of early experiment

Model	Accuracy	Class 0		Class 1	
		Precision	Recall	Precision	Recall
DT–GR	76.42	81.79	49.28	89.06	34.89
DT–IG	73.29	82.04	42.37	83.43	40.01
DT–GI	73.22	82.04	42.25	83.31	40.09
ID3–GR	72.81	82.11	41.64	82.74	39.80
ID3–IG	72.71	81.86	41.24	82.74	39.80
ID3–GI	73.53	82.13	42.87	83.67	40.23
RF–GR	80.00	79.74	85.25	99.09	17.34
RF–IG	80.48	80.18	85.90	99.02	19.60
RF–GI	80.28	80.09	83.54	98.84	19.31
Average	75.86	81.33	51.14	89.10	32.34

an offense, while class 1 means it was a recidivist case. The experimental results are presented in Table 3. The research was conducted using three models: random forest (RF), decision tree (DT), and ID3. Each model was implemented based on different criteria. The three criteria were gain ratio (GR), information gain (IG), and GINI index (GI).

The average accuracy was approximately 75.86%. The average precision of class 0 was 81.33%, while the average precision of class 1 was only 51.14%. The average recall of class 0 was 89.10% which was much higher than the average recall of class 1. The average recall of class 1 was only 32.34%. The results of the experiment indicate that all models were not practical for class 1 prediction because they gave low precision and low recall scores. All models provided too many false-negative instances. In other words, many delinquents were predicted not to be in recidivist cases, but they would commit crimes in the future.

The main reason why all models above gave poor results was that the data were imbalanced. The total number of instances was 5953, but only 20% of them were recidivist cases. All models had not enough data from class 1 to learn. The research proposed to employ an algorithm that was especially designed for imbalanced data. The algorithm focuses on the data sampling process to create a balanced dataset, and the prediction model could then learn effectively from both class 0 and class 1.

5 Proposed Algorithm

The proposed algorithm combines a sampling technique called "Cluster and Over (CaO)" with the standard machine learning algorithms to deal with imbalanced data. Three machine learning techniques, namely decision tree, KNN, and random forest, were applied with CaO. Since the Juvenile Delinquency Dataset was collected by different psychologists, data wrangling might occur. Some clusters of data possibly represented real knowledge; however, it was difficult to draw conclusions from some groups of clusters. Therefore, CaO was designed to perform the oversampling technique for certain clusters.

Cluster and Over (CaO) comprised two main steps. The first step was to perform clustering in the original dataset. Some clusters at this stage still had both members from the majority class and the minority class. Therefore, the second step was to apply an oversampling technique to each cluster. The oversampling technique was continually executed and was terminated when the highest value of Gmean was discovered. Gmean is the squared root of the product of the sensitivity and specificity. This measure tries to maximize the accuracy on each of the two classes while keeping these accuracies balanced [21]. In CaO, Gmean was calculated and compared with the previous Gmean in every oversampling step. If the current Gmean was greater than the previous Gmean, the cluster would be replaced by the oversampling set and Gmean updated. Whenever Gmean reached the highest number, the oversample step was stopped. The algorithm of CaO is shown in Fig. 1.

Refer to the proposed algorithm on line1 and 2, CaO begins working by dividing the dataset D into training set S and test set T. Then, on line 3 and 4, CaO created the first model M1 from this training set and calculated Gmean with the test set to avoid the data distribution bias. The Gmean, obtained from this process, is called initial Gmean. Therefrom, in line 5, the algorithm divides data into clusters to study data concepts in each subgroup. For each clusters $S[i]$, the algorithm calculates the internal imbalance ratio and performs SMOTE with the predefined oversampling ratio IRn. This step is shown on algorithm line 10. Then, CaO created a new classifier from the synthesized sample and calculated the performance by comparing new Gmean with the initial Gmean. On line 16–19, if the performance improves, the oversampling

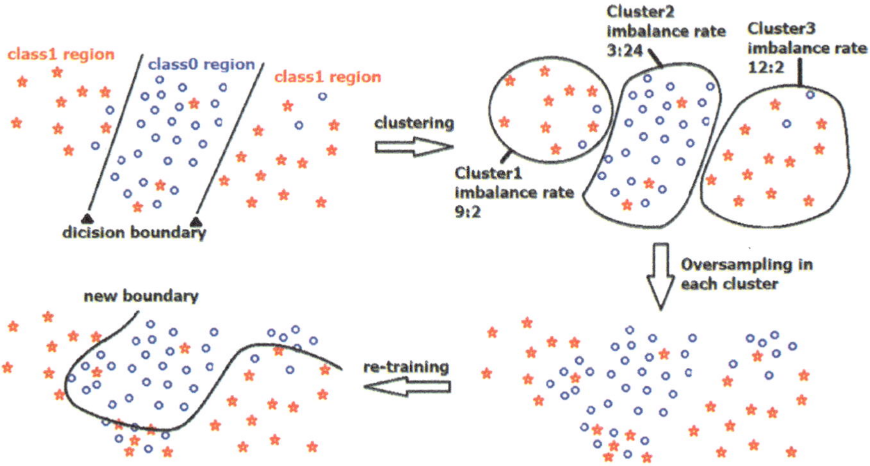

Fig. 1 Cluster and Over (CaO) algorithm

sample will be saved and the initial Gmean will be updated to the next round. After that, the algorithm tried to increase the oversampling ratio and repeat every step again. Finally, on line 23–24, CaO will create the classifier M3 from the highest Gmean dataset and can be used to calculate performance in different areas.

6 Experimental Results

CaO, applies by Python programming language with the Scikit-learn library, yielded good results when the number of clusters was 4 which corresponds to the number obtained from the elbow method. The Gmean value was highest when the ratio of the majority class and the minority class was 1:1. The number of instances in each class when an oversampling technique was applied is shown in Table 5. The first records in Table 4 are the results from the original imbalanced dataset. Upon completion, the second record provided the best results. All clusters had the same number of instances from both the majority class and the minority class.

When CaO was applied with a decision tree and in the early experiment criterion, the values of Gmean were 58.3 percent. Gmean increased by 5.3 (i.e., 58.3–53) and 3.9 (i.e., 58.3–54.4) percent compared with SMOTE and KmeansSMOTE, respectively. When CaO was conducted with KNN, the values of Gmean were 56.8%. Gmean increased by 3.80 (i.e., 56.8–53) and 3.10 (i.e., 56.8–53.7)% compared with SMOTE and KmeanSMOTE, respectively. The experimental results comparing CaO with SMOTE and KmeansSMOTE based on decision tree and KNN are shown in Figs. 2 and 3, respectively.

CaO, described in the field of solving class 1 recall, yielded higher values of recall than the early experiment. The average recall value of class 1 increased by 3.73% when compared with the original imbalanced dataset. Even though the recall values increased the Gmean and the area under the ROC curve (AUC) values had no dramatic changes. Both Gmean and AUC are used to show the effect of the proposed method on the original results. Gmean only increased by 0.69% while AUC decreased by 1.99 percent. The recall of class 1, Gmean, and AUC values are shown in Table 6. The first record in Table 5 is the results from the original imbalanced dataset. The number in parenthesis is the difference between the certain value and the value of the original imbalanced dataset.

CaO could increase the recall of class 1 by 8.46% which was higher than SMOTE and KmeansSMOTE. SMOTE could increase the recall value of class 1 by 5.05%, while the recall value of class 1 was increased by 6.19% in the case of KmeansS-MOTE. The experiment results reveal that CaO was notable because the increase of the recall value had some small effects on Gmean and AUC. The difference of the recall values of class 1, Gmean, and AUC when compared with the original imbalanced dataset are shown in Table 7.

Table 4 Proposed algorithm (CaO)

```
Input : D ← Original imbalanced dataset,
        n ← Number of clusters,
        i ← oversampling clusters,
        Thr ← oversampling threshold,
     Rat ← oversampling ratio,
Output: Notable classifier
--------------------------------------------------------------
S ← Training dataset (some parts of D)
T ← Testing dataset (D-S)
M1 ← Initial classifier constructed from S
initialGmean ← Initial Gmean of T based on M1
{ S[1], S[2], ..., S[n] } ← KmeansClustering(S, n)
foreach S[i];
    IRn ← Imbalanced ratio of S[i]
    Rat ← 0
    while IRn is less than Thr
        SM ← SMOTE(S[i], Rat)
        ST[i] ← S[i] ∪ SM        //ST[i] is a new instance set of cluster i
        Update IRn
        S_temp ← S - S[i] ∪ ST[i]
        M2 ← Modified classifier(S_temp)
        newGmean ← Modified Gmean of T based on M2
        if newGmean is greater than previous Gmean
            Update Gmean
            S ← S_temp
        end if
        Increase resamplingRatio by Rat
    end while
end for
M3 ← Notable classifier(S)
Determine the performance of M3 on D
```

7 Conclusion

The early experiment illustrated that conventional resampling techniques could handle an imbalanced problem. However, the recall value of class 1 from the early experiment was still too low. In other words, there were too many false-negative cases predicted by the early model. However, the proposed algorithm, Cluster and Over (CaO), focused on increasing the value of recall of class 1 without dramatic changes in Gmean and AUC.

Table 5 Number of instances in each cluster

Oversampling cluster	Cluster0	Cluster1	Cluster2	Cluster3
{}	1445+499	2217+367	333+100	230+78
{0, 1, 2, 3}	1445+1445	2217+2217	333+333	230+230
{0}	1445+1445	2217+367	333+100	230+78
{1}	1445+499	2217+2217	333+100	230+78
{2}	1445+499	2217+367	333+333	230+78
{3}	1445+499	2217+367	333+100	230+230
{0, 1}	1445+1445	2217+2217	333+100	230+78
{0, 2}	1445+1445	2217+367	333+333	230+78
{0, 3}	1445+1445	2217+367	333+100	230+230
{1, 2}	1445+499	2217+2217	333+333	230+78
{1, 3}	1445+499	2217+2217	333+100	230+230
{2, 3}	1445+499	2217+367	333+333	230+230
{0, 1, 2}	1445+1445	2217+2217	333+333	230+78
{0, 1, 3}	1445+1445	2217+2217	333+100	230+230
{0, 2, 3}	1445+1445	2217+367	333+333	230+230
{1, 2, 3}	1445+499	2217+2217	333+333	230+230

Remark Representation of $A + B$, A represented majority class, B represented minority class.

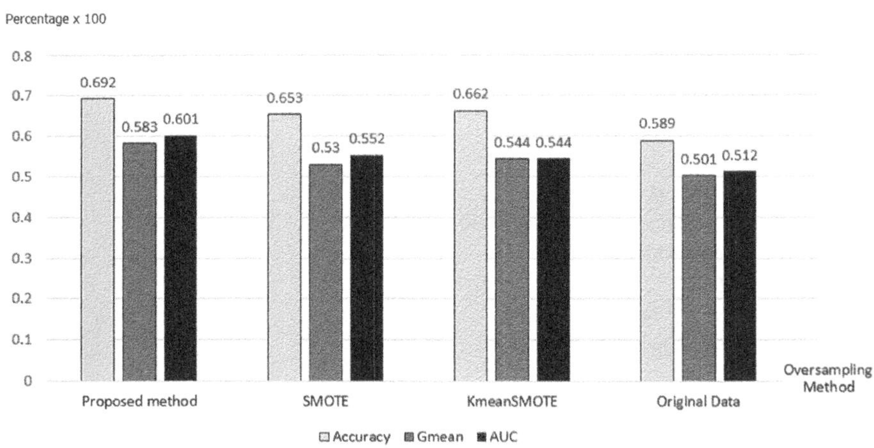

Fig. 2 Experimental results of CaO, SMOTE, and KmeanSMOTE based on decision tree

While SMOTE aims to make the density in areas where the minority concept is unclear. Meanwhile, cluster-based sampling attempts to resolve the imbalanced problem between clusters then tried to identify a concept to the cluster. CaO tries to create a concept in the cluster borderline more clearly, which will allow the classifier to better understand the minority concept in this area.

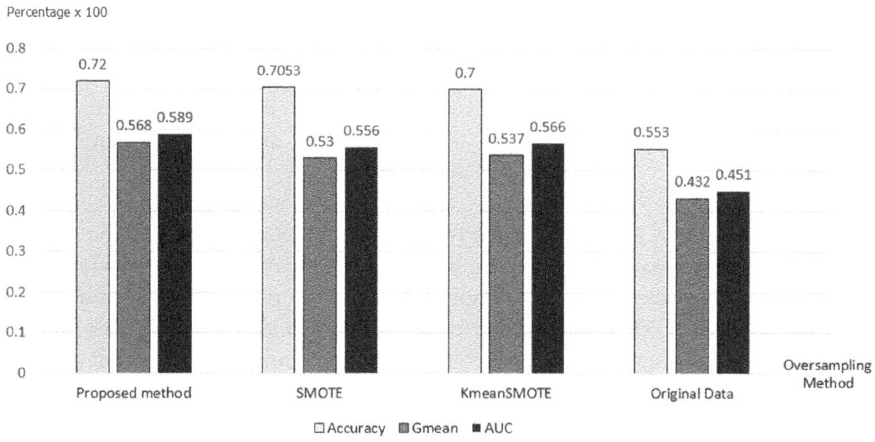

Fig. 3 Experimental results of CaO, SMOTE, and KmeanSMOTE based on KNN

Table 6 Values of recall of class 1, Gmean, and AUC

Oversampling cluster	Recall of class 1	Gmean	AUC
{ }	0.2692	0.4780	0.5589
{0, 1, 2, 3}	0.3384(+0.0692)	0.4911(+0.0131)	0.5256(−0.0333)
{0}	0.2999(+0.0307)	0.4986(+0.0206)	0.5644(+0.0055)
{1}	0.2999(+0.0307)	0.4646(−0.0134)	0.5096(−0.0493)
{2}	0.2600(+0.0092)	0.4768(−0.0012)	0.5568(−0.0021)
{3}	0.2462(−0.0230)	0.4600(−0.0180)	0.5529(−0.0060)
{0, 1}	0.3230(+0.0538)	0.4827(+0.0047)	0.5223(−0.0366)
{0, 2}	0.2999(+0.0307)	0.4886(+0.0106)	0.5480(−0.0109)
{0, 3}	0.3076(+0.0384)	0.5063(+0.0283)	0.5704(+0.0115)
{1, 2}	0.3230(+0.0538)	0.4820(+0.0040)	0.5212(−0.0377)
{1, 3}	0.3230(+0.0538)	0.4879(+0.0099)	0.5300(−0.0289)
{2, 3}	0.2616(+0.0076)	0.4736(−0.0044)	0.5594(+0.0005)
{0, 1, 2}	0.3307(+0.0615)	0.4847(+0.0067)	0.5206(−0.0383)
{0, 1, 3}	0.3538(+0.0846)	0.4951(+0.0171)	0.5234(−0.0355)
{0, 2, 3}	0.2999(+0.0307)	0.4933(+0.0153)	0.5557(−0.0032)
{1, 2, 3}	0.3307(+0.0615)	0.4877(+0.0097)	0.5250(−0.0339)
Average	(+0.0373)	(+0.0069)	(−0.0199)
Maximum	(+0.0846)	(+0.0283)	(+0.0115)
Minimum	(+0.0230)	(−0.0180)	(−0.0493)

Table 7 Difference of recall of class 1, Gmean, and AUC compared with the orginal imbalanced dataset

	CaO	SMOTE	KmeanSMOTE
Recall of class 1	+0.0846	+0.0505	+0.0619
Gmean	+0.0171	+0.0211	+0.0281
AUC	−0.0355	+0.0134	+0.0140

The efficiency of CaO depended on two main parameters, which are the number of clusters and the resampling ratio. Determining too much number of clusters leads to the scattered synthesis of new sample units. On the other hand, the result will not differ from SMOTE if the number of clusters is too small. The proper number of clusters and resampling ratio will synthesize new samples in the right concept area and make the decision better. Future research could adjust these two parameters and determine other parameters that could improve the performance of the proposed algorithm.

Acknowledgements The researchers would like to thank the Department of Computer Science, Faculty of Science, Kasetsart University in Thailand, for their support and advice. Furthermore, the researchers greatly appreciate the assistance of the Department of Juvenile Observation and Protection, Ministry of Justice (Thailand), for their contributions in providing the Juvenile Delinquency Dataset.

References

1. S. Sawanyathipat, *Juvenile offending against property, narcotics and violent crimes in bangkok and peripheral area: a study of self-control theoretical implications*, Ph.D thesis, Mahidol University, Thailand (2011)
2. W. Lholong-in, K. Praiwan, S. Sripathar, P. Buraphan, Repeated offending of jail inmates in Maha Sarakham provincial prison. Rajabhat Maha Sarakham Univ. J. **3**(1), 29–38 (2009)
3. N.V. Chawla, Data mining for imbalanced dataset: an overview, in *Data Mining and Knowledge Discovery Handbook* (Springer, Boston, 2005)
4. H. He, E.A. Garcia, Learning from imbalanced data. IEEE Trans. Knowl. Data Eng. 1263–1284 (2009)
5. M. Kubat, R.C. Holte, S. Matwin, Machine learning for the detection of oil spills in satellite radar images. Mach. Learn. **30**, 195–215 (1998)
6. X.Y. Liu, J. Wu, Z.H. Zhou, Exploratory under sampling for class imbalance learning, in Proceedings of International Conference on Data Mining (2006), pp. 965–969
7. N.V. Chawla, K.W. Bowyer, L.O. Hall, W.P. Kegelmeyer, SMOTE: synthetic minority oversampling technique. J. Artif. Intell. Res. **16**, 321–357 (2002)
8. H. Han, W.Y. Wang, B.H. Mao, Borderline-SMOTE: a new over-sampling method in imbalanced data sets learning, in Proceedings of International Conference on Intelligent Computing (2005), pp. 878–887
9. S. Piri, D. Delen, T. Liu, A synthetic informative minority over-sampling (SIMO), in *Decision Support Systems*
10. A.F. Kallappanamatt, Application of synthetic informative minority over-sampling (SIMO) algorithm leveraging support vector machine (SVM) on small datasets with class imbalance. M.Sc. in Computing (Data Analytics), DIT (2018)

11. S. Wang, L.L. Minku, X. Yao, Resampling-based ensemble methods for online class imbalance learning. Data Eng. **27**(5) (2015)
12. P. Lim, C.K. Goh, K.C., Tan evolutionary cluster-based synthetic oversampling ensemble (ECO-Ensemble) for imbalance learning. Cybernetics **47**(9) (2017)
13. I. Tomek, Two Modifications of CNN. IEEE Trans. Syst. Man Cybern. **6**, 769–772 (1976)
14. M. Kubat, S. Matwin Addressing the curse of imbalanced training sets: one-sided selection, in Proceedings of International Conference on Machine Learning (1997), pp. 179–186
15. X. Tao, Q. Li, W. Guo, C. Ren, Q. He, R. Liu, J. Zou, Adaptive weighted over-sampling for imbalanced datasets based on density peaks clustering with heuristics filtering. Inf. Sci. **519**, 43–73 (2020)
16. Y. Sanguanmak, A. Hanskunatai, DBSM: The combination of DBSCAN and SMOTE for imbalanced data classification, in *13th International Joint Conference on Computer Science and Software Engineering (JCSSE)*, Khon Kaen (2016), pp. 1–5
17. G. Douzas, F. Bacao, F. Last, Improving imbalanced learning through a heuristic oversampling method based on k-mean and SMOTE. Inf. Sci. **465**, 1–20 (2018)
18. S. Wang, L.L. Minku, X. Yao A systematic study of online class imbalance learning with concept drift. Neural Netw. Learning Syst. **29**(10) (2018)
19. A.S. Iwashita, V.H.C.de Albuquerque, J. Paulo Papa, Learning concept drift with ensembles of optimum-path forest-based classifiers. Future Gener. Comput. Syst. **95**, 198–211
20. Y. Sun, M.S. Kamel, Y. Wang, Boosting for learning multiple classes with imbalanced class distribution, Proccedings of International Conference on Data Mining (2006)
21. R. Barandela, J.S. Sánchez, V. Garcıa, E. Rangel, Strategies for learning in class imbalance problems. Pattern Recogn. **36**, 849–851 (2003)

An Experimental Stack Overflow Chatbot Architecture Using NLP Techniques

Rahul Namboodiri, Kanishka Singla, and Priyanka Verma

Abstract In computer programming, it is very common to come across faults or bugs that may result in unintended outcomes or none at all. To get the program to function correctly so as to obtain the desirable results, debugging is vital. In some scenarios, errors are noticeable, obvious, and easy to understand, while in others, with only a cryptic code present, it sometimes becomes tedious to debug. While there are several community question answering (CQA) websites such as Stack Overflow, it is noted that there is a general difficulty in finding the appropriate post pertaining to user's query, especially for beginners. In this paper, we have proposed and implemented a task-specific chatbot model that is also capable of handling general conversations. The bot uses the community question answering website Stack Overflow's posts to be capable of answering programming related queries. For implementation of the proposed model, a custom Stack Overflow general dialog dataset has been used. An intent classifier classifies the user's input, and if the user's query is unique to a task, the task-specific classifier classifies the programming language for which the question is being asked. Long short-term memory (LSTM), a powerful Recurrent Neural Network (RNN) deep learning model, has been utilized to achieve this classification. Additionally, word embeddings enable transformation of the dataset and users' queries into word vectors to facilitate similarity-based fetching. The experimental analysis indicated that the chatbot could identify user motives with 99.15% accuracy and the question programming language with 82.77% accuracy.

Keywords Chatbot · Deep learning · LSTM · Machine learning · TF-IDF · Word embeddings

R. Namboodiri (✉) · K. Singla · P. Verma
Mukesh Patel School of Technology Management and Engineering, Narsee Monjee Institute of Management Studies, Mumbai, Maharashtra, India

P. Verma
e-mail: priyanka.verma@nmims.edu

M. Paprzycki et al. (eds.), *Intelligent Systems, Technologies and Applications*, Advances in Intelligent Systems and Computing 1353,
https://doi.org/10.1007/978-981-16-0730-1_23

1 Introduction

Conversational agents in the form of personal assistants and chatbots have increasingly been implemented as a means of interaction between humans and machines due to the improvement of functionality and increasing the naturalness of human–computer interaction. In 1950, Alan Turing in his paper, Computing Machinery and Intelligence, theorized a test to answer the question "can machines think" by means of substituting it with a test, a game he called the imitation game where to pass, the machine has to be completely indistinguishable from a human in a textual conversation. His work influenced the development of chatbots when the field was in its infancy, with bots increasingly being given frameworks to simulate human speech patterns and human conversation.

Modern chatbots have developed with the overall advancement of natural language processing, with new levels of computational power and the advancement of the field of artificial intelligence to utilize this new found computing power. IBM's Watson became one of the first major milestone bots in 2006, then created to play jeopardy and has now evolved into a complete voice and textual AI platform along with Apple's Siri, Google's Google Now, and Microsoft's Cortana. We increasingly notice a shift from making conversational AI, an emulation of human speaking behavior to combination of it with specific tasks. Modern chatbots really took-off with social media chatting platforms such as Facebook Messenger, Telegram, and WhatsApp that began creating bot extensions in their APIs, allowing for user case-specific customized bots for business purposes. This along with the development of cloud computing has allowed for the creation of highly efficient and accurate bots capable of processing large datasets with high speed allowing for increasingly seamless experiences.

With the advent of the Internet, the amount of data has increased exponentially with users now searching the Internet for query-specific information. Subsequently to cater for the same, a large number of websites have been launched, using community-based questioning and answering solutions with a particular niche topic. Taking the example of Stack Overflow, we found that there are a vast number of CQA posts on a huge range of topics and languages even in its programming language niche, and therefore, finding the desired result became difficult, due to having an absence of knowledge on the encountered problem. We therefore propose a chatbot system as an alternative to simplify the experience of searching through the CQA posts of Stack Overflow using deep learning and allowing the user to have easier access by connecting it to mobile messaging app such as Telegram. This improves accessibility by not requiring the user to be on the said website or have a dedicated app. This way the bot can improve upon the user experience to search through such CQA websites and give them an improved accessibility empowering the capability of the same.

2 Related Work

The field of computation capable of conversational abilities stretch far back to 1950 with Alan Turing's controversial but influential paper [1] where he discusses the philosophical aspects of the capability of machines to think, and devices a test to answer his question called the imitation game, with the goal of the test to see if a computational code can mimic human conversational behavior.

The first known application of a chatbot was in the mid-1960s, when an MIT Professor Joseph Weizenbaum built ELIZA [2] which simulated dialogs with the base concepts being pattern matching and substitution performed on a defined set of keywords. Modeled to act as a therapist, the system identified keywords from a query made by the user and applied matching for retrieval with or without string manipulation. While the techniques used are capable, it did not possess true understanding of the conversation. Another well-known traditional three times Loebner Prize winning bot, A.L.I.C.E [3] was a chatbot system developed by Richard Wallace in 1995. With ELIZA as its inspiration, the bot uses heuristic pattern matching to find the finest match to the user input and respond in the most human-like way possible. Information about conversion patterns is stored under subunits of AIML files. These subunits are called categories and contain the rules for matching and conversion in order to obtain the desired output. While this allowed for a very capable bot, it required manual programming for its rules and essentially the bot conversed due to the extensive efforts in designing over 50,000 categories by the end of the project resulting in its fluid humanistic conversational style.

While the traditional approaches focused heavily on mimicking human conversation, with the advent of the Internet, chatbots started becoming more about providing certain services through the Internet, aiding user experience. In a closely related work, Suta et al. [4] used the highest similarity score to fetch potential answers to queries based on Python programming made by the user, from Stack Overflow website. A corpus of 200,000 Q&A statements acquired from Kaggle was fed to Doc2Vec model whose function is to cumulate all the words in a sentence into multi-dimensional vectors. While the model designed is capable, Doc2Vec's limitation of being unable to handle out-of-vocabulary words and limitations in the dataset results in a validation accuracy of 50%.

In 2019, in an expert recommendation system, Cerezo et al. [5] implemented term frequency—inverse document frequency (TF-IDF) algorithms for classification of sentences from a TF-IDF dataset and used Discord API to link the system to the Pharo ecosystem. The system, although was accepted by most of the respondents, was still expected to be more conversational. This becomes challenging due to the phenomenon of "uncanny valley effect" which states that humanoids that possess human-like characteristics have a tendency to evoke a sense of uneasiness to its users. The goal would be to either improve further to pass this phase or stay behind it to maintain its functionality while still appearing for what it is as a bot providing a service.

Another 2018 work by Petre Anghelescu et al. [6] is centered around the designing of a conversational self-learning chatbot, a bot that learns new information as it is fed into it enabling fair imitation of a human. The designed chatbot is specifically targeting to be more human-like with its ability to put a randomized delay time before replying, and learns answers to previously unknown queries directly from the user by requesting an input and updating its SQL database. With this, it retains a humanistic touch, and its utilization of search engine algorithms ensures appropriate answers for known questions. Nevertheless, it is limited by its need for extensive manual training before being utilizable and its complete incapability of handling unknown information.

Developed in 2017, MILABOT [7] is an award-winning deep reinforcement learning chatbot that responds through both text as well as speech. The conversational system works on the application of neural networks as well as template-based models for retrieval of data from crowd sourced data.

Panitan et al. [8] built an FAQs Chatbot by employing RNN-LSTM to extract responses taking Thai text questions as their database. The system is capable of identifying queries with 86.36% accuracy and providing answers with an accuracy of 93.2%. The use of sequential RNN models allows it to learn an impressive amount of detail allowing it to achieve successful classification and query answering,

Further, the field has increasingly begun to merge with internet of things (IoT) with chatbots capable of controlling devices and thereby allowing Internet-based control of devices. An IoT based office environment was described by Muslih et al. [9] that uses a chatbot as the interface for communication between devices. Telegram bot by Telegram Messenger has been used as the user interface. In a similar work, a home automation system that functions conjointly with a chatbot was developed by Baby et al. [10]. NLP techniques like parsing, tokenization of words, and elimination of stop words were used for preprocessing the data while Raspberry Pi board along with assorted sensors and relays were used to allow for automation of home appliances.

3 Proposed Chatbot Architecture

3.1 System Description

The flowchart (Fig. 1) draws up the general architecture of the chatbot system created. The system requires a server machine for communication between the back-end programming and the Telegram front-end. To generate the bot, Telegram's Bot Father API was used for generation of the API key with built-in functions to help fetch and send data to the bot. The system was designed with portability and accessibility in mind and hereby uses a universal app that is available across all platforms.

When a user inputs a query into the bot, the data is fetched from the Telegram server using the API key to the back-end. This user query is then passed through a binary intent classifier for classification of user's intention being either a basic

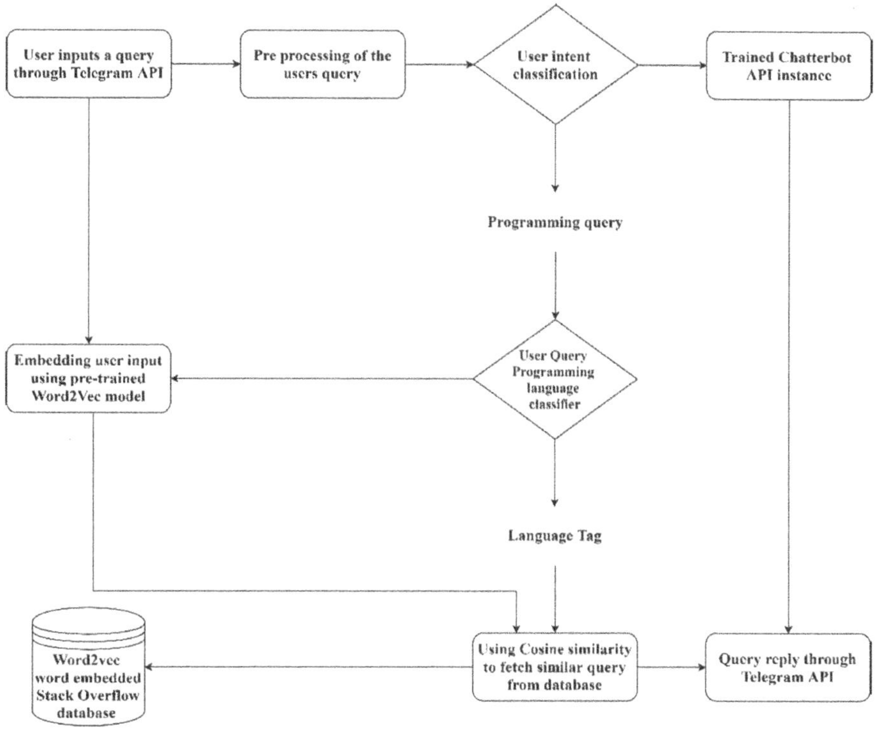

Fig. 1 Generalized architecture of the proposed chatbot system

conversation or a programming query. For a simple conversational dialog, the query is passed to the trained chatterbot API instance which responds to the user's query on Telegram. If the user's query is regarding programming, the programming language classifier classifies the user's query into one of the ten classes available (C#, C++, Java, JavaScript, PHP, Python, R, Ruby, Swift, and Visual Basic). Now, the original dataset has been converted into word embeddings using Word2Vec technique, using Google's pre-trained word embeddings. This is done in such a way that the large dataset is split into its ten component classes, i.e., the programming languages into which the query is to be classified to enable faster searching. Now for programming language query, the user query is also converted to word embeddings, and along with the programming language tag after classification, the two together are used for similarity matching by using cosine similarity. The similar user query then fetched is passed to the Telegram API as a reply to the user.

In our work, we devised two distinct approaches for the classification section—using machine learning algorithms along with TF-IDF for feature vector generation and using a sequential deep learning model with an embedding layer for feature extraction and LSTM layer for utilizing sequential textual data. The techniques used and the stages required have been demonstrated with the help of the flowchart.

3.2 Data and Pre-processing

Implementation of the proposed dually capable model has been carried out by using two custom databases, one having 218,609 conversational dialogs and the other comprising of 2,171,575 Stack Overflow posts along with their corresponding solution post IDs, which were included within Stack Overflow database. For the purpose of classification of user intent, a merger of the above two datasets containing 200,000 dialogs and 200,000 Stack Overflow posts was used. The sizes were limited to prevent class imbalance. The Stack Overflow dataset consists of posts in total of ten programming languages, hence containing ten classes and is available at [11].

Both the databases show the sequence numbers for rows present. Queries put up on the Stack Overflow website are stored under the 'Title' column and have unique identification numbers given by their 'Post ID'. Likewise, the dialog database has a column 'Text' that holds all the conversational sentences. 'Tags' have been assigned to either database that indicates the programming language for Stack Overflow database and the type of communication for the dialog database.

Now, before we begin using the datasets for extraction of any textual feature points, we require the data to be processed. This pre-processing can be performed using many different methodologies and techniques. In our work, we have used the following:

1. **Data cleaning**. Initially, columns of both the datasets containing sentences are integrated after which it is processed to get rid of redundancy which may have arisen due to repeated posts. To achieve this, all the characters in the consolidated dataset are converted to lower-case. Further, cleaning of noisy characters was performed using regular expressions. A regular expression is a string of alphanumeric characters that facilitates convenient identification of patterns and help in management of the same. In our work, a pattern of generally insignificant symbols was created on which employing pattern matching enabled the removal of these symbols from the consolidated data resulting in a clean dataset.
2. **Removal of stop words**. This stage allows us to eliminate words that are of least significance, but are the most repetitive across the corpus. Python's Natural Language Toolkit (NLTK) library stores these stop words in 16 different languages and helps in filtering the same out of the corpus with ease. This is necessary for avoiding weighting common grammatical words with high significance, preventing skewing of information.
3. **Tokenization**. Tokenization, an important pre-processing stage of natural language processing was carried out by splitting each sentence into its individual word components. Tokenization enables us to use bag-of-words approach and related models such as n-gram, skip-gram, and TF-IDF for feature extraction purposes.

3.3 *Classification*

For the purpose of classification, we have experimented two distinct approaches. Firstly, in our work, we have proposed a combination of TF-IDF for textual feature extraction and passing the TF-IDF feature vector through machine learning classifiers for the purpose of intent and programming language classification. The second approach uses recurrent neural networks (RNN) in the form of long short-term memory networks (LSTM) with four layers in total. The four layers being—an input layer in the form of an embedding layer, a bi-directional LSTM layer, a dense hidden layer, and a dense output layer. The embedding layer generates the word embedding features required further by the network for better classification.

First approach. The machine learning-based classification system can be split further into two more stages—feature extraction and classification with machine learning algorithms.

Feature extraction

For facilitating analysis, complications within the data need to be removed for which non-essential features should be eliminated under a procedure termed as feature extraction. To obtain the necessary feature vectors, we used the approach of TF-IDF (term frequency-inverse document frequency). This information retrieval technique allows weighting of words based upon the frequency of their occurrence in a corpus. This has been utilized for down weighting the common words across the entire corpus while giving importance to words common in the document but uncommon across the corpus. This results in feature vectors with weights more representative of the different information present within. It has been derived by [12] as

$$\text{Term Frequency } (t, d) = \log(1 + \text{frequency}(t, d)) \tag{1}$$

$$\text{Inverse Document Frequency } (t, D) = \log\left(\frac{|D|}{1 + \{d \int D : t \int d\}}\right) \tag{2}$$

$$\text{TFIDF } (t, d, D) = \text{TF}(t, d) \times \text{IDF}(t, D) \tag{3}$$

where t is denoting the terms, d is denoting each document, and D denoting a collection of documents. Here, term frequency calculates the number of times each word appeared within each of the document. This is done to get the local frequency of the words. Inverse document frequency is calculating a ratio between the document space D with the total number of times a term t has appeared in all documents. 1 is added to avoid a zero-division case. We can clearly see here that this will upweight the values of terms rare across all documents but frequent within its own document. The product of the two yields the TF-IDF score.

Table 1 Comparison of accuracies obtained upon application of machine learning classifiers to both, intent and programming classifiers

S. No.	Machine learning classifier	Accuracy (intent) (%)	Accuracy (programming) (%)
1	Logistic regression	98.66	79.42
2	K-nearest neighbors	76.93	60.02
3	Random forest	88.27	75.267
4	Support vector machines	99.05	80.293

Classification

The TF-IDF feature vectors acquired from both the datasets, i.e., conversational as well as posts, were passed in turn to different machine learning classifiers for an assessment of their performance. Logistic regression, support vector machines, K-nearest neighbors, and random forest were the algorithms implemented in our work. The training and validation split were maintained at 80–20% throughout all classifiers.

Table 1 shows the classifiers along with the accuracies obtained upon their application. Hyperparameter tuning was avoided, and default values were taken as provided in the Scikit-learn library for Python, due to mounting computational cost resulting in extremely high wall times on 10th Gen Intel i5 10,210 U processors with over 3 h of training time per classifier in general.

Second approach. For the deep learning approach, recurrent neural networks (RNN) in the form of long short-term memory (LSTM) model was used due to the design of RNN architecture allowing it to make use of and learn sequential data, with LSTM allowing it to learn longer sequences of inputs due to its three-gated structure.

Before passing the tokenized textual data to the designed sequential model, we require certain further stages of pre-processing. Every single word in the corpus vocabulary must first be converted into an integer token, and a corresponding dictionary must be created mapping the integer tokens with their corresponding vocabulary words. Further, a selection of a smaller vocabulary on basis of repetition throughout the corpus is selected for reduction of less consequential data points. This vocabulary size has been varied from 5000 to 100,000 for the purpose of hyperparameter tuning. Further, we need to replace the text present in the dataset with its corresponding integer token and replace out-of-vocabulary words with a common token. Additionally, for having a fixed input to the LSTM's input layer, we fix a word length of 200 words per subsection of the corpus, and padding with out-of-vocabulary token is performed if the number of words in the subsection is shorter than the same.

The designed four-layer sequential model (Fig. 2) consists firstly an embedding layer as the input layer, with its parameters as vocabulary size of the dataset and

Fig. 2 Generalized representation of the four-layer model with layer detail format as (input, output)

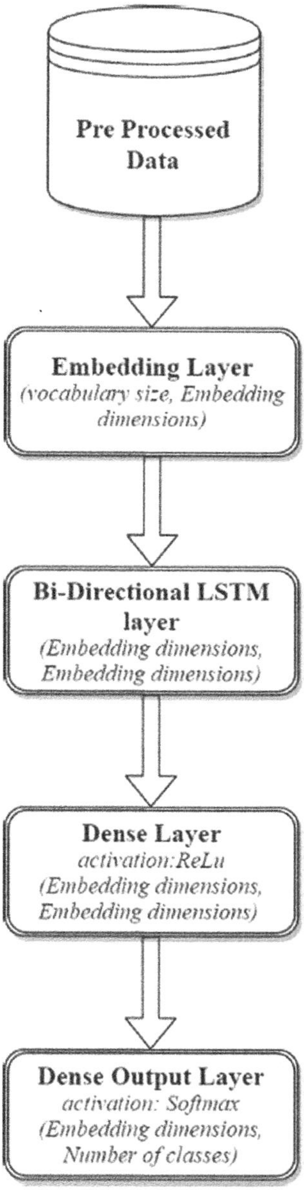

the dimensionality of embeddings produced as the output. The two parameters here majorly define the complexity of the model and have been variably tuned to produce the best results. The second layer, a bi-directional LSTM layer, takes the dimensionality of the word embeddings produced by the input layer as its input and uses its capability to process sequential data and its longer memory compared to a simple

RNN model due to the LSTM architecture containing three-gated units compared to the single gated unit on a simple RNN structure and two on the gated recurrent unit (GRU).

The third layer, a dense connecting layer is added with its input size equal to the embedding dimensions with the activation function as rectified linear unit (ReLu). This was done to reduce the computational cost due to computational limitations as ReLu which is much less computationally expensive than training with tanh activation function.

The final layer, a dense layer, is added as the output layer with its layer neuron count equal to the number of classes (2 for intent classification and 10 for programming language classification). The activation function is set to SoftMax for providing outputs as probabilistic values for each class.

Hyperparameter tuning

For enhancing the performance of models without overfitting or maximized losses, selecting an optimal combination of parameters plays a crucial role. Precise hyperparameter tuning assures an overall desirable behavior of the algorithms. We carried out optimization for both the classifiers (intent and programming) till the highest possible accuracies were reached. Following are the parameters tuned in the proposed method:

1. Vocabulary size: Assessment of performance was done keeping vocabulary size fixed at 5,000 and 10,000, and peak value with minimal loss was obtained at 10,000 words for intent classifier. In the case of the programming classifier, due to the size of the dataset, a vocabulary of 15,000, 30,000, 50,000, and 100,000 was tested, and 100,000 was taken as the optimal value. The starting vocabulary size was taken corresponding to the size of the dataset with the Stack Overflow dataset being much larger, hence requiring a larger size for better performance.
2. Embedding dimensions: Dimensions of embedded vectors were adjusted to 64, 100, and 200 with 100 being the optimum in case of intent classifier. For programming classifier, embedding dimension was set to 32 because of the large computational cost making higher sizes unfeasible due to high wall times for training.
3. Epochs: For both classifiers, training was performed up until the tenth epoch due to the peak accuracies appearing much before the same. After assessment, the peak value was obtained at third epoch for intent classification with the corresponding embedding dimensions and vocabulary, while for programming classifier it was obtained at fourth epoch. Training was performed using Nvidia Mx 250 GPU with 2 Gb of dedicated memory using TensorFlow 2.0 and CUDA DNN 10.0.0.

Here in our work, the manual parameter tuning approach was taken, and after tuning the parameters to their optimal values, maximum accuracy of 99.15% was obtained during intent classification and 82.77% accuracy was obtained in the case of programming language classification. Graphical representation of loss and accuracy after hyperparameter tuning for both classifications are as shown below (Figs. 3 and 4).

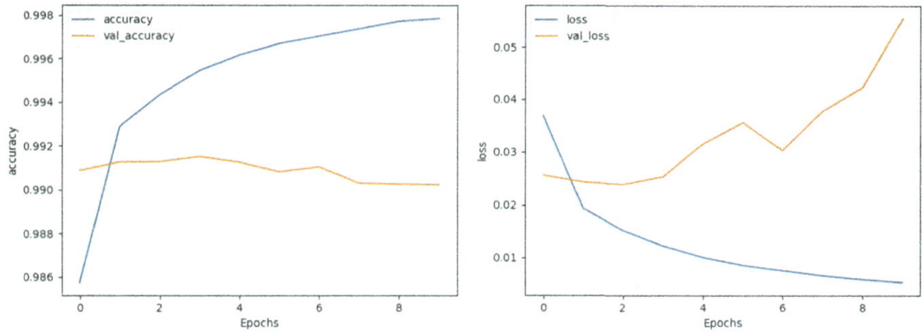

Fig. 3 Test versus validation accuracies (left) and test versus validation losses (right) for vocabulary size of 10,000 words with embedding dimensions 100 for intent classification

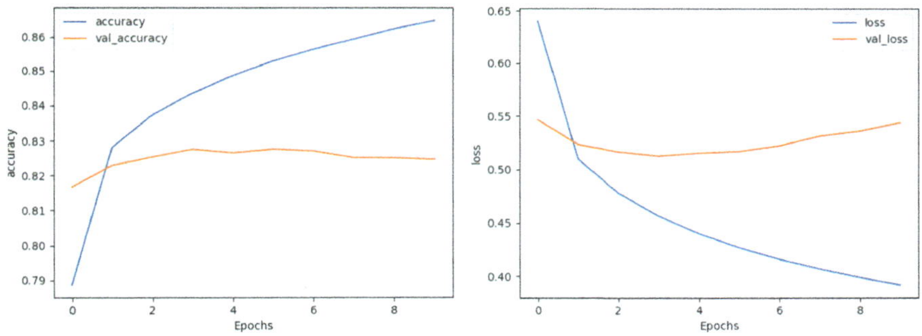

Fig. 4 Test versus validation accuracies (left) and test versus validation losses (right) for vocabulary size of 100,000 words with embedding dimensions 32 for programming classification

Further, a tabulation of all the accuracies and loss values for intent and programming classifiers are as follows (Tables 2 and 3).

Word Embeddings

For fetching of similar queries from the programming database in our, we have converted the original Stack Overflow database into word vectors by using the method

Table 2 Obtained accuracy and loss values for all the sets of hyperparameter settings used for intent classification

S. No.	Parameters (vocabulary, embedding dimensions)	Validation accuracy (%)	Validation loss
1	(5000, 100)	98.895	0.0302
2	(5000, 200)	99.15	0.0235
3	(10,000, 100)	99.13	0.0238
4	(10,000, 200)	99.15	0.0232

Table 3 Obtained accuracy and loss values for all the sets of hyperparameter settings used for programming language classification

S. No.	Parameters (vocabulary, embedding dimensions)	Validation accuracy (%)	Validation loss
1	(15,000, 32)	81.022	0.5241
2	(30,000, 32)	82.44	0.5191
3	(50,000, 32)	82.528	0.5170
4	(100,000, 32)	82.77	0.5144

of Word2Vec word embeddings. This has been done in such a way that the dataset is split into sections on basis of the classes, i.e., programming languages for which the post had been created in the dataset. This is done so as to allow for utilization of the programming language classifier's output for faster searching through the dataset for a similar question, that is, the system will only search through the section belonging to the predicted language by the programming language classifier. This is to avoid bottlenecks resulting due to having to search over large datasets.

For performing the same, we have used Word2Vec embeddings pre-trained by Google on part of the Google News dataset. The embeddings generated were trained with a vocabulary of over three million words and phrases with a dimensionality set to 300. This model was trained using the continuous bag-of-words (CBOW) approach for obtaining word embeddings. We have used the same instead of training on our own custom dataset due to the sheer size of the Google dataset allowing us to avoid the out-of-vocabulary word problem suffered by Word2Vec models. This is necessary due to the possibility of variable phrasings and possibly uncommon words used by the user.

3.4 *Chatterbot API*

When the user's intent has been classified as a conversational dialog, the message is passed to the chatterbot API. Chatterbot API is a machine learning-based chatbot API capable of learning by either creating an untrained instance and learning from user interaction or by training it on any conversational dataset. The fact that the API is language independent allows for chatbot instances to be trained in any language based on user input or the provided dataset. For our work, we have used the Ubuntu dialog corpus for training the chatbot instance. The API after training on the dataset handles user's input by using machine learning to match user's input with a similar known input and then give a corresponding known response with a confidence value based on the matching level of the user's input and the known input by the bot upon training.

3.5 Telegram API

Finally, for the user interface, we have used Telegram's Bot Father API for creation of our Telegram-based bot. The interfacing with Telegram's API is rather direct, simple, and intuitive with a simple API key allowing us for communication with Telegram server. The information stored includes the user's name, the message ID, the text of the message, and other essential details such as the message timestamp. With update fetching and communication being quite easy and with no barriers for creating third-party bot applications, we picked Telegram for its ease of use and the fact that it allows our bot to be platform independent, portable, and due to Telegram being a widely used app it allows the bot to not require a dedicated app improving accessibility for users and increase in bot usage in general.

4 Demonstration and Discussion

The designed architecture requires a Web-connected server system for connecting the back-end software with the Telegram Bot API front-end, and the user end is multi-platform due to Telegram's versatility. The bot's capability is of two general functionalities, and they have been demonstrated in the images below (Fig. 6). The first is its capability to handle general conversations, which works through the chatterbot API, while the second demonstrates the bot replying to programming queries. This setup performs admirably with its limited database, and the classification and similarity-based fetching performance is good barring novel queries absent from the dataset. The bot's failings with the current design lie in its incapability to provide a response through Internet-based search for such novel queries. Further, it is currently incapable of using the full benefit of a CQA website such as Stack Overflow due to its inability to add new posts (Fig. 5).

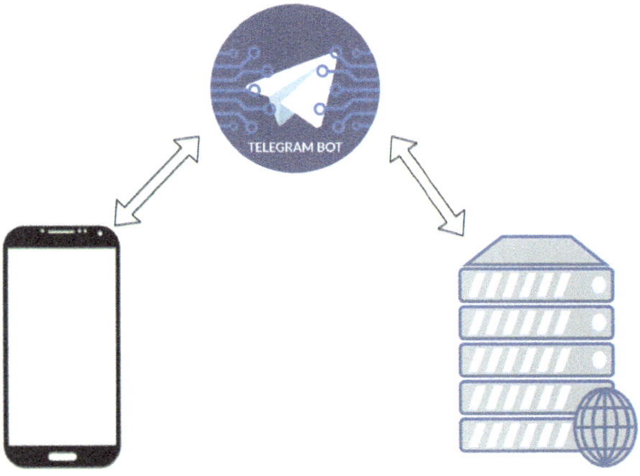

Fig. 5 Representative image of the interactions between user end and server end of the bot along with the Telegram API

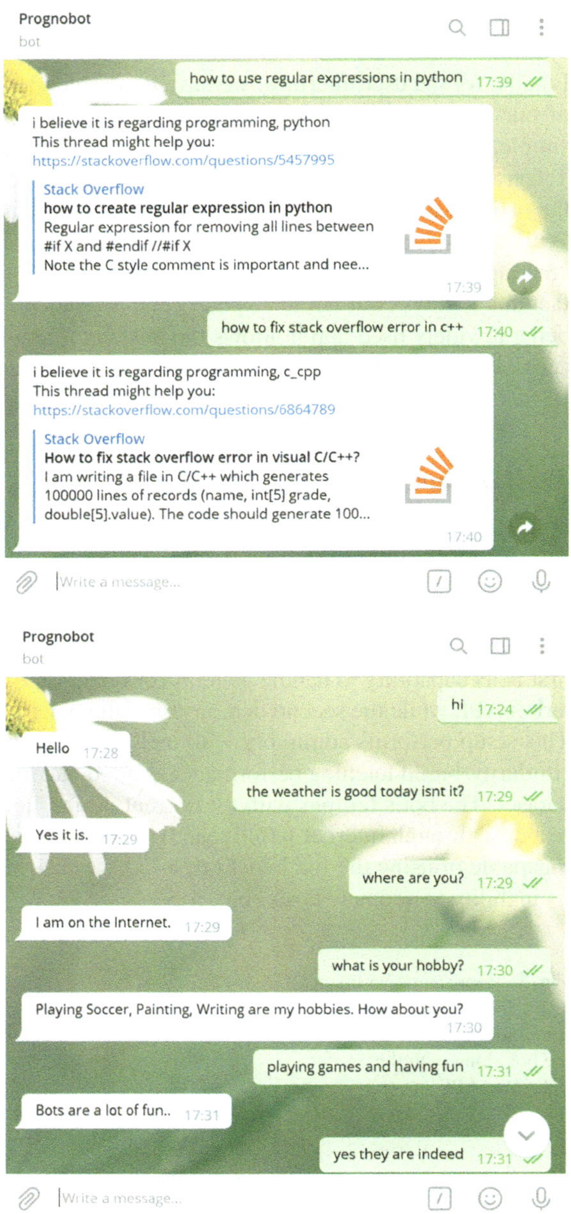

Fig. 6 Demonstration of the chatbot architecture on the user end—left demonstrates programming query answering while right demonstrates general conversational ability

The two distinct approaches to classification add a sense of versatility allowing this model to be tuned on both methods and select the best one for the application and its corresponding dataset. Here, the sequential LSTM-based model performed much better providing an intent classification accuracy of 99.15% and programming language classification accuracy of 82.77%.

5 Conclusion

In this research paper, we have proposed and implemented a chatbot architecture for a task-specific chatbot capable of handling basic human conversations as well as its specified task. We have built it with the capability to answer programming-related queries using community question answering website Stack Overflow's posts. The chatbot framework proposed consists of two datasets, one for chatbot's capability to hold human conversation and a custom Stack Overflow dataset. User's input is classified using an intent classifier, and if the user's query is task-specific, the task-specific classifier classifies which programming language the question has been asked for. This classification is achieved using a recurrent neural network (RNN) as a long short-term memory (LSTM) deep learning model. Further, word embeddings are used to convert the dataset and user's questions into word vectors to allow for similarity-based fetching. The experimental results show that the chatbot could classify user intent with 99.15% accuracy and query programming language with 82.77% accuracy.

6 Future Scope

The model designed needs to be tested further by performing user analysis study to see how the chatbot performs compared and to note how well it is capable to handle conversations and answer user queries appropriately as this would be a true metric for performance in terms of helpfulness, accuracy, and in total the capability of the chatbot itself. The chatbot architecture designed is versatile and has been made task-specific for answering Stack Overflow questions. With appropriate datasets and modifications, the architecture can work for completely different tasks such as answering queries regarding menus for a restaurant, for a product-search bot on an e-commerce website and other commercial uses. The bot architecture defined here would be more capable if connected to the company's datasets by means of an API access key, allowing the bot access to live updates of the data and increasing possible features that can be added. For example, with Stack Overflow, additional functionality of creation of posts by the user, fetching answers from a particular post, and so on can be added. Also providing the bot search functionality through search engine API's would allow for the bot to handle out-of-data queries by performing a search on the internet.

For classification, the LSTM network can be made deeper with additional layers for increased accuracy if the computational power is feasible. Also, dropout layers may be added to tackle the overfitting problem. Furthermore, state-of-the-art and more complex methods such as attention-based LSTM models may be used to improve classification accuracies. The dialog corpus for chatterbot API can be switched with larger conversational datasets for more life-like responses.

References

1. A.M. Turing, Computing machinery and intelligence, in *Parsing the Turing Test*, ed. by R. Epstein, G. Roberts, G. Beber (Springer, Dordrecht, 2009)
2. J. Weizenbaum, ELIZA—a computer program for the study of natural language communication between man and machine. Commun. ACM **9**(1), 36–45 (1966)
3. R.S. Wallace, The anatomy of ALICE, in *Parsing the Turing Test* (Springer, Netherlands, 2009), pp. 181–210
4. P. Suta, et al., Matching question and answer using similarity: an experiment with stack overflow, in *2018 IEEE International WIE Conference on Electrical and Computer Engineering (WIECON-ECE)* (IEEE, 2018)
5. J. Cerezo, et al., Building an expert recommender chatbot, in *2019 IEEE/ACM 1st International Workshop on Bots in Software Engineering (BotSE)* (IEEE, 2019)
6. P. Anghelescu, S.V. Nicolaescu, Chatbot application using search engines and teaching methods, in *2018 10th International Conference on Electronics, Computers and Artificial Intelligence (ECAI)* (IEEE, 2018)
7. I.V. Serban, et al., A deep reinforcement learning chatbot (2017). arXiv preprint arXiv:1709.02349
8. P. Muangkammuen, N. Intiruk, K.R. Saikaew, Automated thai-faq chatbot using RNN-LSTM, in *2018 22nd International Computer Science and Engineering Conference (ICSEC)* (2018)
9. M. Muslih, et al., Developing smart workspace based IOT with artificial intelligence using telegram chatbot, in *2018 International Conference on Computing, Engineering, and Design (ICCED)* (IEEE, 2018)
10. C.J. Baby, F.A. Khan, J.N. Swathi, Home automation using IoT and a chatbot using natural language processing, in *2017 Innovations in Power and Advanced Computing Technologies (i-PACT)*, 21 April 2017. IEEE, pp. 1–6
11. https://www.kaggle.com/zethione/stack-overflow-bot-dataset
12. A. Aizawa, An information-theoretic perspective of tf–idf measures. Inf. Process. Manage. **39**(1), 45–65 (2003)

Empirical Analysis of Performance of MT Systems and Its Metrics for English to Bengali: A Black Box-Based Approach

Goutam Datta, Nisheeth Joshi, and Kusum Gupta

Abstract There are numerous use cases of machine translation (MT) systems. Therefore, it has become very important to evaluate the performance of MT which can help researchers design a robust and reliable machine translation system. Although there are number of automatic MT evaluation metrics available nowadays, most of these fail to produce correct score. In this paper, we shall describe the most recent framework in MT industry that is Neural Machine Translation, and our approach is to test the performances of most popular translators—Google Translate and Microsoft's Bing translator. We use language pairs English and Bengali for our detail analysis. Experiments are performed to translate English to Bengali. Bengali is a resource poor and one of the most widely spoken language in Indian subcontinent. Unlike glass box approach, where the performance of the MT systems are tried to enhance by adjusting various hyper parameters, we will be using black box approach, i.e., evaluating the performance of the already built systems. Beside performance analysis of the two translators, our main focus is to evaluate the performance of one of the very popular automatic evaluators Bilingual Under Study (BLEU) and some other automatic evaluation metrics in our work. This paper aims to measure the performance of BLEU and other automatic metrics during English to Bengali translation process by conducting survey with the help of questionnaires distributed among twenty people having moderate to high linguistic expertise on both the languages. Their responses are collected, and mean score is calculated which we have considered as human score (human judgement). BLEU and scores generated by other automatic metrics are used and compared their performance with human-generated score. A correlation is measured between human score and other metrics with Pearson correlation coefficient. Finally, some important observations are reported for this language pairs.

G. Datta (✉)
School of Computer Science, University of Petroleum and Energy Studies, Dehradun, India

N. Joshi · K. Gupta
Department of Mathematical and Computer Science, Banasthali Vidyapeeth, Vanasthali, Rajasthan, India
e-mail: jnisheeth@banasthali.in

K. Gupta
e-mail: kusumgupta@banasthali.in

M. Paprzycki et al. (eds.), *Intelligent Systems, Technologies and Applications*, Advances in Intelligent Systems and Computing 1353,
https://doi.org/10.1007/978-981-16-0730-1_24

Keywords Neural machine translation · Cosine similarity · F-measure · Precision · BLEU

1 Introduction

MT translates source language to target language automatically. Various attempts have been made in the past to automate the translation process. There are many challenges in MT design. Some of the linguistic challenges are language understanding, language generation and language pair mapping [1]. Translation industry has witnessed different phases of MT since its inception. But none of translator developed so far are full proof in terms of adequacy and fluency. Even numerous researches have taken place to automate the evaluation process of MT system. Human evaluation is considered to be the best but extremely time consuming. This paper attempts to measure the performance of one of the most popular automatic MT evaluation metric BLEU and some other automatic metrics. The performance of BLEU and other automatic metrics are measured by comparing the scores produced for the same set of hypothesis by BLEU and other automatic metrics against human-generated score (human judgment). Many researchers claimed that often BLEU scores are misleading and hence can't be relied completely [2]. A detailed statistical analysis has been presented for the score produced by automatic metrics and human judgment. We have used popular translators—Google Translate and Microsoft's Bing translator. The language pairs we have selected English and Bengali. Bengali is one of the low resource languages. Bengali is very widely spoken language in Indian subcontinent and much material on Bengali is not available on net. Hence, we found developing better MT for English to Bengali is very necessary. Previously, many researchers have already worked in this direction to measure the performance of automatic evaluation metrics for different language pairs. Because correct automatic evaluation of MT systems helps researcher to design robust and better MT systems. Such type of meta-evaluation, i.e., finding correlation between automatic evaluation, and human score has already been done before and some important observations are reported by researchers [3]. Here our approach is to do such meta evaluation for English→Bengali language pair.

The remaining part of the paper is organized as follows. Phases of Machine Translation (MT) systems (which we have used in later section of this paper) and some work on MT evaluation metrics are introduced in Sect. 2. Section 3 briefly describes the linguistic differences between the two languages English and Bengali. Section 4 briefs some evaluation metrics we have used in our work. Methodology is described in Sect. 5. Section 6 describes data and our experiment. Section 7 includes detail analysis and discussion. Finally, we conclude our work in Sect. 8.

2 Phases of MT Systems and Some Related Work on Evaluation Metrics

The first statistical machine translation (SMT) system was developed by a group of researchers from IBM in 1980s. It combined two models, one is language model and the other is translation model. Early SMT models ignored the linguistic aspects, although it tried to incorporate linguistic aspect in language or in translation model later on [4]. SMT systems which is mainly based on probabilistic-based approach select the most probable word y given a set of words x_i where $i = 1$ to n can be represented as $P(y|x_i)$. In SMT, phrase-based SMT outperforms word-based SMT [5].

There was a novel framework in MT system which introduced kernel regression technique. In the kernel-based regression, the source language string is mapped to a vector in to an input feature space. This vector is then mapped to a vector in the output feature space with the help of suitable regression function. Finally, output feature space vector is mapped to target language. Kernel-based method in MT system produced encouraging results for French–English language pair [6]. Subsequently, neural network was introduced in the language model and researcher reported significant improvement in the performance over n-gram model on two corpora [7]. The performance of NMT is better than traditional phrase-based SMT when training data is large. In terms of fluency, NMT outperforms SMT. On the other hand, in terms of adequacy SMT outperforms NMT. Hence, researchers combined these two approaches and claimed to significantly improve the performance for Chinese to English translation task [8]. Under linguistic consideration, NMT uses an end-to-end encoder decoder approach to automate source to target translation task. The encoder takes the variable length input string and converts into fixed length string. Decoder uses this fixed length string to generate output. The encoder–decoder contains two recurrent neural networks (RNNs). There is one attention mechanism which is used to align target with the source sentence. The performance of NMT degrades with larger sentence size and with larger unknown words [9–13]. In NMT, word-level training demands exact matching and this problem is nicely addressed by the researchers and claimed to achieve significant improvement on Chinese → English and English →German datasets [14].

Some researchers even claimed that NMT's encoder part captures few syntactic information during training at the source side and can perform well without having prior linguistic information for some language pairs [15].

Contrast to this, some researchers claim prior linguistic information improves NMT model for other language pairs [15]. Despite being the fact, NMT outperforms SMT but it's performance degrades for the low resource languages. Even the language pairs where there is much syntactic differences, morphologically rich and don't have much data for such types of languages NMT's performance drops [16]. Bengali being a low resource language and we have selected this language in our study.

NMT has some major drawbacks like handling rare words, large training time, handling unknown words, domain mismatch problem, handling large sentences, etc.

Evaluating translations generated by MT systems are more challenging tasks. A lot of research has been done in this direction to find the correlation between human and auto-generated score and how to maximize the correlation between human and automatic evaluation.

In another paper, researchers used various metrics such as BLEU, F-measure, Word Error Rate (WER). They used SVM-based classifier and claimed to improve the correlation between SVM and human score compared to other metrics in the MT evaluation process [17]. In other paper, researchers proposed a pairwise neural network-based approach in MT evaluation process and claimed to improve the correlation with human score [18]. It has also been found that metrics that have performed well in evaluating adequacy also performed well in evaluating fluency [19].

In MT evaluation, with word embedding and Earth Movers Distance (EMD), researchers designed automatic MT system which can evaluate on the basis of word meaning. Researchers claimed their method achieved highest correlation with human judgment [19].

3 Linguistic Features of English and Bengali

English and Bengali are two different types of languages. There is phonetic difference also. Because of phonetic difference transliteration is very important in many natural language processing applications. Transliteration is a process of transcribing a source language exactly as it's phonetic to target language [20]. There are many linguistic differences between English and Bengali. Some of the prominent differences are highlighted here. In few cases, similarity may be noticed with English and other Indian languages such as Hindi also [21].

Sentence structure or word ordering: Linguistically, English is considered to be subject verb object (SVO). On the other hand, Bengali is considered to be subject object verb (SOV), though not strictly, for example, "John goes to Park". Here, John is subject, goes is verb and Park is object. The same sentence in Bengali is জন পার্কে যায়(John park a jaye), which is subject object verb. Though if we say in Bengali জন যায় পার্কে(John jaye park a) is still not wrong. But in BLEU, it fails to capture because there is order mismatch among words.

Article level: There are two types of article in English, definite (the) and indefinite (a, an). In Bengali, there is no definite article. However, indefinite article (a, an) is represented in Bengali as একটি(ekti) [21].

Verb, adverb sequence: In English, verb precedes adverb always. For example, he walks slowly, whereas in Bengali, adverb precedes verb. The Bengali translation of the sentence will be " সে আস্তে আস্তে চলে" (Se aste aste chale).

Above listed are only a few differences in linguistic features between these two languages. Translation and evaluation task become difficult because of these linguistic differences also.

4 Some Evaluation Metrics

We evaluated our results generated by two MTs using following automatic evaluation metrics:

BLEU: The BLEU score varies from 0 to 1. When candidate and reference translations match exactly, it gives highest score 1. BLEU computes geometric average of modified precision score using n-grams and max up to N and positive weights w_n summing to one [22].

Precision: Precision is calculated as $\mathrm{P} = m/w_t$, where m is the total number of words in the candidate translation which exactly matches with the reference translation words. w_t is the number of words in the candidate translation.

Recall: Recall is calculated as $\mathrm{R} = m/w_r$, where m is the total number of words in the candidate translation which exactly matches with reference translation and w_r is the total number of words in the reference translation.

F-measure: F-measure is calculated with the help of following 2PR/P + R, where P is the precision and R is recall.

Cosine similarity: Cosine similarity (CS) is used to measure the similarity between documents. It actually measures the cosine of the angle between two vectors. In our case, cosine of the MT outputs is compared with reference translation. The cosine similarity of the sentences represented as vectors can be represented as follows:

$$\mathrm{CS}(d_1 \cdot d_2) = d_1 \cdot d_2/||d_1|| \times ||d_2||$$

where d_1 and d_2 are two vectors we need to compare, whereas $||d_1||$ and $||d_2||$ are the product of the Euclidean norms of the two vectors. Actually, it's the length of the two vectors. When the cosine value is 0 that means two vectors are orthogonal and similarity is null. The cosine value 1 indicates two vectors are coinciding, pointing to the same direction and the angle between them is 0, which indicates match is maximum. We have used above-mentioned metrics in our work. Although these are used in some popular automatic evaluation metrics including BLEU.

5 Methodology

In our work, we have used English–Bengali parallel corpus taken from IEEE data port SuPara 0.8 M [23].

We have selected randomly few English sentences and its corresponding Bengali from the corpus. We have translated these English sentences to Bengali using Google Translate and Microsoft's Bing translator. The performance of these two translators has been tested with automatic metrics BLEU and other such as precision, recall, f-measure and cosine similarity. BLEU and other metrics scores have been computed considering the hypothesis (Translation output) produced by the individual translators and reference translations (English counterpart of parallel corpus, i.e., Bengali). The scores generated by BLEU and other metrics for different English sentences are compared with the scores generated for the same sentences by human expert (human judgment). We have conducted a survey based on the questioners to generate human score. The survey was conducted from twenty people who have moderate to high linguistic expertise on these two languages. Mainly, the questioners' query were formed that address adequacy and fluency of the translation. Adequacy and fluency were measured based on scale value ranging from 0 to 5. Adequacy ensures the meaning of source and translated output same or not, completeness between source and target sentence, no distortion between source and target sentence etc. Similarly, fluency ensures syntactical correctness of the source and target language. In adequacy when all meanings are correct between source and target language we assign score 5, most of the meaning correct score is 4, much meaning score is 3, little meaning score is 2 and for none score is 1. In case of fluency, flaw less language score is 5, good language score is 4, non-native language score is 3, dis-fluent language score is 2 and incomprehensible score is 1. Researchers claimed that in MT evaluation even human generated scores are not consistent and not completely reliable and obvious reason automatic evaluation measures are more unreliable [24]. We have calculated the mean of scores for individual sentences given by all twenty human experts. The survey report is provided in Fig. 1. It is the graphical representation of scores given by twenty people having moderate to high linguistic expertise on these two languages. They have given the score in the metric scale ranging from 0 to 5 for all five sentences given in the Table 1. 0 represents least score and 5 represents highest score.

Similar survey was conducted with the help of questioners from the same group of people for Bing translator (Fig. 2). Finally, mean value of individual sentence's score as given by twenty different people were computed to generate human score. Figure 2 is the graphical representation of scores given by the same twenty people.

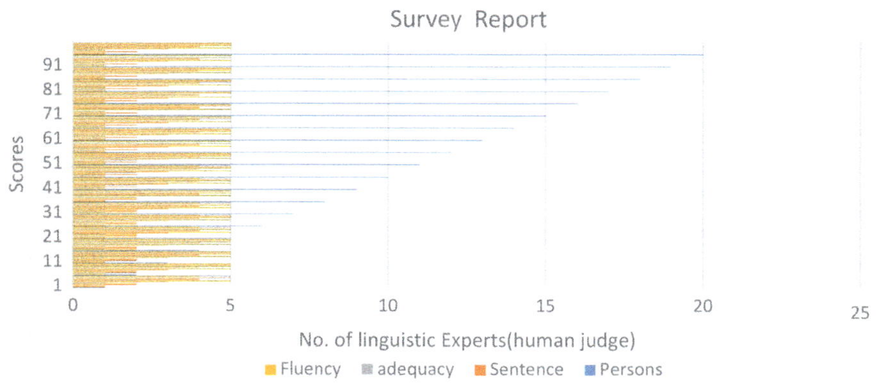

Fig. 1 Survey report of translations generated by Google Translate

Table 1 BLEU and other scores are computed and compared with manual score for the translations produced by Google Translate and Microsoft's Bing translator

English sentence	Translation in Bengali	Scores	Translation in Bengali	Scores
She likes red roses	R.T=সে লাল গোলাপ পছন্দ করে। Google Translate= তিনি লাল গোলাপ পছন্দ।	BLEU=32.51% Precision=3/4=75% Recall=3/5=60% F-measure=66.66% Cosine similarity=67.11% Human Score=23%	R.T=সে লাল গোলাপ পছন্দ করে। Bing Translator= সে লাল গোলাপ পছন্দ করে।	BLEU=100% Precision=5/5=100 Recall=5/5=100% F-measure=100 Cosine similarity=100% Human Score=100
It makes our body firm	R.T=ইহা আমাদের শরীরকে দৃঢ় করে। Google Translate= এটি আমাদের শরীরকে দৃ makes◌় করে তোলে।	BLEU=15.91% Precision=3/5=60% Recall=3/5=60% F-measure=60 Cosine similarity=54.74% Human Score=20%	R.T=ইহা আমাদের শরীরকে দৃঢ় করে। Bing Translator= এটা আমাদের শরীরকে দৃঢ় করে তোলে।	BLEU=19.30 Precision=4/5=80% Recall=4/5=80% F-measure=80% Cosine Similarity=72.99% Human score=98%
I love my daughter	R.T=আমি আমার মেয়েকে ভালবাসি। Google Translate= আমি আমার কন্যাকে ভালবাসি.	BLEU= 37% Precision=3/4=75% Recall=3/4=75% F-measure=75% Cosine Similarity=75% Human Score=100%	R.T=আমি আমার মেয়েকে ভালবাসি। Bing Translator= আমি আমার মেয়েকে ভালবাসি।	BLEU=100 Precision=4/4=100% Recall=4/4=100% F-measure=100% Cosine Similarity=100% Human score=100%
Physical exercise is very necessary to preserve good health	R.T=ভাল স্বাস্থ্য রক্ষার জন্যে শারীরিক ব্যায়াম অত্যাবশ্যক। Google Translate= শারীরিক অনুশীলন সুস্বাস্থ্য রক্ষার জন্য অত্যন্ত প্রয়োজনীয়।	BLEU=9.46% Precision=3/7=42.85% Recall=3/7=42.85% F-measure=42.85% Cosine Similarity=28.57% Human Score=98%	R.T=ভাল স্বাস্থ্য রক্ষার জন্যে শারীরিক ব্যায়াম অত্যাবশ্যক। Bing Translator= সুস্বাস্থ্য বজায় রাখতে শারীরিক ব্যায়াম অত্যন্ত প্রয়োজনীয়।	BLEU=9.15% Precision=2/7=28.57% Recall=2/7=28.57% F-measure=28.57% Cosine Similarity=28.57% Human score=99%
Mass awareness should be created	R.T=গণসচেতনতা সৃষ্টি করা উচিত। Google Translate= জনসচেতনতা তৈরি করতে হবে।	BLEU=19.88% Precision=0/4=0% Recall=0/4=0% F-measure=0% Cosine Similarity=0% Human score=99%	R.T=গণসচেতনতা সৃষ্টি করা উচিত। Bing Translator= জনসচেতনতা তৈরি করা উচিত।	BLEU=35.93% Precision=2/4=50% Recall=2/4=50% F-measure=50% Cosine similarity=50% Human score=97%

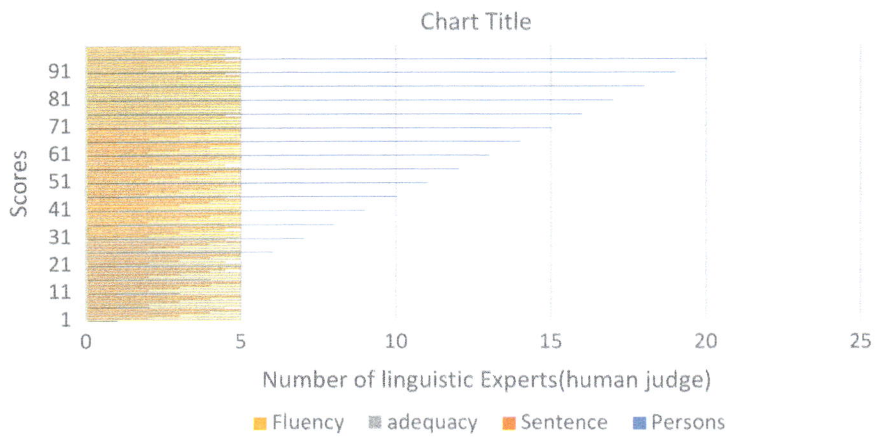

Fig. 2 Survey report of translations generated by Microsoft's Bing translator

They have given the scores in the metric scale ranging from 0 to 5 for all five same sentences. The value 0 represents least score, and 5 represents highest score. The score value was assigned based on the adequacy and fluency of translations.

BLEU and other scores like precision, recall, F-measure, cosine similarity are calculated for all translations generated by two translators. Finally, human generated score from survey and automatic evaluations scores are compared. A detail comparative analysis in terms of the correlation between these two types (automatic and human generated) of scores is reported for this English → Bengali language pair.

6 Data and Experiments

To evaluate how the two translators and automatic performance evaluators generate scores for selected sentences are compared with human generated score. Human judgements are considered to be the best. Hence, we have set human score as a standard reference and against that rest of the metrics are compared. For this experimentation we have used data sets taken from SuPara 0.8 M. In this English-Bengali corpora we randomly selected five different sentences and translated with two different translators—Google Translate and Microsoft's Bing translate. Automatic evaluation BLEU is used to measure the performance of the translators. BLEU generates score by comparing hypothesis, i.e., candidate translation and reference translation. As mentioned before, we have considered few set of sentences from parallel corpora's English and it's corresponding Bengali as reference translation. For the first sentence "She likes red roses", Google Translate translates this as "তিনি লাল গোলাপ পছন্দ।" (tini lal golap pachanda), which is not a complete Bengali sentence. BLEU score for this translation is 32.51% and human score is 23%, whereas other scores like precision is 75%, recall is 60%, F-measure is 66.66% and cosine similarity is 67.11%. For the same sentence, translation produced by Bing is "সে লাল গোলাপ পছন্দ করে।", (se lal golap pachanda karey") is a correct translation because it is adequate and fluent. BLEU score for this is 100% and human score is also 100%. Rest of the measures like precision, recall, f-measure, and cosine similarity are also 100% (Table 1). Similarly for rest of the sentences, translations generated by two translators and their manual and all other automatic scores are represented in Table 1. Translations generated by two different translators and their human-generated scores and automatic evaluations are represented in graphical form in Fig. 3. We have also represented the different scores generated by automatic and human judgement (human score) for two different translators, viz. Google and Bing are represented graphically in Fig. 4 and Fig. 5 respectively.

From the graphical observation, we can see that there is very little and no correlation across different scores produced by automatic metrics and human-generated score. Automatic scores are compared against human judgement with Pearson's correlation.

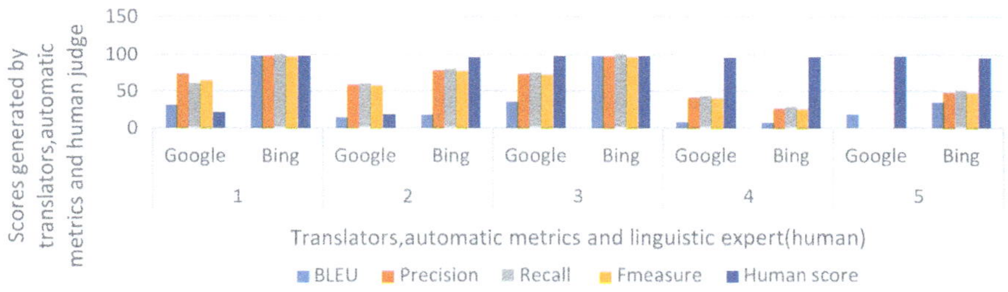

Fig. 3 Manual and other scores of translations generated by Google and Bing. Abscissa represents translators, and ordinate represents scores

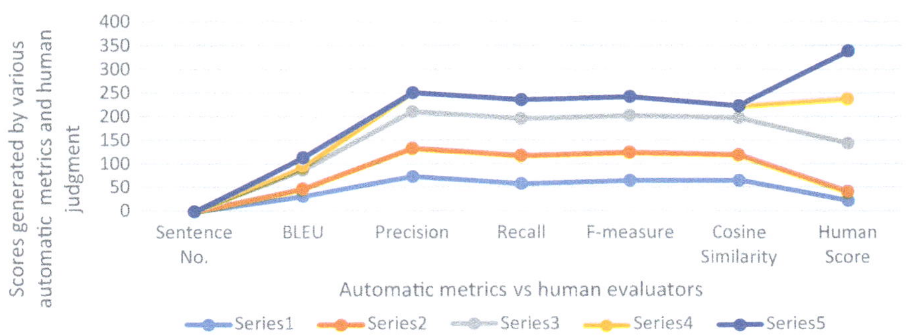

Fig. 4 Performance evaluation of Google Translate with the help of different automatic and human-generated scores

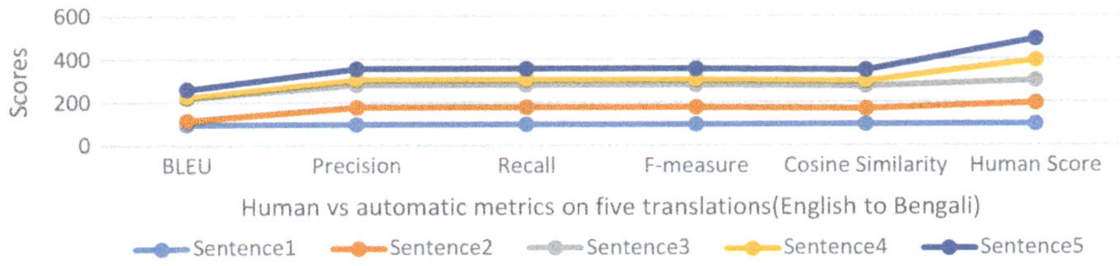

Fig. 5 Performance evaluation of Microsoft's Bing translator with the help of various automatic and human-generated scores

Researchers observed that poor quality translations are even more challenging for human being (human evaluator) to score correctly. And even automatic metrics perform poorly when try to score poor quality translations [19].

We have computed and observed correlation between various automatic metrics and human judgement with the help of Pearson correlation coefficient. Pearson correlation coefficient values range from −1 to +1. When two variables increase or decrease simultaneously, then coefficient is positive. When one increases and other decreases, then coefficient is negative. Table 2 shows the Pearson coefficient to detect correlation between BLEU and human judgment when Google Translate is used. Interestingly, we have found that for our selected set of sentences, all automatic metrics and human judgments are negatively correlated and there is no strong correlation found (Fig. 6).

Table 2 Pearson correlation between various metrics and human judgement for Google translations

Metric type	Human judgement (score)	Pearson coefficient
BLEU	Human score	−0.07292
Precision	Human score	−0.48446
Recall	Human score	−0.38557
F-measure	Human Score	−0.4337
Cosine similarity	Human Score	−0.45784

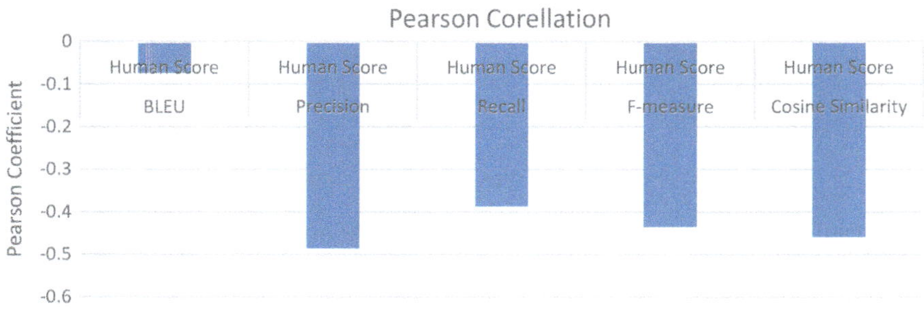

Fig. 6 Pearson correlation of Google translate

Table 3 Pearson correlation between automatic scores and human judgement for Bing translator

Metric type	Human judgement (score)	Pearson coefficient
BLEU	Human score	0.703637
Precision	Human score	0.555976
Recall	Human score	0.555976
F-measure	Human Score	0.555976
Cosine similarity	Human Score	0.59572

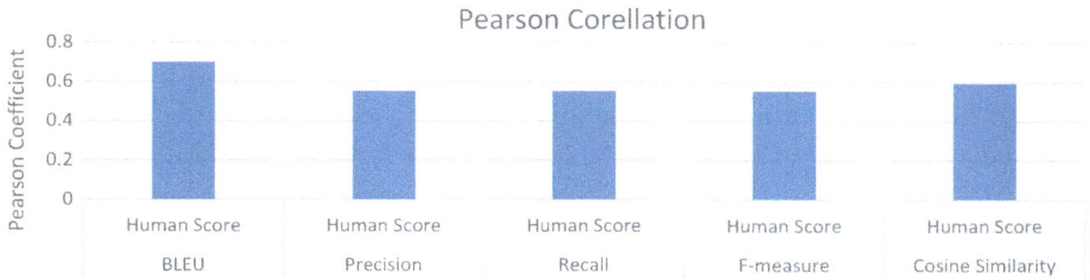

Fig. 7 Pearson correlation graph between human-generated score and automatic metrics

We have also computed the correlation between automatic evaluation and human judgment for translations generated by Microsoft's Bing translator. In this case, also there is no promising correlation found between human score and automatic metrics for our set of five sentences. Except BLEU, having correlation around 70.36%, rest all have correlation ranging between 55.59 and 59.57% only (Table 3). Figure 7 represents the correlation score graph for the translations generated by Bing Microsoft translator.

7 Analysis and Discussion

Measuring the quality of MT may require different tasks such as assessing the output generated by MT system may be any application specific, i.e., absolute evaluation. It may compare the different systems and if there is any change incorporated what is the impact of this change (relative evaluation). Finally, when the system is in learning stage, how parameter tuning helps improve its performance. Depending on the importance of the task human-generated scores may also be required because ultimately human judgment is considered to be the best [25]. From the above experiments and various automatic evaluation scores, we can see that there is very little or no correlation between human and automatic scores. Specifically for BLEU which is very popular among different automatic evaluations, failed to produce correct score in many situations. Since BLEU is primarily dependent on n-gram approach and compares candidate translation with the given reference translation and searches for exact match without considering the semantic of the match which causes its poor performance. So the obvious approach is making candidate translation compared with more number of reference translation which can increase its performance up to certain extent. Referring Table 1 and considering two different translators and considering first sentence "She likes red roses", Google translates this as " 'তিনি লাল গোলাপ পছন্দ।" (tini lal golap pachando) which is not a complete and meaningful translation. It may be fluent but not adequate. Human judgment for this is only 23%, whereas automatic evaluations such as BLEU, precision, recall, f-measure, cosine similarity are 32.51%, 75%, 60%, 66.66%, 67.11%, respectively. Bing translates this sentence as "সে লাল গোলাপ পছন্দ করে।" (se lal golap pachanda kore) is an excellent translation and exactly matches with reference translation word by word. It is adequate and also fluent. Human judgment for this translation is 100%. Automatic evaluations such as BLEU, precision, recall, f-measure and cosine similarity are 100%, 100%, 100%, 100% and 100%, respectively. Second sentence is "It makes our body firm". Google translates this as " এটি আমাদের শরীরকে দৃ makes় করে তোলে।" (Ati amader sharir k দৃ makes়korey toley) which is a worst translation. It even did not find proper match for the word "firm" and translated in Bengali as " "দৃmakes" which does not make any sense in Bengali and might have treated as unknown word. Human score for this translation is 20%. Automatic evaluation scores generated by BLEU, precision, recall, f-measure and cosine similarity are 15.91%, 60%, 60%, 60% and 54.74%, respectively. However, Bing translates this sentence as " এটা আমাদের শরীরকে দৃঢ় করে তোলে।" (Ata amader shorir k diiro korey toley) which is a good translation. Human score for this translation is 98%. Automatic scores such as BLEU, precision, recall, f-measure and cosine similarity are 19.30%, 80%, 80%, 80% and 72.99%, respectively. Taking third sentence, "I love my daughter", google translates this as "আমি আমার কন্যাকে ভালবাসি" (Ami amar kanya k bhalobashi), which is a good translation. This translation is adequate and fluent both. Human judgment for this translation is 100%. However, automatic evaluation such as BLEU, precision, recall, f-measure and cosine similarity are 37%, 75%, 75%, 75%, and 75%, respectively. Second MT, Bing also translates this sentence correctly and returns as "আমি আমার মেয়েকে ভালবাসি।" (Ami amar meye k bhalobashi) which is an excellent translation. This translation exactly matches with reference translation. Human score and all other automatic evaluations are 100% for this translation. For the fourth sentence "Physical exercise is very necessary to preserve good health". Google translates this as " শারীরিক অনুশীলন সুস্বাস্থ্য রক্ষার জন্য অত্যন্ত প্রয়োজনীয়।" (saririk anushilan su sasthya rakhaar janya proyajanio). This is a good translation generated by google. It's adequate and fluent. Human score for this translation is 98%, whereas automatic evaluations BLEU, precision, recall, f-measure and cosine similarity return scores 9.46%, 42.85%, 42.85%, 42.85% and 28.57%, respectively. Bing translates this as " সুস্বাস্থ্য বজায় রাখতে শারীরিক ব্যায়াম অত্যন্ত প্রয়োজনীয়।" (su sasthaya bajay rakhtey sharirik bayam atyanta proyojanio). This is also a very good translation. It is fully adequate and fluent. Human judgment (human score) for this translation is 99%. However, BLEU and other metrics such as precision, recall, f-measure, cosine similarity are 9.15%, 28.57%, 28.57%, 28.57%, and 28.57%, respectively. N-gram-based BLEU did not find exact word-based match and wrongly scored. Last sentence is "Mass awareness should be created". Google translates this as " জনসচেতনতা তৈরি করতে হবে।" (Janasachetanata taire kartey habey). This translation is correct and human score for this translation is 99%. BLEU and other scores for this translation are 19.88%, 0%, 0%, 0% and 0%, respectively. BLEU and other metrics scored incorrectly. Bing translates this as " জনসচেতনতা তৈরি করা উচিত।" (Janashachetanata tairey kara uchit). This translation is adequate and fluent, and human judgment is 97%. BLEU and other metrics could not score correctly. BLEU score is only 35.93%. Precision, recall, f-measure and cosine similarity are all scored same which is only 50%. From the above translations and their scores, we observe there is very little or in some cases no correlation between automatic matrices and human-generated score. Variations in scores imply evaluating translations generated by MT which is more challenging when it is automatic.

We now analyze the performance of two translators Google and Bing with their translations where there is significant difference noticed.

In our first sentence "She likes red roses", Google could not generate adequate translation. Gold standard human judgment scored it only 32.51%, whereas for the same sentence Bing generated excellent translation and human generated score is 100%. For the second sentence "It makes our body firm", Google translate failed to generate correct translation. Human score for this translation is only 20%. However, Bing translates this correctly. Translation is adequate and fluent. Human score for this translation is 98%. Google translate and Microsoft's Bing both are though NMT and for our sample sentences Bing performed better. But it does not mean that for all other sample sentences and always Bing would outperform. There are several factors behind NMT's performance. Some of the possible reasons are the datasets it uses. The framework it implements and many other [8, 10, 14]. Researchers even found performance of MT systems can be enhanced by combining the outputs of different MT and even combining SMT and NMT [8, 26].

8 Conclusions

We can conclude by saying automatic metrics including most popular one, BLEU failed to achieve satisfactory correlation with manual score in the English Bengali language pair. BLEU and other metrics are still not completely reliable. Researchers have explored machine learning-based approaches such as artificial neural network (ANN) and support vector machine (SVM) in auto-MT evaluation process and claimed to have enhanced the correlation with human judgement for different language pairs [17, 18]. Recently proposed approach, using word embedding claimed to achieve high correlation coefficient for the language pairs where grammar, differs much, viz. English → German, German → English and English → Turkish, Turkish → English [27].

From the above survey and analysis, it is evident that although machine learning and other approaches enhanced the correlation with human judgment, but it is for some selected language pairs only. And as the point mentioned before based on the purpose and importance human judgment (human generated score) is still very much needed and can't be avoided.

Automatic metrics perform well and their scores are more reliable in NMT as compared to traditional SMT [19]. In future work, we will try to exploit machine learning-based approach in automatic evaluation and verify the performance on English→Bengali language pair.

References

1. B.J. Dorr, P.W. Jordan, J.W. Benoit, A survey of current paradigms in machine translation. Adv. Comput. **49**, 1–68, Elsevier (1999)
2. R. Ananthakrishnan, P. Bhattacharyya, M. Sasikumar, R.M. Shah, *Some Issues in Automatic Evaluation of English-Hindi MT: More Blues for BLEU*
3. C. Callison-Burch, C. Fordyce, P. Koehn, C. Monz, J. Schroeder, (Meta-) evaluation of machine translation, in *Proceedings of the Second Workshop on Statistical Machine Translation* (Pragua, Association for Computational linguistic, 2007), pp. 138–158
4. P.F. Brown, S.A. Della, V.J. Pietra, D. Pietra, R.L. Mercer, The Mathematics of statistical machine translation: parameter estimation. Association Comput. Linguist. **19**(2), 1993 (1993)
5. P. Koehn, *Statistical Machine Translation* (Cambridge publisher, 2010)
6. Z. Wang, J. Shawe-Taylor, S. Szedmak, Kernel regression based machine translation, in *Proceedings of NAACL HLT 2007, Companion Volume* (Rochester, NY, Association for Computational Linguistics, 2007), pp. 185–188
7. Y. Bengio, R. Ducharme, P. Vincent, C. Jauvin, A neural probabilistic language model. J. Mach. Learn. Res. **3**, 1137–1155 (2003)
8. L. Zhou, W. Hu, J. Zhang, C. Zong, *Neural System Combination for Machine Translation.* arXiv:1704.06393v1 [cs.CL] 21 Apr 2017
9. M.-T. Luong, H. Pham, C.D. Manning, Effective approaches to attention-based neural machine translation, in *Proceedings of the 2015 Conference on Empirical Methods in Natural Language Processing* (Association for Computational Linguistics, Lisbon, Portugal, 17–21 September 2015). c©2015
10. M. Wang, J. Xie, Z. Tan, J. Su, D. Xiong, C. Bian, Neural machine translation with decoding-history enhanced attention, in *Proceedings of the 27th International Conference on Computational Linguistics* (Santa Fe, New Mexico, USA, 20–26 Aug 2018), pp. 1464–1473
11. D. Bahdanau, K. Hyun, C.Y. Bengio, *Neural Machine Translation by Jointly Learning to Align and Translate.* arXiv:1409.0473v7 [cs.CL] 19 May 2016
12. K. Cho, B. van Merrienboer, D. Bahdanau, Y. Bengio, *On the Properties of Neural Machine Translation: Encoder–Decoder Approaches.* arXiv:1409.1259v2 [cs.CL] 7 Oct 2014
13. X. Shi, I. Padhi, K. Knight, Does string based neural MT learn source syntax?, in *Proceedings of the 2016 Conference on Empirical Methods in Natural Language Processing* (Association for Computational Linguistics, Austin, Texas, 1–5 Nov 2016), pp. 1526–1534. c©2016
14. Y. Wu, M. Schuster, Z. Chen, Q.V. Le, M. Norouzi, *Google's Neural Machine Translation System: Bridging the Gap between Human and Machine Translation.* arXiv:1609.08144v2 [cs.CL] 8 Oct 2016

15. R. Sennrichand, B. Haddow, *Linguistic Input Features Improve Neural Machine Translation* (School of Informatics, University of Edinburgh)
16. A. Das, P. Yerra, K. Kumar, S. Sarkar, A study of attention-based neural machine translation models on Indian languages, in *Proceedings of the 6th Workshop on South and Southeast Asian Natural Language Processing* (Osaka, Japan, 11–17 Dec 2016)
17. A. Kulesza, S.M. Shieber, A learning approach to improving sentence-level MT evaluation, in *Proceedings of the 10th International Conference on Theoretical and Methodological Issues in Machine Translation* (Baltimore, MD, 4–6 Oct 2004
18. F. Guzmán, S. Joty, L. Màrquez, P. Nakov, *Machine Translation Evaluation with Neural Networks* (Elsevier, Computer Speech and Language, 2017).
19. M. Fomicheva, L. Specia, Taking MT evaluation metrics extremes: beyond correlation with human judgment. Comput. Linguist. **45** (2019)
20. T. Dasgupta, M. Sinha, A. Basu, Resource creation and development of an English-Bengali back transliteration system. Int. J. Knowl. Intell. Eng. Syst. **19**, 35–46 (2015)
21. R. Balyan, N. Chatterjee, Factor-based evaluation for English to Hindi MT outputs. Lang Res. Eval. **52**, 969–996 (2018)
22. K. Papineni, S. Roukos, T. Ward, W.-J. Zhu, BLEU: a method for automatic evaluation of machine translation, in *Proceedings of the 40th Annual Meeting of the Association for Computational Linguistics (ACL)* (Philadelphia, July 2002), pp. 311–318
23. M.A.A. Mumin, M.H. Seddiqui, M.Z. Iqbal, M.J. Islam, SUPara0.8M: A Balanced English-Bengali Parallel Corpus, IEEE Dataport, 2018 (2020)
24. J.P. Turian, L. Shen, I. Dan Melamed, *Evaluation of Machine Translation and it's Evaluation* (2003)
25. N. Cancedda, M. Dymetman, G. Foster, C. Goutte, A statistical machine translation primer, in *Learning Machine Translation* (PHI Learning Pvt. Ltd., 2010)
26. E. Matusov, G. Leusch, H. Ney, Learning to combine machine translation systems, in *Learning Machine Translation* (PHI learning Pvt. Ltd., 2010)
27. H. Echizen'ya, K. Araki, E. Hovy, Word embedding-based automatic MT evaluation metric using word position information, in *Proceedings of NAACL-HLT 2019*, pp. 1874–1883 (2019)

A Spot Rainfall Prediction During Cyclones by Using Time Series Analysis Model

J. Senthil Kumar, V. Venkataraman, S. Meganathan, P. Thiruvasagam, N. Venkatanathan, and Kannan Krithivasan

Abstract Forecast models of climate and rainfall are exceptionally nonlinear and complex, which require definite models to acquire precise predictions. Weather forecasting is an application to predict the atmospheric changes in a particular location, but unexpected changes in the weather condition is a major threat for the forecasts. Numerous tools and procedures have been made to predict weather forecasting, but still more accurate methods are needed to predict weather forecasting in right time to avoid the devastation of properties and loss of lives. In this work, time series analysis models with scatter plots and smooth lines have been used to predict the forthcoming rainfall. This research article focuses on correlating the cyclones and observed rainfalls using statistical as well as data analytical model. The rainfall data during the cyclone period in the Nagapattinam station (10°76′ N and 79°85′ E) of Bay of Bengal during *North-East monsoon season* for the period 2015–2019 are extracted from the National Oceanic and Atmospheric Administration (NOAA) of the World Meteorological Organization. There were 14 cyclones occurred during the five-year period in this region, and these cyclones have been analyzed using statistical as well as data analytical model. This model is verified, validated with training and testing datasets, and the results are found to be encouraging.

Keywords Cyclone · Rainfall prediction · Scatter plot · Time Series · Statistical and analytical model

---

J. Senthil Kumar · S. Meganathan
Department of Computer Science and Engineering, SRC, SASTRA Deemed University, Kumbakonam, Tamil Nadu, India
e-mail: senthilkumarcse@src.sastra.edu

V. Venkataraman (✉) · K. Krithivasan
Department of Mathematics, SASTRA Deemed University, Thanjavur, TamilNadu, India
e-mail: mathvvr@maths.sastra.edu

P. Thiruvasagam
Department of Chemistry and Biosciences, SRC, SASTRA Deemed University, Thanjavur, India

N. Venkatanathan
School of Electrical and Electronics Engineering, SASTRA Deemed University, Thanjavur, India

M. Paprzycki et al. (eds.), *Intelligent Systems, Technologies and Applications*, Advances in Intelligent Systems and Computing 1353,
https://doi.org/10.1007/978-981-16-0730-1_25

1 Introduction

Nowadays, analyzing huge amounts of data have become a vital need to discover hidden information in various fields. With advanced technology, it is possible to analyze the data and to get answers from it almost immediately. Data mining is a strategy that helps in extracting relevant and meaningful information from the huge volume of data. Weather science is important for human being since the people are living in environments that get affected often by weather. Weather forecasting is part of meteorology to predict how the atmosphere will change at a given location and time and are made by collecting data related to weather like fog, rainfall, temperature, wind, cyclone, etc., Since data cannot be collected from the future, it is one of the most challenging problems to predict weather well in advance by using the available set of data. Synoptic weather forecasting is the common method used by professional meteorologists which involves analyzing a lot of weather observations over a wide area and period. Specific forecast of precipitation in a suitable area accurately is significant for the nations flourishing with an agro-based economy [1]. Monthly precipitation plays a significant role in the planning and management of agricultural scheme as well as water resources systems.

With the advent of computers application of data mining techniques in meteorology, various hidden facts about the weather and weather changes can be brought out. To predict the rainfall occurrence 24 h ahead and 48 h ahead on sample stations located along east coast of India, data mining techniques are used [2, 3]. A few studies have been made regarding the weather conditions and predictions for Indian subcontinent. They have successfully used the data mining techniques to predict the occurrence of rainy day and the quantitative of rain fall for certain stations in peninsular India using in situ data.

Many tools and techniques are available to predict the rainfall, but still there is a need for more accurate technique to predict the rainfall to save properties from destruction and lives from loss. There are various statistical approaches which have been used for the rainfall prediction, and among them one of the methods is statistical downscaling method which includes multiple linear regression. Time series is basically a sequence measurement data taken in chronological order within a certain time. Time series model can be actually an integration of random and deterministic components. If random components are eliminated, then the deterministic components can be easily modeled. A fuzzy-entropy-neuro-based expert system suggests the usage of three techniques, viz. fuzzy set, entropy and artificial neural network (ANN). Based on these techniques, a novel ISMR time series forecasting model is designed to deal with the dynamic nature of the Indian summer monsoon rainfall. The Box–Jenkins time series seasonal ARIMA (Auto-Regressive Integrated Moving Average) model is used for the prediction of rainfall on monthly scales. To attain precise rainfall forecast, a modern method termed fuzzy time series method is used to analyze the rainfall [4]. Time series models are appropriate tools for forecasting in semi-arid climates [5]. Improved multilayer perception neural network is planned on intellectual tool for predicting rainfall time series [6]. The time series rainfall data was decomposed into stochastic trend, seasonal variations and random movements. The yearly rainfall data showed in decreasing trend [7]. Precipitation is a final result of the number of a complex atmospheric process which varies both in space and time. Hence, time series prediction is also significant. The objective of this study is to develop a model which comprises this intelligent behavior. The variables defining weather conditions like temperature (maximum or minimum), rainfall, cyclone formation and movement, etc., vary continuously with time, forming time series of each parameter which can be used to develop a forecasting model that uses this time series data.

2 Data Used

2.1 Study Area

Nagapattinam (10°76′ N and 79°85′ E) is located in east coast of Tamil Nadu state of South India.

2.2 Data Collection

The rainfall data for the cyclone period during north-east monsoon season in the month of October, November and December for the period 2015–2019 are extracted from the National Oceanic and Atmospheric Administration (NOAA) of the World Meteorological Organization. Totally, 92 days of northeast (NE) monsoon have been selected and the parameters are collected because of their operational importance. Daily rainfall data for the years 2015–2017 are considered as training data for developing this model. Data preprocessing steps have been applied on the raw data, and thus they have become machine readable. The normal rainfall in Nagapattinam region is 975.4 mm (38.40 in.), but the variation from year to year is appreciable. The statistical model has been built using this three-year training dataset, and this model is applied with test dataset, the results obtained are considerably good. Similarly, the cyclone track data are collected from IMD for the same period.

2.3 Software Used: MS_EXCEL

Scatter plot model in **MS-Excel** software is used to find the best fit of a time series for past values in order to make forecasts.

3 Methodology

A time series is defined as a set of observations arranged chronologically, i.e., a sequence of observations usually ordered in time. The principal aim of this time series analysis is to describe the history of movements in time of some variables at a particular site.

The objective is to generate data having properties of the observed historical record. To compute properties of a historical record, it is broken into separate components and analyzed individually to understand the casual mechanism of different components. Once properties of these components are understood, these can be generated with similar properties and combined together to give a generated future time series. Analysis of a continuously recorded rainfall time series is performed by transforming the continuous series into a discrete time series of finite time interval.

3.1 Scatter Plot

The graphs discussed so far are useful for visualizing individual time series. They are also useful to explore relationships between time series. This scatter plot helps researchers to visualize the relationship between the variables. Time series forecasting is a process of using a model to generate prediction for future event based on past experiences.

3.2 Rainfall

The historical weather dataset are collected from NOAA website for making rainfall patterns over the northeast monsoon months of Nagapattinam station in October, November and December. These datasets are stored and given as inputs for further processes. Figure 1 shows the data flow diagram for proposed rainfall prediction.

3.3 Cyclone

Regional Specialised Meteorological Centre (RSMC)—Tropical Cyclones, New Delhi, functions in Cyclone Warning Division, Indian Meteorological Department (IMD), New Delhi. This center has the responsibility of monitoring and predicting cyclonic disturbances over the north Indian Ocean and issuing tropical weather outlook and tropical cyclone advisories for the benefit of the World Meteorological Organisation (WMO)/Economic and Social Cooperation for Asia and the Pacific

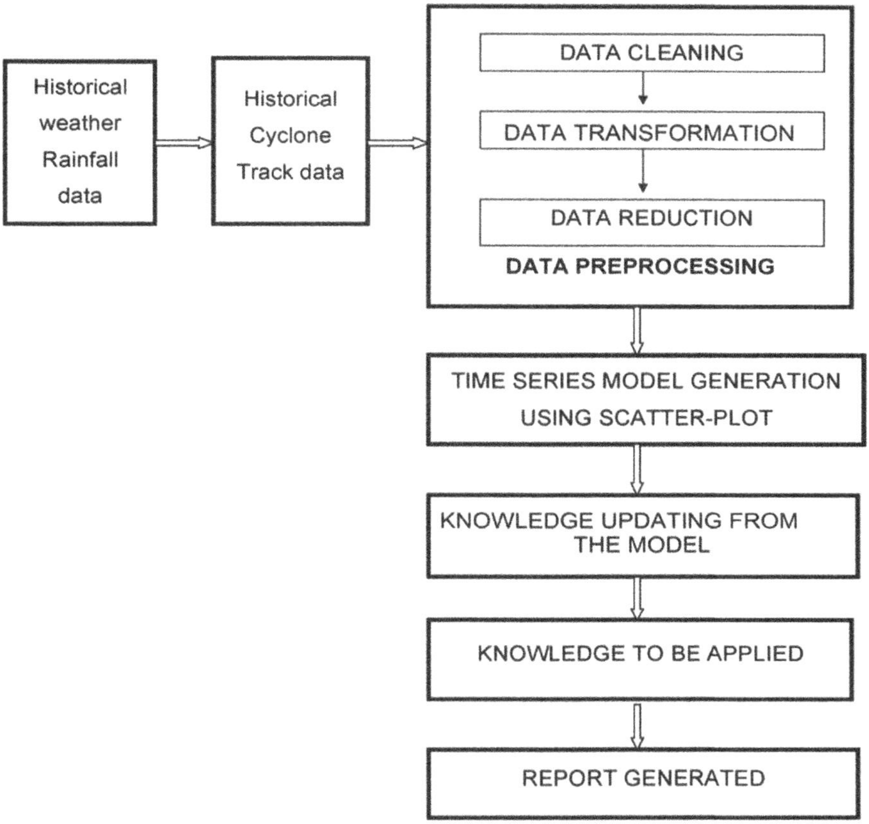

Fig. 1 Data flow diagram for rainfall prediction

(ESCAP) panel member countries. The best track cyclone data of the northeast monsoon for the year 2015–2019 has been used for this research work.

3.4 *Pre-processing Steps*

Data pre-processing is an information mining strategy that includes changing raw information into an understandable format. Real-world data is often incomplete, inconsistent, and or lacking in certain behaviors or trends and is likely to contain many errors. Information pre-processing is a demonstrated strategy for settling such issues. Data pre-processing prepares raw data for further processing. Tasks in data pre-processing consist of four steps, the first step is data cleaning which consists of filling missing values, smoothening noisy data, identifying or removing outliers, and resolving inconsistencies. The second step is the data integration process which functions as to integrate multiple databases, data cubes or files. The third step is the data transformation process which helps to normalize and aggregate the results. The fourth step is the data reduction process and is used to reduce the volume but produces the same or similar analytical results.

4 Results and Discussion

The five years (2015–2019) Northeast (NE) monsoon (October, November and December) rainfall data (in mm) of Nagapattinam coastal station is shown in Table 1. In this table, three months rainfalls data are listed by chronological order (day-1 to day-92). We applied scatter plot technique with smooth lines for the data shown in Table 1, and the time series output is plotted as a graph shown in Fig. 2.

In Fig. 2, it could be observed that the time series with peak curve refers the highest rainfall period of NE monsoon for the year 2015. Highest rainfall was observed during the period 8th November 2015–11th November 2015. During the peak rainfall period 8th November 2015–11th November 2015 (day-38 to day-41), the cyclone (BOB-12) formed in the Bay of Bengal with the distance of 421 km from the Nagapattinam coastal station (Table 2). It is observed that the cyclone track comes from 43 km closer to the Nagapattinam station. It is also observed that prior days of the cyclogenesis (day-37 and day-38), the rainfall at the Nagapattinam station is high (190 mm), when the cyclone moves closer to the station, the rainfall decreases gradually (120 mm).

In the Nagapattinam station, the historical monsoon average rainfall of the cyclone period (day-38, day-39, day-40 and day-41) is 15.0 mm. Due to the cyclone formation, the average rainfall of the four days (day-38, day-39, day-40 and day-41) is 90.6 mm. It is observed that due to the formation of cyclone the average rainfall of the Nagapattinam station increased by 5 times higher than the historical monsoon average rainfall during the days. It has been derived that the rainfall trend = (0.042)x +14.08 where "x" is the day and "y" is the expected rain on the day in mm. It helps us to find out the rainfall of the particular day of the specific monsoon season.

The time series output for northeast (NE) monsoon of year 2016 is plotted and is given in Fig. 3. Highest rainfall was observed during the period day 31 to day 34. In the year 2016 during the NE monsoon period (92 days), there were four cyclones, the first cyclone BOB-06-KYANT formed in the Bay-Of-Bengal (BOB) day-21. On comparing the rainfall data (Table 1) with the cyclones which are formed during the NE monsoon of year 2016, it is observed that the rainfall changes due to cyclone formation. It is noticed that the rainfall is not significant during the cyclone period (day-21 to day-27), the reason is the track of cyclone BOB-06-KYANT is too far from the coastal station and formed at a distance greater than 988 km from Nagapattinam coastal station.

Comparing the rainfall data (Table 1) with cyclone BOB-07 (Table 2) formed in the Bay of Bengal, it is observed that during the period (day-33 to day-37) the Nagapattinam station received the significant rainfall of 135 mm when the cyclone (BOB-07) track was closer to Nagapattinam station in the distance between 703 and

Table 1 Daily rainfall data (year 2015–year 2019)

Nagappattinam station

	Year 2015—daily precipitation (MM)			Year 2016—daily precipitation (MM)			Year 2017—daily precipitation (MM)			Year 2018—daily precipitation (MM)			Year 2019—daily precipitation (MM)		
Date/day	Oct	Nov	Dec	Oct	Nov	Dec	Oct	Nov	Dec	Oct	Nov	Dec	Oct	Nov	Dec
1	4.1	0.0	16.0	0.0	5.1	21.1	0.0	37.1	13.0	0.0	24.9	0.0	0.5	0.0	89.9
2	4.1	0.0	72.9	0.0	0.3	14.0	0.0	70.1	93.0	6.1	141.0	0.0	0.0	0.0	4.1
3	0.0	45.0	0.0	0.0	136.9	2.0	0.0	51.1	104.9	16.0	46.0	7.1	0.0	0.0	22.1
4	1.0	0.0	88.9	0.0	0.0	0.3	0.0	93.0	2.0	45.0	0.5	23.1	0.0	0.0	0.0
5	7.1	7.9	36.1	2.0	0.0	0.0	4.1	17.0	0.0	42.9	0.5	0.0	0.0	0.0	0.0
6	41.9	30.0	23.1	0.0	2.0	0.0	0.0	48.0	1.0	53.1	0.0	6.1	0.0	0.0	0.0
7	10.9	194.1	17.0	0.0	0.0	0.0	0.0	39.9	0.0	0.0	0.0	0.0	0.0	0.0	3.0
8	0.0	82.0	106.9	0.0	0.0	0.0	0.0	71.1	0.0	24.9	76.0	0.0	0.0	0.0	37.1
9	0.0	142.0	46.0	0.0	0.0	0.0	0.0	0.0	0.0	3.1	25.9	0.0	0.0	0.0	0.0
10	0.0	5.1	2.0	0.0	0.0	0.0	0.0	7.1	0.0	0.0	0.0	0.0	0.0	0.0	0.0
11	0.0	18.0	50.0	0.0	0.0	0.0	0.0	85.1	0.0	0.5	0.0	0.0	0.0	0.0	0.0
12	0.0	0.0	2.0	22.1	0.0	0.0	0.5	0.8	0.0	0.0	0.0	0.0	0.0	0.0	0.0
13	0.0	2.0	0.0	0.0	11.9	0.0	0.0	2.0	0.0	0.0	0.0	0.0	0.0	2.0	6.1
14	0.0	22.1	0.0	0.0	2.0	0.0	6.1	0.0	0.0	0.0	0.0	0.0	5.1	0.0	62.0
15	0.0	72.9	0.0	0.0	3.0	0.0	2.0	0.3	2.0	1.0	17.0	0.0	15.0	1.0	1.0
16	0.0	41.9	0.0	0.0	6.1	0.0	0.0	0.0	0.0	27.9	64.0	0.0	16.0	8.9	3.0
17	1.0	0.5	0.0	0.0	0.3	0.0	1.0	0.0	0.0	0.0	0.0	0.0	52.1	1.0	9.9

(continued)

Table 1 (continued)

Nagappattinam station															
	Year 2015—daily precipitation (MM)			Year 2016—daily precipitation (MM)			Year 2017—daily precipitation (MM)			Year 2018—daily precipitation (MM)			Year 2019—daily precipitation (MM)		
Date/day	Oct	Nov	Dec	Oct	Nov	Dec	Oct	Nov	Dec	Oct	Nov	Dec	Oct	Nov	Dec
18	3.0	0.0	0.3	0.0	33.0	0.0	0.0	0.0	0.0	1.0	0.0	0.0	20.1	7.1	0.0
19	13.0	10.9	0.0	0.0	0.0	0.0	0.0	6.1	4.1	1.0	0.0	0.0	0.0	14.0	0.0
20	0.3	0.0	0.0	0.0	0.3	0.5	0.0	0.0	14.0	0.0	0.0	0.0	17.0	9.9	2.0
21	0.0	47.0	0.0	0.0	0.0	0.3	5.1	8.9	8.9	5.1	0.0	0.3	18.0	2.0	0.3
22	0.0	32.0	0.0	0.0	0.0	0.0	47.0	10.9	14.0	33.0	0.0	5.1	49.0	39.1	0.0
23	0.0	103.1	0.0	0.8	0.0	0.0	0.0	0.0	0.0	0.0	29.0	11.9	0.3	82.0	0.0
24	0.0	11.9	0.0	0.0	0.0	0.0	0.0	0.0	0.0	0.0	166.9	7.1	7.1	15.0	0.0
25	0.5	0.0	0.0	0.0	0.0	0.0	0.0	0.0	0.0	3.1	3.1	0.0	0.0	0.8	138.9
26	0.0	0.0	0.0	0.0	0.0	0.0	6.1	22.1	0.0	0.0	0.0	0.0	0.5	3.1	2.0
27	0.0	10.9	0.0	0.0	0.0	1.0	7.9	57.9	0.0	0.0	0.0	0.0	1.0	8.9	0.0
28	2.0	8.9	0.0	0.0	0.0	33.0	22.1	14.0	0.0	0.0	0.0	0.0	54.1	43.9	0.3
29	7.9	0.0	0.0	0.0	0.0	0.3	3.0	7.1	0.0	0.0	1.0	0.0	9.9	56.9	0.0
30	19.1	24.9	0.0	1.0	0.0	0.0	48.0	41.9	0.0	1.0	17.0	4.1	43.9	63.0	1.0
31	41.9		0.0	7.1		0.0	95.0		0.0	7.9		0.0	53.1		0.0

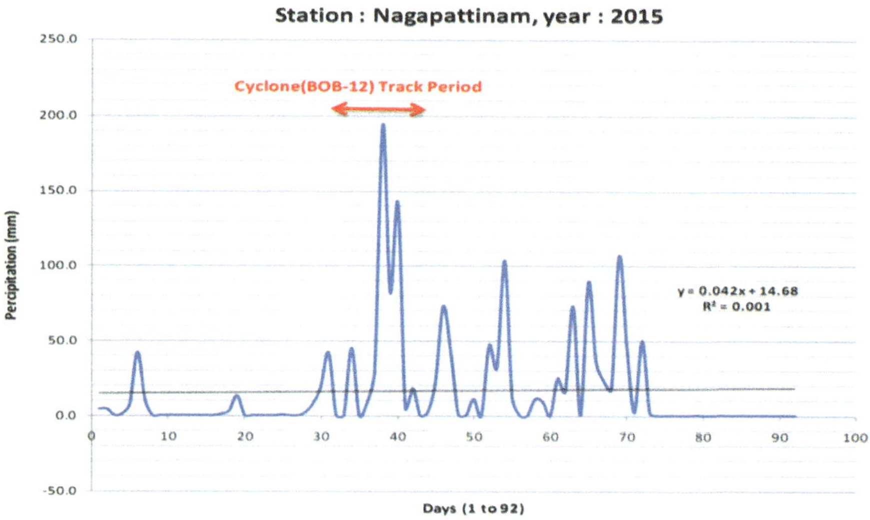

Fig. 2 Time series of observed rainfall during NE monsoon (2015)

1761 km. Thus, it is clear that the reason for getting significant rainfall in Nagapattinam station is due to the closer movement of the cyclone. In Nagapattinam station, the historical monsoon average rainfall of the cyclone period (day-33, day-34, day-35 and day-36) is 16.6 mm. Due to the cyclone (BOB-07) formation, the average rainfall of the four days (day-33, day-34, day-35 and day-36) is 67.46 mm. It is observed that due to the formation of cyclone the average rainfall of the Nagapattinam station has received 27.8 mm. It is 67.46% higher as compared with historical NE monsoon average rainfall of the particular days.

Similarly on comparison the cyclone BOB-08-NADA (Table 2) formed in the BOB region in the period 2016 (day-60 to day-62) the cyclone BOB-08-NADA track at a distance between 229 and 965 km, but the station received minimum rainfall. Even though the cyclone was closer to the station, the station received minimum rainfall. It may be due to lack of climatological parameter. Finally, rainfall data were compared with cyclone BOB-09-VARDAH (Table 2) in the BOB region in the period 2016 (day-66 to day-74). The distance between cyclone track and station Nagapattinam is 229 to 1248 km, but there was no significant rainfall and it may be due to other climatological parameter.

Time series output for Northeast (NE) monsoon of year 2017 is given in Fig. 4. In this year 2017 during the NE Monsoon period (92 days), there were five cyclones formed in the BOB.

On comparing the rainfall data (Table 1) with cyclone of BOB-06, it was observed that during the cyclone track of BOB-06 (Table 2) period (day-9 and day-10), the study station received less rainfall. The reason is the distance of cyclone track of BOB-06 (day-9 and day-10) is at a distance greater than 1249 km which is far away from the Nagapattinam coastal station.

Table 2 Cyclone track data during the year 2015 to 2019 (NE monsoon period)

Station: Nagapattinam				Latitude: 10.767			Longitude: 79.844	
Year	Cyclone No/name	Date		Cyclone origin		Cyclone movement distance (km) *	No. days sustained	Conversion of cyclone life period into monsoon days
		From	To	Latitude range (Degree)	Longitude range (Degree)			
2015	BOB-12	11-08-2015	11-10-2015	10.7–12.4	83.7–79.3	43–421	3	Day-39, Day-40, Day-41
2016	BOB-06-KYANT	22/10/2016	27/10/2016	13.5–15.4	88.5–83.0	618–1502	7	Day-21 to Day-27
	BOB-07	11-02-2016	11-06-2016	12.7–21.6	88.0–91.9	703–1761	5	Day-33, Day-34, Day-35, Day-36, Day-37
	BOB-08-NADA	29/11/2016	12-01-2016	6.5–10.8	87.5–79.7	16–965	3	Day-60, Day-61, Day-62
	BOB-09-VARDAH	12-06-2016	13/12/2016	8.5–12.5	91.0–78.0	264–1248	8	Day-66 to Day-74
2017	BOB-06	10-09-2017	10-10-2017	22.4–23.8	88.4–86.5	1581–1591	2	Day-9, Day-10
	BOB-07	19/10/2017	21/10/2017	16.5–24.3	86.5–87.4	1249–1460	3	Day-19, Day-20, Day-21
	BOB-08	15/11/2017	17/11/2017	15.0–19.5	83.0–86.3	581–1192	3	Day-46, Day-47, Day-48

(continued)

Table 2 (continued)

Station: Nagapattinam				Latitude: 10.767			Longitude: 79.844	
	BOB-09-OCKHI	29/11/2017	12-05-2017	6.5–19.2	81.8–71.9	527–1267	7	Day-60 to Day-66
	BOB-10	12-06-2017	12-09-2017	8.5–19.5	88.5–86.7	870–1217	4	Day-67, Day-68, Day-69, Day-70
2018	BOB-12-TITLI	08/10/2018	10-12-2018	14–22.1	88.8–87.5	846–1501	5	Day-8, Day-9, Day-10, Day-11,Day-12
	BOB-13 -GAJA	11-10-2018	19/11/2018	13.5–10.5	92.5–65.0	30–1384	9	Day-41 TO Day-49
	BOB-14-PHETHAI	13/12/2018	18/12/2018	6.5–17.5	88.7–82.5	444–1082	5	Day-73 to Day-78
2019	BOB-03-BUL BUL	11-05-2019	11-11-2019	13.1–23.1	87.9–91.9	1071–1517	7	Day-36 to Day-42

(*) with respect to Nagapattinam station

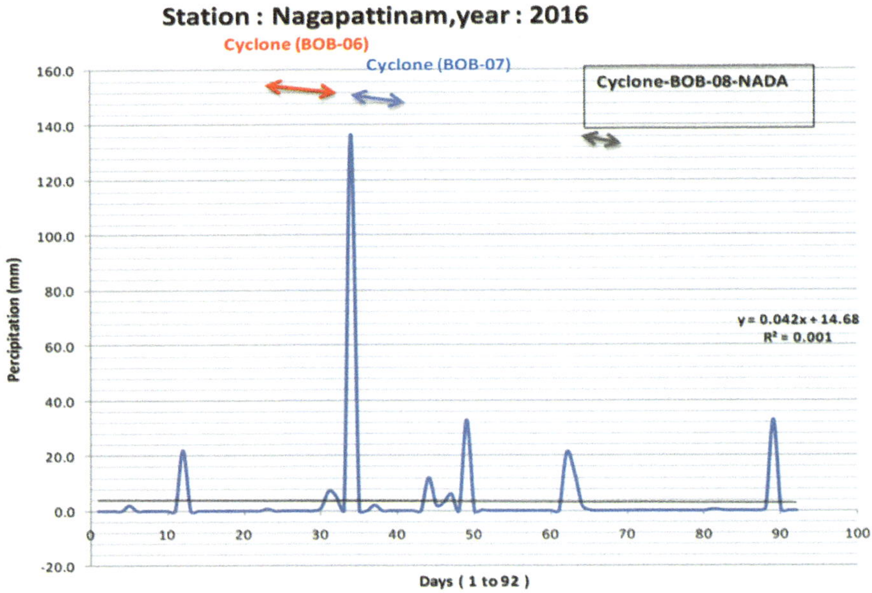

Fig. 3 Time series of observed rainfall during NE monsoon (2016)

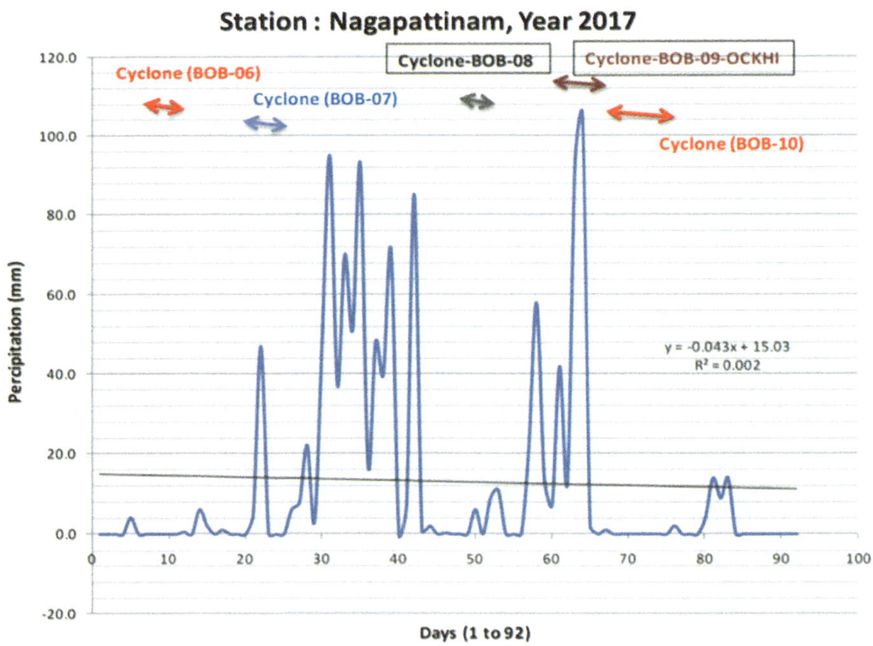

Fig. 4 Time series of observed rainfall during NE monsoon (2017)

The second cyclone BOB-07 formed in BOB at a distance in the range 1249–1581 km, during this period (day-19, day-20 and day 21) the Nagapattinam station received the rainfall of 50 mm. In the Nagapattinam station, the historical monsoon average rainfall of the cyclone period (day-19, day-20 and day 21) is 13.2 mm. Due to the cyclone (BOB-07) formation, the average rainfall of the three days (day-19, day-20 and day 21) is 17.4 mm. It is 31.81% higher when compared with historical NE monsoon average rainfall of the particular days.

The third cyclone BOB-08 (day-46, day-47 and day-48) track movement was at a distance 581–1192 km with respect to Nagapattinam station. Prior to this cyclone period, the station received the maximum rainfall of 84 mm. It was noticed that during the actual cyclone track period, the study station not received enough rainfall but earlier to the cyclone period, the station received enough rainfall.

The fourth cyclone BOB-09-OCKHI (Table 2) formed in the BOB region at a distance 482–1267 km to the Nagapattinam station. During the cyclone track (day-60 to day-66), the study station received maximum rainfall of 106 mm. In the Nagapattinam station, the historical monsoon average rainfall of the cyclone period (day-60 to day-66) is 15.7 mm. Due to the cyclone formation, the average rainfall for the seven days (day-60 to day-66) is 37.4 mm. It is observed that due to the formation of cyclone, the average rainfall of the Nagapattinam station increased by double when compared with the historical monsoon average rainfall of the particular day.

The last cyclone of this NE monsoon was BOB-10 (Table 2) and move at track distance of 870–1217 km. The station received minimum rainfall. (3 mm).

For this proposed study, the three-year (2015, 2016 and 2017) cyclone track data and the daily rainfall data of the Nagapattinam station were taken and considered as training data set in order to train our proposed time series model. Using this model, for the years 2018 and 2019 daily rainfall data and corresponding cyclone track data during the NE monsoon period were tested. Our proposed model found that the trained data cyclone BOB-12-2015 (Table 2) track-data (43–421 km) fit with BOB-13-GAJA-2018 (Table 2) cyclone track-data (30–1300 km) based on the movement distance some how was similar with respect to Nagapattinam station (In Fig. 5, the rainfall days are day-41 to day-50).

In calculating the average rainfall of the Nagapattinam station, 12-day rainfall data (training and test data set) were considered. This includes actual cyclone period, pre-cyclone period and post-cyclone period. The average rainfall of cyclone BOB-12-2015 and cyclone BOB-13-GAJA-2018 have been calculated as 45.7 mm and 15.2 mm, respectively.

From the observations, the proposed model concludes that any cyclone moves at a distance of 30–1300 km to the Tamil Nadu coastal station of BOB, the station will receive average rainfall in between 15.2 and 45.7 mm. Also it shows that the historical average rainfall of the 12-day period of the NE monsoon is 16.4 mm. Thus, our predicted result has been validated.

In another scenario, our model cyclone BOB-09-OCKHI-2017 was similar with BOB-12-TITLI-2018, the track data of both cyclones moved between 846 and 1501 km. Here, in calculating the average rainfall, only 10-day rainfall data (actual cyclone, pre-cyclone and post-cyclone periods) have been considered. The cyclone

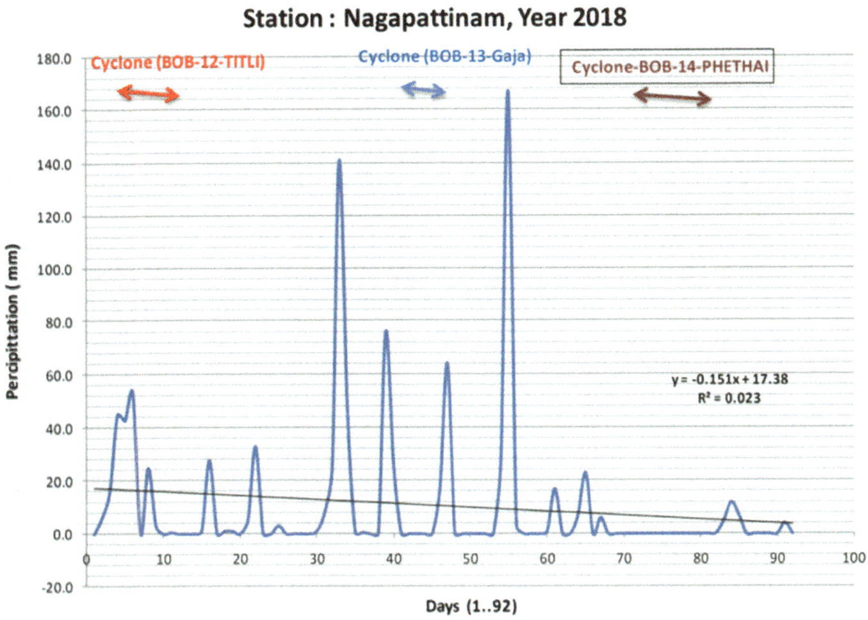

Fig. 5 Time series of observed rainfall during NE monsoon (2018)

BOB-12-TITLI-2018 (Table 2) and its track period with its corresponding rainfall data (day-8 to day-12) are shown in Fig. 5. From the observation, the proposed model has indicated that any cyclone moves with the distance from 846 to 1501 km to Tamil Nadu coastal station of BOB, the station will receive average rainfall in between 19.2 and 35.6 mm. The rainfall data and their time series model are as shown in Fig. 6.

5 Conclusion and Future Work

In the proposed time series model, five year NE monsoon cyclone track data and rainfall data have been analyzed. The rainfall at Nagapattinam station was predicted using this model during the cyclone track period. In the year 2017, during NE monsoon period, there were five cyclones in BOB region. During this cyclone track period, the Nagapattinam station got significant rainfall, when compared with the rainfall of other years of NE monsoon not forming of any cyclone. It is observed that when the cyclone movement was close to the particular coastal station, in most of the cases the station rainfall was positively correlated. In few cases, rainfall was not positively correlated, for example, when the cyclone BOB-08-NADA was closer to our study station, the station received only minimum rainfall. In calculating the rainfall received the station, the cyclone period, pre-cyclone and post-cyclone period were taken into account and more accurate results were observed. The proposed model

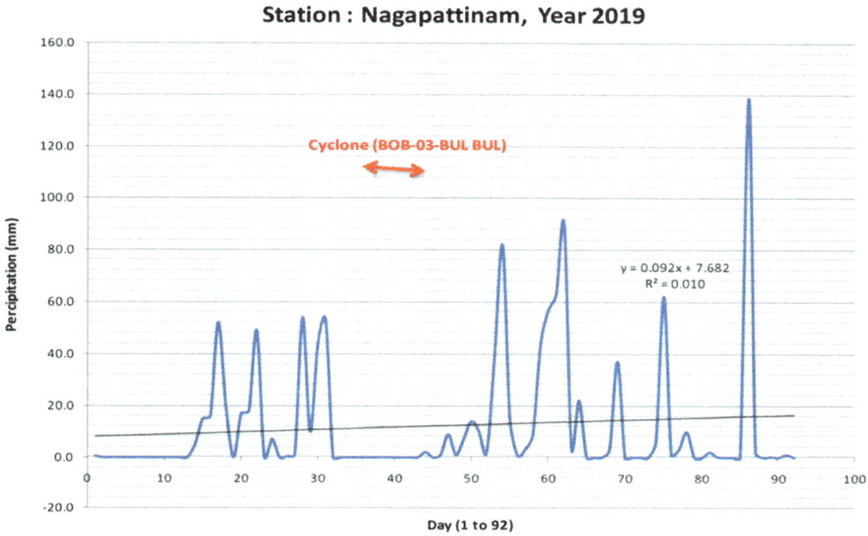

Fig. 6 Time series of observed rainfall during NE monsoon (2018)

concluded that when the cyclone track is at a distance in the range of 30–1300 km to Tamil Nadu coastal station of BOB, the station will receive average rainfall at the range of 15.2–45.7 mm. The proposed model also indicated that when the cyclone track is at a distance 846–1501 km to Tamil Nadu costal station of BOB, the station will receive an average rainfall at the range of 19.2–35.6 mm. It was planned to apply this model by considering the other parameters like wind speed, wind direction, air pressure and cyclone distance.

Acknowledgements The authors of this paper thank the scientists of India Meteorological Department Chennai, Tamil Nadu, India, and scientists of NOAA, WMO, USA; for having provided the factual meteorological data. The authors fervently thank Dr. S. VaidyaSubramaniyam, Vice Chancellor and Dr. S. Swaminathan, Dean (Planning and Development) of SASTRA Deemed University, for having given permission to utilize the DST-FIST sponsored Discrete Mathematics Laboratory at Srinivasa Ramanujan Centre, Kumbakonam, to carry out this research work. The authors would like to thank Y. E. A. Raj, Former Deputy Director General of Meteorology, RMC, Chennai, for his valuable comments and suggestions. Authors are grateful to the nameless reviewers for their valuable clarifications to enhance the quality of the paper.

References

1. M. Swapna, N. Sudhakar, A hybrid model for rainfall prediction using both parametrized and time series models. Int. J. Pure Appl. Math. **119**(14), 1549–1556 (2018)
2. T.R. Sivaramakrishnan, S. Meganathan, Data mining as a tool for precipitation prediction. Arch. Sci. **65**(3), 8 (2012)

3. T.R. Sivaramakrishnan, S. Meganathan, Association rule mining and classifier approach for quantitative spot rainfall prediction. J. Theor. Appl. Inf. Technol. **34**(2), 173–177 (2011)
4. T.O. Olatayo, A.I. Taiwo, Statistical modelling and prediction of rainfall time series data. Global J. Comp. Sci. Technol. **14**(1), 1–10 (2014)
5. M. Valipour, How much meteorological information is necessary to achieve reliable accuracy for rainfall estimations? Agriculture **6**(4), 53, 1–9 (2016)
6. N. Divya prabha, P. Radha, Prediction of weather and rainfall forecasting using classification techniques. Int. Res. J. Eng. Technol. (IRJET) **06**(02), 803–806 (2019)
7. Md. Abu Shahin, Md. Monimul Huq, F. Hasan, Md. Ayub Ali, A.B.M. Shawkat Ali, Time series analysis, modeling and forecasting of climate variable rainfall: a case study of Rajshahi district in Bangladesh. Int. J. Adv. Res. Rev. IJARR **1**(1), 71–81 (2016)

Multivariate Variational Mode Decomposition based Analysis on Stock Sectors

Silpa Balagopal, Vijay Krishna Menon, E. A. Gopalakrishnan, and K. P. Soman

Abstract The price of a single stock is seldom independent. It has been known to brokers and fund managers, that, they heavily influence each other. Portfolios are built, on the premise of minimizing such dependencies between stocks. There have been several efforts to quantify these dependencies, predominantly using conventional statistics and correlations. Most approaches use single independent variable-based approaches to compare two stock price signals at a time. In this paper, we analyze stock sectors with multiple stocks in each, presenting a multivariate case. We apply two multivariate analysis techniques, namely multivariate variational mode decomposition (MVMD) and spatio-temporal intrinsic mode decomposition (STIMD), to analyze stock sectors based on their individual stock day-wise price series. Sector-wise data is downloaded from Google Finance. Furthermore, we quantify the dependence of a sector on each of its constituent stock, by decomposing this multivariate signal and selectively reconstructing it multiple times, excluding each one from the reconstruction process. The error of reconstruction in each case serves as a measure of how much other stocks depend on the one that was excluded.

Keywords Mode Decomposition · Reconstruction Error · Multivariate VMD · STIMD · Stock Sectors · Portfolio Analysis

1 Introduction

Stock market is a place where trading of stocks happens. Stock price indices are a metric used to measure the performance of the stock market compared to an accepted but trending normal. Stock market consists of stock sectors, which is a group of companies active in the same economic sector, e.g., energy, infrastructure, telecommunications, banking, etc. Every sector has its own local performance indices.

S. Balagopal (✉) · V. K. Menon · E. A. Gopalakrishnan · K. P. Soman
Centre for Computational Engineering and Networking, Amrita Vishwa Vidyapeetham, Coimbatore, India

M. Paprzycki et al. (eds.), *Intelligent Systems, Technologies and Applications*, Advances in Intelligent Systems and Computing 1353,
https://doi.org/10.1007/978-981-16-0730-1_26

Stock market dynamics is an active area for multi-disciplinary research, for various applications including price predictions, market crash predictions [1] and correlation between stock markets [2, 3]. Research in this area relies on textual data like news articles, financial report and numerical data like price series and indices [4, 5]. For example, stock market crash prediction has been done using sentiment analysis (NLP) on news feed as well as using crash coefficients from stock prices.

Stock price prediction techniques have been improving over the years, from using statistical models like ARIMA [6], to application of machine learning (ML) and deep learning algorithms like CNNs, RNNs and LSTMs [7, 8]. It further went onto data-driven techniques such as DMD [9, 10], which is a non-orthogonal spatio-temporal algorithm.

It is really difficult to predict stock prices accurately as stock markets are very sensitive in nature [11], which means that it changes due to a variety of reasons that could be social, political and/or economical changes. The effect of these changes could be direct or indirect; the indirect changes are harder to identify [12, 13].

In order to better understand such changes, mode decompositions can be utilized. The modes or bases that we arrive upon after the decomposition process hold vital and fundamental information that constitute the input data [14].

Mode decompositions have been used for applications including quantum mechanics. From Fourier transforms to Hilbert transforms and to wavelets, decompositions have found importance in signal processing. A signal can be decomposed into multiple signals, each in a different frequency band, ordered according to its dominance in the original signal; hence, we get frequency modes.

Collating closing value of stock prices day after day gives us such a signal which can be decomposed into its maximum varying to minimum varying frequency modes. This helps us to better understand and correlate the governing factors behind price changing dynamics [15]. In more common terms, the influence a particular stock has on a sector can be better analyzed and quantified by group (multivariate) decomposition of all prices signals within the sector.

2 Multivariate Variational Mode Decomposition

Multivariate variational mode decomposition (MVMD) [16] algorithm was found by Naveed ur Rehman and Hania Aftab in 2019. It is a generalization of the variational mode decomposition (VMD) [17] algorithm for multivariate data. The objective of MVMD is to draw out a predefined number of modes, $uk(t)$ from the input data $x(t)$. These modes, each would have c number of channels, i.e., $x(t) = [x_1(t), x_2(t), \ldots x_C(t)]$.

$$x(t) = \sum_{k=1}^{K} u_k(t) \tag{1}$$

where $u_k(t) = [u_1(t), u_2(t), \ldots u_C(t)]$

MVMD decomposes the input data into frequency modes.

3 Spatio-Temporal Intrinsic Mode Decomposition

Spatio-temporal intrinsic mode decomposition (STIMD) [18] was introduced in 2018 by Seth M. Hirsh, Bingni W. Brunton and J. Nathan Kutzz. STIMD decomposes the mixed time series to intrinsic mode functions (IMFs)

The input X is decomposed to mode matrix S and its coefficient matrix B.

$$x_i(t) = \sum_{j=1}^{r} b_{i,j} s_j(t) \tag{2}$$

In matrix form,

$$X = BS \tag{3}$$

The reconstruction can be seen Fig. 1.

4 Data Description

Twenty-six stocks were downloaded from Google Finance. They comprised of five stocks from five sectors each, making a total of twenty five. The selected sectors include banking, IT, pharmaceutical, automotive and media (Table 1).

The stocks in each sector were joined based on the dates. The dataset has a dimension of $5 \times n$ where n is the number of data points available. The dataset specifications for each sector are detailed in Table 2.

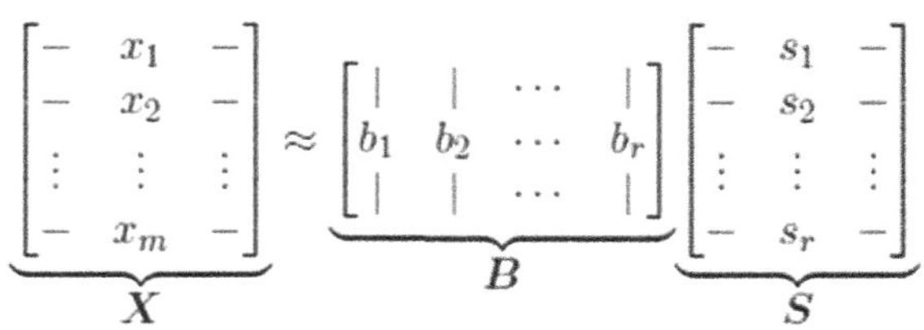

Fig. 1 Reconstruction logic in STIMD

5 Reconstruction Using STIMD

The multivariate stock data for a sector (with five price series each) was decomposed using STIMD [18]. The process yielded two matrices, namely the coefficient matrix, B and the mode matrix, S. For the reconstruction process, the two matrices were multiplied. Unfortunately, the reconstruction process gave us some uninterpretable results, wherein we could not compare the outputs with the original stock signals. Hence, a second technique was employed. The plot of the reconstruction of the stock prices using STIMD can be seen in Fig. 2.

6 Reconstruction Using Multivariate VMD

The second technique used for decomposition was MVMD [16]. Multivariate VMD computes 'n' modes, with 'n' channels each, which reflects all the combined information within the multivariate signal (the independent price signals of each constituent stock in the sector). The first input can be reconstructed by adding the first channels of all modes;similarly, other stocks were reconstructed. The modes are arranged as per the information captured by them from the original data; The first mode had most information, followed by the second so on and so forth till the final mode that will have the least information (Figs. 3, 4 and 5).

Table 1 Stocks in each sector

Banking	IT	Pharma	Auto	Media
YES	WIPRO	SUN	TATA MOTORS	ZEE
AXIS	HCL	BIOCON	BAJAJ	BALAJI
HDFC	INFOSYS	CIPLA	MAHINDRA	DISHTV
SBI	TATA ELXSI	LUPIN	MARUTI	PVR
ICICI	TCS	PIRAMAL	MRF/HEROMOTOCO	SUNTV

Table 2 Sector-wise data

Sector	No. of points	Start date	End date
Banking	3478	12th July 2005	28th August 2019
IT	3756	25th August 2008	29th November 2019
Pharmaceutical	3852	7th April 2004	29th November 2019
Automotive	3997	9th July 2003	29th November 2019
Media	3087	18th April 2007	29th November 2019
Inter-sector—major	3040	18th April 2007	28th July 2019
Inter-sector—minor	3025	18th April 2007	28th August 2019

In order to quantify the dependency, stocks are sequentially excluded from the sector, one at time and the same decompose reconstruct cycle is repeated for every stock in the sector. The MAPE error of the reconstructed signal over the original signal is computed and acts as a metric to measure to cross compare the performance in each case.

Figure 6 shows the stocks B, C, D and E being excluded one by one . The error in reconstruction, which was calculated using MAPE, was between A' & A, B' & B, C'& C, D' & D and E' & E.

The computational experiment indirectly measures the dependence and correlation, the other stocks have, on stock A. The average error of reconstruction is considered a quantification of this dependence. The experiment is extended to reconstruct all stocks in the same manner, for a given sector of five stocks.

The error of reconstruction of all the stock when excluding, say, stock A is collated and averaged, to quantify the importance of stock A with in the whole sector. This will aid financial players to understand and quantify the non-obvious dependencies between stocks in a sector, especially when building sensitive portfolios.

Fig. 2 Reconstructed and input stock prices using STIMD

Fig. 3 Decomposition of stock prices using MVMD for one sector of five stocks

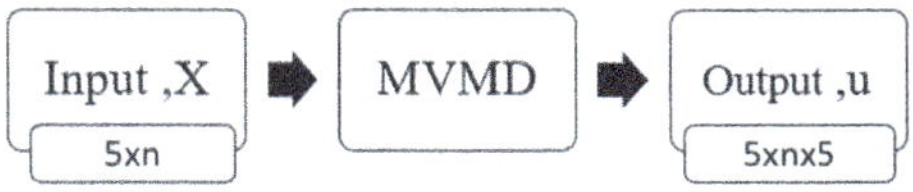

7 Results

Stock prices were decomposed and then reconstructed, by excluding one stock at a time. The results are tabulated in Sect. 7.1.

We have tabulated the mean absolute percentage error (MAPE) over all reconstruction of each stock excluding the stock in question. The rows show the stock which was excluded, and the columns indicate the stock that was reconstructed . For example, the reconstruction error of Axis Bank when Yes Bank was excluded can be observed in the column of Axis Bank intersecting the row of Yes Bank.

The first set of results is for the reconstruction using only first mode of every decomposition and is labeled as 'single channel'. The second set was the error noted while adding all the modes of every decomposition and is labeled as 'sum of all channels'.

In all the experiments, the error can be seen as following an order among the group. This can be seen if the values are read column-wise.

7.1 Reconstruction Error for Single Modes

In the banking sector, the error is highest when Axis Bank is excluded in the data for reconstruction. It is the lowest when State Bank of India is excluded. In the IT sector, the error is highest when HCL is excluded in the data for reconstruction. It is the lowest when TATA ELXSI is excluded. In the pharmaceutical sector, the error is highest when PIRAMAL is excluded in the data for reconstruction. It is the lowest when CIPLA is excluded. In the automotive sector, an order of influence could not be found. MRF being a tyre company was hence replaced with HEROMOTOCO stock. However, on choosing MRF, the error showed a proper order, and the stock of MARUTI showed highest value. In the media sector, the error is highest when SUN TV is excluded in the data for reconstruction. It is the lowest when PVR is excluded. In

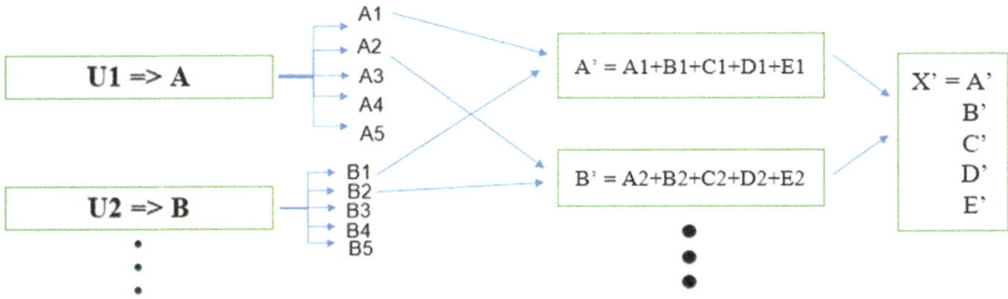

Fig. 4 Reconstruction of stock prices using MVMD for one sector of five stocks. The input matrix 'X' contains the data from one sector, with five stock price series for n days. In order to reconstruct the original signal, the output channels from each 'u' (output) is linearly combined

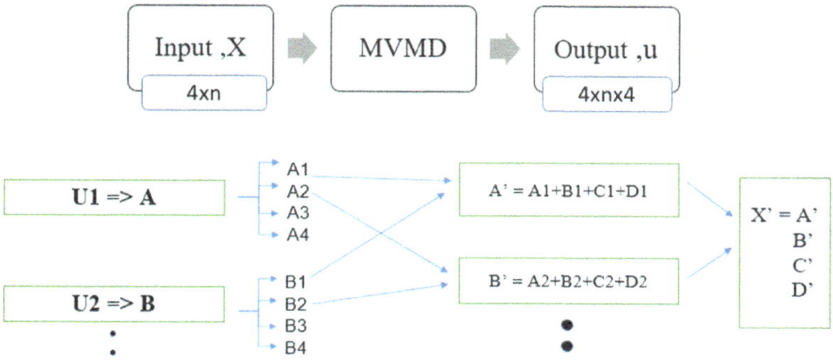

Fig. 5 Reconstruction of stock prices using MVMD excluding one stock price from a dataset

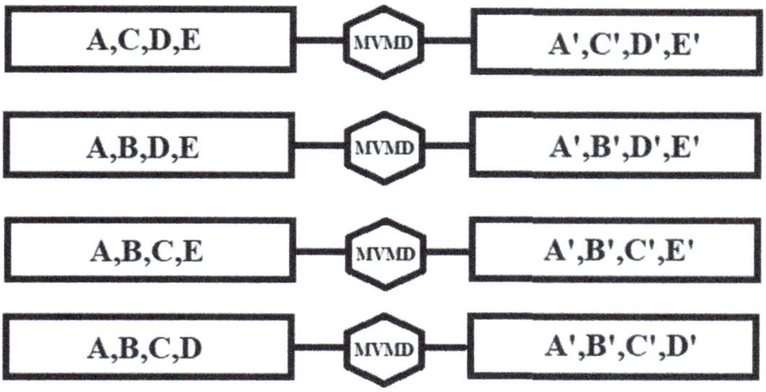

Fig. 6 Pictorial representation of the process when stock A was reconstructed by excluding all the other stocks one at a time

the inter-sector—major group, we have companies that showed maximum error from each sector. The error is highest when PIRAMAL is excluded in the data for reconstruction. It is the lowest when MARUTI is excluded (Tables 3, 4, 5, 6, 7, 8, 9 and 10).

In the inter-sector—minor group, we have companies that showed lowest error from each sector. The error is highest when CIPLA is excluded in the data for reconstruction. It is the lowest when PVR is excluded.

Table 3 Banking sector (single channel)

Company	YES	AXIS	HDFC	SBIN	ICICI
YES	–	1.345	0.939	1.227	1.340
AXIS	1.700	–	1.011	1.327	1.453
HDFC	1.557	1.336	–	1.216	1.332
SBIN	1.176	1.022	0.722	–	1.008
ICICI	1.542	1.324	0.926	1.205	–

7.2 Reconstruction Error for Sum of All Modes

The results of reconstruction when channels from all the modes are added follow the same pattern as for single mode, except for media and inter-sector—minor. Another important point to note is that the error value considerably reduces when compared to the single mode reconstruction.

In the banking sector, the error is highest when Axis Bank is excluded in the data for reconstruction. It is the lowest when State Bank of India is excluded (Table 11).

In the IT sector, the error is highest when HCL is excluded in the data for reconstruction. It is the lowest when TATA ELXSI is excluded (Table 12).

In the pharmaceutical sector, the error is highest when PIRAMAL is excluded in the data for reconstruction. It is the lowest when CIPLA is excluded (Tables 13, 14 15, 16 and 17).

In inter-sector—minor, TATA ELXSI has the largest error, and PVR has the lowest, which is different from the single mode reconstruction. In media sector, a proper order could not be established. The automotive sector in single channel reconstruction showed that some combinations fail to give a proper order.

The differences in media and inter-sector—minor datasets, need more analysis. It could be due to some specific dynamics of the five stocks.

In all the above experiments, we see that the stock prices have an effect on other stock prices, since their presence in the group reduces the error in reconstruction of other's stock prices. This can be an influence of company dynamics directly or indirectly. This correlation is in most cases clearly visible in the value of the error in reconstruction.

Table 4 IT sector (single channel)

Company	WIPRO	HCL	INFOSYS	TATA ELXSI	TCS
WIPRO	–	0.865	0.713	0.971	0.7
HCL	1.235	–	1.238	1.735	1.3
INFOSYS	1.146	1.368	–	1.584	1.2
TATA ELXSI	0.793	0.919	0.758	–	0.8
TCS	0.901	1.058	0.877	1.196	–

Table 5 Pharmaceutical sector (single channel)

Company	SUN	BIOCON	CIPLA	LUPIN	PIRAMAL
SUN	–	1.194	1.070	1.089	1.28
BIOCON	1.092	–	1.080	1.103	1.30
CIPLA	0.905	0.990	–	0.903	1.07
LUPIN	1.091	1.212	1.080	–	1.30
PIRAMAL	1.096	1.215	1.085	1.108	–

Table 6 Automotive-1 sector (single channel)

Company	TATA MOTORS	BAJAJ	MAHINDRA	MARUTI	MRF
TATA MOTORS	–	1.743	1.789	1.699	1.854
BAJAJ	2.126	–	1.789	1.699	1.853
MAHINDRA	2.126	1.743	–	1.699	1.854
MARUTI	2.126	1.743	1.789	–	1.854
MRF	2.007	1.626	1.686	1.599	–

Table 7 Automotive-2 sector (single channel)

Company	TATA MOTORS	BAJAJ	MAHINDRA	MARUTI	HEROMOTO
TATA MOTORS	–	1.727	1.777	1.688	1.556
BAJAJ	2.112	–	1.777	1.688	1.556
MAHINDRA	2.112	1.727	–	1.688	1.556
MARUTI	2.115	1.730	1.779	–	1.559
HEROMOTOCO	2.110	1.725	1.775	1.687	–

Table 8 Media sector (single channel)

Company	ZEE	BALAJI	DISH TV	PVR	SUN TV
ZEE	–	2.006	2.034	1.659	1.84
BALAJI	1.575	–	2.081	1.618	1.798
DISH TV	1.559	1.958	–	1.616	1.795
PVR	1.505	1.869	1.974	–	1.712
SUN TV	1.651	2.063	2.199	1.71	–

Table 9 Inter-sector—major (single channel)

Company	SUN	HCL	AXIS	MARUTI	PIRAMAL
SUN	–	1.754	1.994	1.589	1.81
HCL	2.2	–	1.993	1.588	1.81
AXIS	2.2	1.754	–	1.588	1.81
MARUTI	2.156	1.718	1.949	–	1.77
PIRAMAL	2.201	1.755	1.994	1.589	–

Table 10 Inter-sector—minor (single channel)

Company	CIPLA	HEROMOTOCO	PVR	SBIN	TATA ELXSI
CIPLA	–	1.524	2.062	1.898	2.094
HEROMOTOCO	1.241	–	1.83	1.699	1.857
PVR	1.032	1.127	–	1.408	1.528
SBIN	1.385	1.524	2.062	–	1.094
TATA ELXSI	1.385	1.524	2.062	1.564	–

Table 11 Banking sector (all channels)

Company	YES	AXIS	HDFC	SBIN	ICICI
YES	–	0.100	0.058	0.081	0.088
AXIS	0.143	–	0.077	0.108	0.117
HDFC	0.103	0.098	–	0.078	0.085
SBIN	0.039	0.058	0.022	–	0.036
ICICI	0.10	0.096	0.055	0.077	–

Table 12 IT sector (all channels)

Company	WIPRO	HCL	INFOSYS	TATA ELXSI	TCS
WIPRO	–	0.019	0.024	0.043	0.025
HCL	0.149	–	0.151	0.228	0.154
INFOSYS	0.113	0.132	–	0.174	0.117
TATA ELXSI	0.023	0.028	0.029	–	0.03
TCS	0.036	0.044	0.041	0.067	–

8 Conclusion

Analyzing stock price signals within a sector, treating them as a multivariate signal, using MVMD effectively allow us to study and quantify their interdependencies. The stocks in a sector were decomposed and reconstructed with one of them selectively

Table 13 Pharmaceutical sector (all channels)

Company	SUN	BIOCON	CIPLA	LUPIN	PIRAMAL
SUN	–	0.086	0.082	0.089	0.099
BIOCON	0.089	–	0.084	0.092	0.102
CIPLA	0.052	0.053	–	0.053	0.054
LUPIN	0.091	0.098	0.085	–	0.104
PIRAMAL	0.099	0.108	0.094	0.102	–

excluded in each sector; exclusion was repeated for every stock in the multivariate signal (sector). We have used the error of reconstruction, of each individual stock price signal, as a measure of the influence of the missing stock on them.

The experiments were conducted in two categories, one, where only the dominant mode channel was used in the reconstruction process and a second category which used all the modes for reconstruction. Empirically, both categories have insights that clearly indicate the degree of influence between stocks. As expected, the second method gives smaller errors at the same time the pattern of influence is maintained as in the first approach.

We thus come to the conclusion that correlation does exist among stock sectors. This is not only inside a stock sector but outside them as well. The extent of influence of the stocks can be seen in the error in reconstruction while that stock is not included in the input. A higher error indicates that the information of that stock is vital in arriving at the stock price of the reconstructed stock. Another interesting inference is that, in both inter-sector experiments, pharmaceutical sector showed maximum influence. The future work could explore the possibilities of dynamics of a particular

Table 14 Automotive sector (all channels)

Company	TATA MOTORS	BAJAJ	MAHINDRA	MARUTI	HEROMOTO
TATA MOTORS	–	0.337	0.332	0.324	0.302
BAJAJ	0.400	–	0.330	0.322	0.298
MAHINDRA	0.401	0.337	–	0.323	0.299
MARUTI	0.413	0.348	0.341	–	0.308
HEROMOTOCO	0.400	0.338	0.330	0.322	–

Table 15 Media sector (all channels)

Company	ZEE	BALAJI	DISH TV	PVR	SUN TV
ZEE	–	0.286	0.301	0.236	0.25
BALAJI	0.198	–	0.272	0.214	0.226
DISH TV	0.199	0.259	–	0.215	0.226
PVR	0.202	0.263	0.283	–	0.23
SUN TV	0.217	0.282	0.294	0.145	–

Table 16 Inter-sector—major (all channels)

Company	SUN	HCL	AXIS	MARUTI	PIRAMAL
SUN	–	0.311	0.294	0.294	0.328
HCL	0.384	–	0.295	0.295	0.328
AXIS	0.386	0.312	–	0.296	0.330
MARUTI	0.382	0.305	0.296	–	0.324
PIRAMAL	0.383	0.310	0.260	0.294	–

Table 17 Inter-sector—minor (all channels)

Company	CIPLA	HERO	PVR	SBIN	TATA ELXSI
CIPLA	–	0.351	0.457	0.423	0.487
HERO	0.185	–	0.258	0.52	0.301
PVR	0.098	0.110	–	0.15	0.160
SBIN	0.303	0.357	0.138	–	0.495
TATA ELXSI	0.310	0.364	0.442	1.48	–

stock with different groups of stocks. It can also explore the results of this paper on the global market as well.

References

1. R. Zhao, Quantifying the correlation of media coverage and stock price crash risk: a panel study from China. Phys. A Stat. Mech. Appl. **537**, 122378 (2020)
2. E. Simiyu, J. Korir, G. Laiboni, Dynamic linkages of stock market sectors: evidence from the Kenyan stock market. J. Fin. Econ. **8**(1), 33–9 (2020)
3. J.W. Yu, W.J. Xie, Z.Q. Jiang, Early warning model based on correlated networks in global crude oil markets. Phys. A Stat. Mech. Appl. **490**, 1335–1343 (2018)
4. V.K. Menon, N.C. Vasireddy, S.A. Jami, V.T.N. Pedamallu, V. Sureshkumar, K.P. Soman, Bulk price forecasting using spark over NSE data set, in *International Conference on Data Mining and Big Data* (Springer, Cham, 2016), pp. 137–146
5. T.H. Nguyen, K. Shirai, J. Velcin, Sentiment analysis on social media for stock movement prediction. Expert Syst. Appl. **42**(24), 9603–9611 (2015)
6. G.P. Zhang, Time series forecasting using a hybrid ARIMA and neural network model. Neurocomputing **50**, 159–175 (2003)
7. M. Hiransha, E.A. Gopalakrishnan, V.K. Menon, K.P. Soman, NSE stock market prediction using deep-learning models. Proc. Comput. Sci. **132**, 1351–1362 (2018)
8. S. Selvin, R. Vinayakumar, E.A. Gopalakrishnan, V.K. Menon, K.P. Soman, Stock price prediction using LSTM, RNN and CNN-sliding window model, in *2017 International Conference on Advances in Computing, Communications and Informatics (ICACCI)* (IEEE, 2017), pp. 1643–1647
9. D.P. Kuttichira, E.A. Gopalakrishnan, V.K. Menon, K.P. Soman, Stock price prediction using dynamic mode decomposition, in *2017 International Conference on Advances in Computing, Communications and Informatics (ICACCI)* (IEEE, 2017), pp. 55–60
10. N.A. Unnithan, E.A. Gopalakrishnan, V.K. Menon, K.P. Soman, A data-driven model approach for daywise stock prediction, in *Emerging Research in Electronics, Computer Science and Technology* (Springer, Singapore, 2019), pp. 149–158
11. R. Zhang, X. Xian, H. Fang, The early-warning system of stock market crises with investor sentiment: evidence from China. Int. J. Fin. Econ. **24**(1), 361–369 (2019)
12. C. Liu, Y. Zheng, Q. Zhao, C. Wang, Financial stability and real estate price fluctuation in China. Phys. A Stat. Mech. Appl. **540**, 122980 (2020)
13. L. Zhou, L. Qiu, C. Gu, H. Yang, Immediate causality network of stock markets. EPL (Europhys. Lett.) **121**(4), 48002 (2018)
14. J. Mann, J.N. Kutz, Dynamic mode decomposition for financial trading strategies. Quant. Fin. **16**(11), 1643–1655 (2016)

15. I.H. El-adaway, G.G. Ali, I.S. Abotaleb, H.M. Barber Jr., Studying the relationship between stock prices of publicly traded us construction companies and gross domestic product: preliminary step toward construction-economy Nexus. J. Construct. Eng. Manage. **146**(1), 04019087 (2020)
16. N.U. Rehman, H. Aftab, Multivariate variational mode decomposition. IEEE Trans. Signal Process. **67**(23), 6039–6052 (2019).https://doi.org/10.1109/TSP.2019.2951223
17. K. Dragomiretskiy, D. Zosso, Variational mode decompositio. IEEE Trans. Signal Process. **62**(3), 531–544 (2014). https://doi.org/10.1109/TSP.2013.2288675
18. S.M. Hirsh, B.W. Brunton, J. Nathan Kutz, Data-driven spatiotemporal modal decomposition for time frequency analysis (2018). arXiv:1806.08739

Case-Based Expert System for Smart Air Conditioner with Adaptive Thermoregulatory Comfort

Akshaya Sundaram, Hamsini Ravishankar, Uma Subbiah, Nalinadevi Kadiresan, and R. Karthika

Abstract With the onset of the Internet revolution and development in hardware, IoT has become an integral part of our lives, optimising various aspects of day-to-day activities. An important role of IoT is played in providing thermal comfort through smart heating, ventilation, and air conditioning (HVAC) systems. The smart AC proposed in this paper aims to create a pleasant thermal environment with minimal human intervention, according to various personal and environmental parameters using a case-based expert system. The novelty of this system is a case-based expert system that has been implemented with querying and analytics to identify and set the temperature of the AC. Based on users' prior settings, preferences, and ambient conditions, the knowledge base will continue to expand. Moreover, if the user overrides the actuated temperature with a preference of his/her own, this will be stored as a new case in the case database. This smart AC system also considers the theory that resulted from studies on the variation in human temperature caused by human thermoregulation. Integrating all of these factors, the most similar case is extracted from the case database using the k-NN regression similarity criteria. The system has been optimised to minimise the number of re-computations with respect to actuation and the number of retrievals performed on the knowledge base. Equipped with cloud-based data utilisation, behavioural query analytics, and historical query analytics, this futuristic system possesses the ability to automatically perform optimal actuation on the air conditioner, making it more convenient and user friendly than its present-day counterparts.

A. Sundaram · H. Ravishankar · U. Subbiah · N. Kadiresan (✉)
Department of Computer Science and Engineering, Amrita School of Engineering, Coimbatore
Amrita Vishwa Vidyapeetham, Coimbatore, India
e-mail: k_nalinadevi@cb.amrita.edu

R. Karthika
Department of Electronics and Communication, Amrita School of Engineering, Coimbatore
Amrita Vishwa Vidyapeetham, Coimbatore, India
e-mail: r_karthika@cb.amrita.edu

M. Paprzycki et al. (eds.), *Intelligent Systems, Technologies and Applications*,
Advances in Intelligent Systems and Computing 1353,
https://doi.org/10.1007/978-981-16-0730-1_27

Keywords IoT · Case based reasoning · Expert system · Smart AC · k-NN similarity measure · Knowledge base · Cloud analytics

1 Introduction

Internet of things (IoT) is a domain that enables the interconnection of everyday objects through the Internet, to enable increased sync between them and comfort in our lives. IoT has been used in air conditioning systems both on a localised (in one or two rooms) and on a more globalised (throughout a building) scale. This paper proposes a smart air conditioning system that is driven by a case-based expert system. A NoSQL case database stores the set of previously encountered cases, along with the user approved action decisions. The knowledge base consists of a set of rules that dictates the action decision on the air conditioner. In addition, this knowledge base is constantly updated and continues to grow, gaining more information with each case. In this way, the case database and the knowledge base work together to be dynamic and adaptable to human thermal requirements.

With careful consideration of the available protocols, this system is designed to incorporate the Message Queuing Telemetry Transport (MQTT) publisher-subscriber architecture, along with Transmission Control Protocol (TCP) and IPv4. Lower in the stack, our system uses MAC as the data link layer protocol and IEEE 802.15.4—Open sensor network standard—in the physical layer. In order to achieve maximal thermal comfort, this system uses five sensors to capture relevant data, while also considering the human thermoregulation patterns and the natural variation of body temperature throughout a day. These sensors used have been selected after an extensive study on the factors that help provide optimal thermal comfort. Additional factors that have been studied and shown to have an influence on the human body temperature include location, age, gender, health status, height, and weight which are collected from the user at the time of registration in the system. These are all integrated, and an interactive graphic user interface is provided for easy human interfacing. The performance of our system is measured by the feedback that is obtained from the user after actuation.

This paper contains the following sections. Section 2 details an extensive literature survey of the current and significant research conducted in this field. Section 3 details the design of our system. Section 4 is an explanation of the working of the case-based expert application along with the query format, case format, case matching, knowledge base, and the feedback evaluation. The proposed system is concluded in Sect. 5.

2 Literature Survey

2.1 Theory of Human Thermoregulation

The human body temperature changes continuously throughout the night. Zakri [1] suggests that the body temperature directly correlates with the sleep quality of a person. The details of the thermoregulation are tabulated in [1] and can be used as a reference to train the IoT-based air conditioner proposed in our system. Since human beings are endotherms, the body temperature changes in accordance with the surroundings, to regulate the temperature and maintain it at a relatively constant number [2].

Moreover, [3] shows that there exists a relation between the approximate time a person wakes up (his sleep habits—early bird or night owl) and the temperature variations during one's sleep. The phases of sleep and various stages also affect the way the body temperature is maintained. Kelly [4] introduces the concept of a multitude of factors affecting our body temperature like fitness and ageing. Further extension of our project can be to incorporate this myriad of factors with the AC to provide the ultimate optimal temperature for comfort.

2.2 IoT Hardware and Software Requirements

IoT has a wide variety of applications, some of the more successful being smart cities, smart homes, and smart buildings. In all of these models, a variety of software (protocols, cloud integration services, etc.) and hardware (sensors, base microcontroller, and connectivity components) combinations are possible. The optimal combination ultimately depends on the application and environment of operation. Abouzar et al. [5] propose a flexible layered approach of selecting the hardware enabling technologies and software platforms to address the different design challenges, to develop the smart cities. This hierarchical architecture is effective at connecting the things and people in the environment with the cloud services. An energy-efficient, low-cost alternative to the traditional home automation is made possible through the use of an IoT prototype called IoT@HoMe [6].

Their choice of hardware (a base NodeMCU) along with software (adafruit IO, Google Assistant) and the MQTT protocol provides an option for low-cost prototypes. Another low-cost solution in terms of a building automation system uses an Arduino Uno with the required sensors (LM35) and connectivity components (HC05—Bluetooth component) [7]. A more extensive survey of the options available for software and protocol choices along with their functions and potential challenges is discussed in [8].

2.3 IoT in HVAC Systems

The highest IoT-based consumers of energy are heating systems [9], in which IoT is used to reduce the excess wastage of energy and increase the subsequent efficiency of the resulting systems. The analysis stage of the system is done using an extreme learning machine, and there has been a reported decrease in the greenhouse gas production levels of these systems. The approach detailed here can be effectively adapted to our approach too, in order to reduce the carbon emissions of our system. The enhancement of the IoT analytics phase using predicted mean vote (PMV) along with data collected from the sensors to provide satisfactory thermal comfort is described in [10].

The use of IoT and analytics to overcome problems faced in providing customisable services in HVAC systems is detailed in [11]. Extending the local HVAC systems into a global solution (throughout a building), while optimising energy consumption and providing thermal comfort, can be done using model predictive control (MPC) [12]. This energy reduction is further optimised in [13] using self-powered wireless sensors (SPWS). The success of this motivates the use of SPWS in our proposed system, while considering energy requirements. An alternative method to reduce energy consumption is by using time-series predictive models, based on deep learning, which can lead to a reduction of 20–40% in energy requirements as reported in [14].

2.4 Expert Systems and Case-Based Reasoning in IoT

IoT has many applications in the field of health care, for instance, the prevention of adverse effects of elderly people falling [15]. The use of expert systems in health care for creating smart homes that aid elderly people by monitoring activities of people in the smart home and suggesting behavioural changes using clinician-based machine learning is promising [16]. The extension of a case-based system into the IoT domain to pervade the medical health care cycle of diagnosis, monitoring, treatment, and rehabilitation, using IoT for monitoring along with a decision system, is seen in [17].

The use of CBR paired with a hierarchical knowledge-based indexing is seen in [18]. Retrieval of cases in [18] also depends on features, which depend on domain knowledge. A proposal is also made to include artificial neural networks on top of the CBR system to optimise the adaptation of known cases from the knowledge base to the new input case. [19] talks about an expert system for concept design of economical car (ESCDEC) using the traditional four-step case-based reasoning process—retrieve, reuse, revise, and retain. There is ease at which solutions for problems in specific scenarios are attributed to the stronger resemblance of this method to the natural thinking process (cerebration), along with the continual increase in the knowledge base, as new cases are added to the system. The adaptation of case-based expert analytics to adjust the behaviour of the AC is an alternative to traditional and hybrid machine learning analytical strategies, with the hybrid models proven to have better leverage over the individual models. Hybrid strategies like using the SVM-KNN as a component classifier with AdaBoost ensemble have outperformed the use of its individual counterparts on a mammogram classification [20].

3 The Architecture

The human-thermo AC expert application is unique for its query handling strategies that are obtained directly from the inferential analytics fed into the knowledge base from the cloud and a case database of previously decided cases. The knowledge base in the expert system is a data store of the theories and logical decisions to ensure the thermal comfort of the individual. It also derives its knowledge from the behavioural and historical analytics derived from the cloud. The queries are aimed to give the user an informed decision that would help maintain the satisfied thermal comfort of everyone in the room with the smart AC installed. The primary stakeholders of such a system are the smart AC manufacturers, the cloud service providers, and the users.

3.1 IoT Architecture of Proposed Smart Air Conditioning System

The human thermoregulated case expert AC has five sensors in the physical layer—humidity, CO sensor, a temperature sensor, a body worn sensor, and a sensor to measure the carbon dioxide levels. The sensors communicate with the Arduino microcontroller. The microcontroller is embedded in the AC, and it wirelessly communicates with the rest of the system. Based on the inputs from the sensor, simple analysis is done in the edge or cloud. The inference drawn from this analytics is fed to the knowledge base of the case-based expert application. The messaging protocol we have used in the application layer is MQTT, which is a publisher-subscriber architecture using a broker in the constrained IoT environment which is similar to other messaging protocols that exist in the client–server model. TCP is a standard specifying how a network connection should be built and maintained for computer programs to share data. TCP works with the Internet protocol (IP), which defines how computers are sending data packets to each other.

The MAC allows MAC frames to be transmitted using the physical channel. In addition to the data service, an administration interface is provided, and the physical channel and network beaconing are managed. It also checks the validation of the frame, manages nodes, and assures time-slots. Finally, it provides secure services with hook points [21]. Basically, the physical layer (PHY) provides the data transmis-

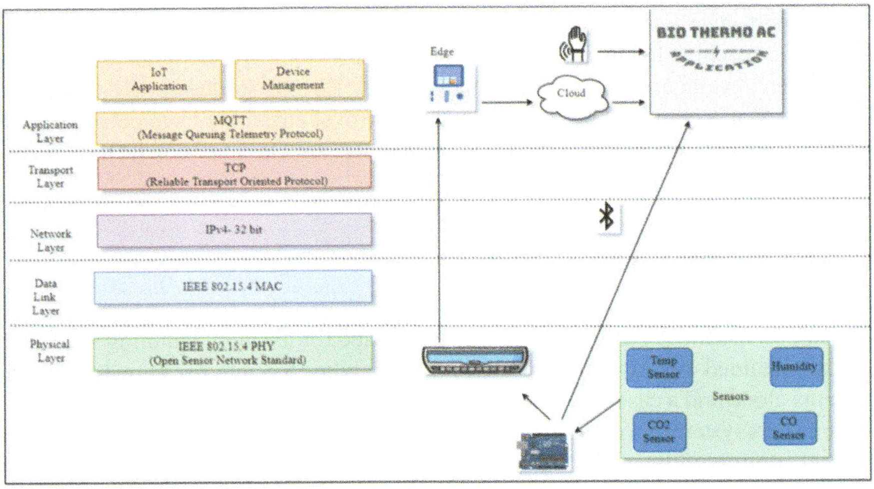

Fig. 1 System architecture of smart AC for the thermal comfort of an individual

sion facility, as well as the interface to the physical layer management organisation, which provides access to any layer management function and maintains a knowledge database on relevant personal area networks, as mentioned in Fig. 1.

3.2 *Device Management*

Device management is very fundamental to ensuring and supporting the security, connectivity, and health of any virtual or physical device. Therefore, it becomes a necessity to discuss how the device management requirements would be satisfied by any IoT-based system, as it is primarily a connection of physical and hardware devices. The four major IoT device management requirements are provisioning and authentication, configuration and control, monitoring and diagnostics, and software maintenance and updates [22].

Provisioning and Authentication All smart ACs would be given a unique identification using mixed characters and numbers. Moreover, the sensors embedded in the ACs would also be given a unique enrolment ID. The customers would also be given a unique ID. At the time of purchase, the AC and customer ID would be mapped and enrolled into a central database system for authorisation of customer services. The customers would also be given encrypted credentials to connect to the external expert system application with which they could query and interact with it. Enrolled customers are only authorised to use the services of the expert application. In the case that any other member would like to have their own access to a smart AC purchased, then they would also have to enrol independently.

Configuration and Control The facility to control, configure and monitor the smart AC is established after framing security policies on data and remote control levels. There are many configuration management activity models that could be used to ensure remote configuration and control. Configuration management systems create a sustainable chain among the smart ACs that would ensure configuring it with the latest updates to improve performance, functionality, and security [23].

Monitoring and Diagnostics Any expert system application bugs or the smart AC operational breaches or system security invasion could be detected by a central monitoring system that would constantly monitor the system and network. In-person monitoring could be replaced with cloud-hosted monitoring software that would alert and provide insights on unforeseen problems in the operations of the smart AC and the case-based expert application. Diagnostics in the use of application and the AC would also be logged or reported into the database dedicated to record diagnostics. Such a database would help to also provide inputs or fix a common security vulnerability for any new software maintenance and update that is needed.

Software Maintenance and Update Remote updates to the expert application would ensure to keep the system up to date with the latest security measures and customer service standards. The knowledge base of the case-based expert software also needs to be maintained and updated. Such updates would allow the system to be a promise for long-term investments. Updates will be remotely communicated to the system, when there is a persistent connection established. Software must be maintained and supported with new feature enhancements or new set of query handling by the expert system to keep the customers engaged in the system.

3.3 Challenges in the System

The operations of the smart AC are dependent on the cooperation between the services of the cloud, AC manufacturer, and the application. The expert system needs to use the Internet services to keep updating its knowledge base. It also comes along with the other disadvantages of using an expert system like the lack of creative decisions or the lack of adaptability to changing situations, unless the knowledge base is explicitly updated.

4 Case-Based Expert Application for Thermoregulation Comfort

The working of the case-based expert application for smart AC actuation is depicted by Fig. 2. Query is the sensor reading based on an event triggered by change in body temperature, captured by the body worn sensor. When the body temperature changes, then it raises an event that takes the sensor readings as a query. The sensor values about

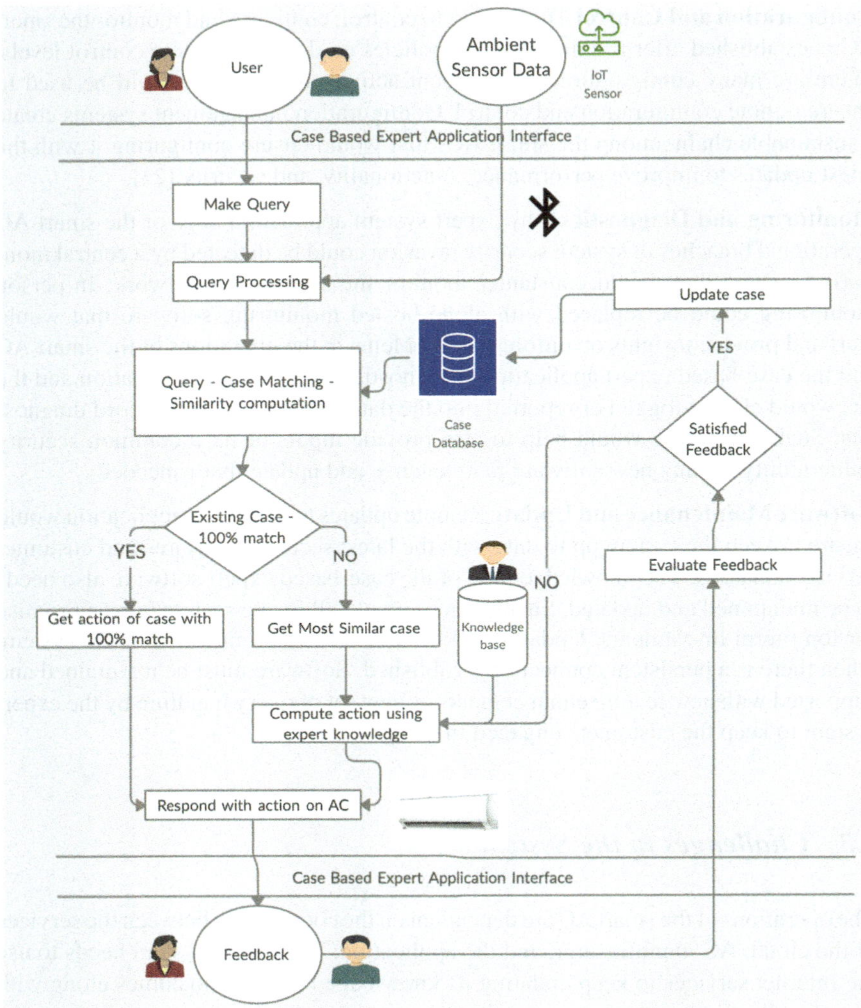

Fig. 2 Case-based expert application for thermal comfort—flow diagram

the ambient data of the room and the human body temperature is communicated to the application through Bluetooth. The query is then parsed, translated, and encapsulated with the ambient sensor values to the case format and executed to find a match with the already existing cases in the case database. There is a similarity algorithm that aims to match the existing cases with the newly encountered one. The K-nearest neighbour regression is used to calculate the numerical target, which is the AC temperature, based on the Euclidean distance function. A typical AC operates in the range of 18–30 °C.

For two cases to be 100% similar, the query parameter values and case parameter values need to fall within an established deviation. For example, if we allow a deviation of 5 for ambient temperature, 2 for body temperature, 20 for humidity, 10 for CO, and 10 for CO_2 sensors, then all cases with the individual parameters falling within one standard deviation of the above values will be 100% similar. At this step, the next decision is based on two different instances. In the instance, where the system has already registered a similar case in its database, then the same action would be actuated on the smart AC. But, if there is not a 100% match, then the most similar case is retrieved, and the action is computed using the case parameters and the inferences drawn from the knowledge base. In both the scenarios, regular feedback is collected from the user on how comfortable he/she is with the thermal tuning of the AC. Such a feedback evaluation would ensure a continued learning of the case-based expert system.

The feedback survey is then evaluated and given a 'satisfactory' score. If the score is beyond a reasonable threshold, then it means that the thermal comfort is achieved and the case along with the decision is updated into the database. But, if the feedback is negative, then a new action is re-computed from the expert knowledge base and actuated on the AC. This step continues until a satisfactory feedback is formed.

4.1 Query Format

Query is the reading of the sensor, based on an event caused by a shift in the values of the sensor. The application allows the processing of probabilistic queries and spatial queries related to parameters like temperature and fan movement, respectively. The query format consists of attributes like CO sensor, CO_2 sensor, humidity sensor, ambient temperature, and body worn temperature sensor readings.

4.2 Case Format

A case is a dynamic sequence of events or parameters with the decision made. The JSON format is used to represent a case in the system. MongoDB, a NoSQL database, is taken as the case database for its flexibility to store each case as a document. Figure 3 represents a case instance in JSON format, along with the parameters and the decision to be made.

4.3 Case Matching

The Euclidean distance measure of k-NN regression algorithm is used for matching the retrieved cases with the incoming query. Before case matching, the incoming

Fig. 3 JSON Format of case instance


```
{"case based":{
        "case ID":"123",
        "userID":"001",
        "Temperature":"26.3",
        "Humidity":"0.55",
        "CO sensor":"25",
        "CO2 sensor":"10",
        "human temperature":"37",
        "feedback_metric": "12",
        "number of people":"2",
        "retreival_index":"4",
        "decision":"22"
    }
}
```


query is processed and converted to the JSON format. The k-NN uses only the common labels between the case and query formats like the sensor values to compute the k-NN regression value. The set of cases are then retrieved, and the matching is checked. Two parameter values are same if the values fall within small deviations from one another. If there is a 100% match, then the action decision corresponding to the retrieved case is actuated on the AC. If there is no 100% match, then the prediction of the k-NN regression is actuated after comparing and reviewing it with the expert knowledge base.

4.4 Knowledge Base

There is only one knowledge base, which is a word document. The contents of the document are HVAC theory, human thermoregulatory theory, and the inferences fed from the cloud analytics. The feedback collected by the application is also taken in the knowledge base, which is used to understand the user preference. Although the former two theories are only updated if there follows a significant finding in that field, the cloud analytics in the knowledge base is updated every time the application gets connected to the Internet, and the feedback is updated every time it is collected. The action is decided by combining the facts of a particular case with the knowledge contained in the knowledge base to produce a recommendation.

A production rule is a part of IF statement (a condition) and a part of THEN statement (an action). IF condition THEN action (conclusion) is the syntax of the production rule. IF (hr= avg_sleep_time-2) THEN (actuate the AC to higher temperature) is an example of a rule in the system. The process of inferencing shifts from the facts of the case to a conclusion. Thus, the approach is guided by the information available in the working memory and by the assumptions which can be met. The inference engine tries to align the condition (IF) part of each law in the knowledge base with the facts or case in the working memory at present [24].

The prediction of k-NN regression is additionally verified by the knowledge base before it is taken for actuation. In case there is an inconsistency between the two, then a small deviation is introduced in the value predicted by the k-NN regression using the production rules in the knowledge base. Consequently, the number of retrievals will be reduced through which optimisation is achieved. The cases which have been retrieved the most are retained in the case database, and the most unused cases, indicated by the retrieval_index parameter, are discarded. Behavioural query and historical query analytics is done on the data stored in the cloud. The inferences from these analytics are fed into the knowledge database of the case-based expert system.

Data Utilisation in Google Cloud Platform (BigQuery) The behavioural queries to understand the sleeping patterns for a particular month are shown in Fig. 4 and

Fig. 4 Behavioural query to understand sleeping pattern in a month

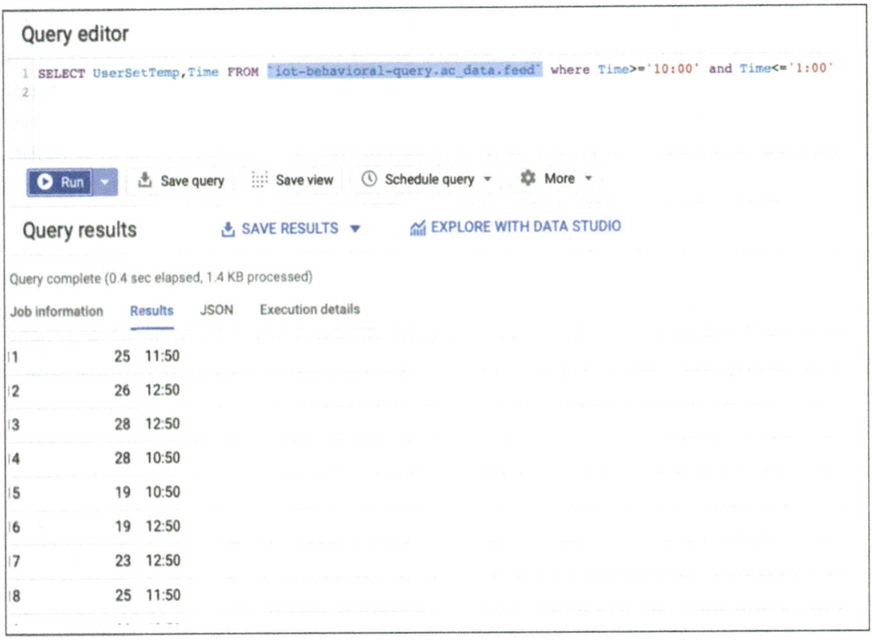

Fig. 5 Behavioural query to understand temperature versus time

to understand what temperature the user is setting at a particular time interval in the night, say from 10:00 PM to 1:00 AM, is shown in Fig. 5.

Figures 6 and 7 show the data utilisation by querying for historical analysis of finding the maximum temperature during the summer season (March–May) and finding the temperature at a particular place, where 3 is the encoded numeral for the city 'Coimbatore', for each month in the past one year, respectively. Such historical surveys help in understanding the levels of utilisation at different phases of time.

4.5 Feedback

Feedback is collected 15 min after the first actuation. To accommodate user convenience, we have kept the complexity of the feedback form to a minimum. An in-depth design for the feedback form, for seasonal surveys, can be integrated into the system at later stages. A set of three questions related to thermal comfort are prompted to the user as in Fig. 8. The user ranks each of these questions on a scale of 0 to 5 (0—very bad, 5—very good). The outcome of the three questions determines whether the case is updated in the case database or if it is sent to the expert knowledge system to recompute another action. If the cumulative of the three questions is less than 10,

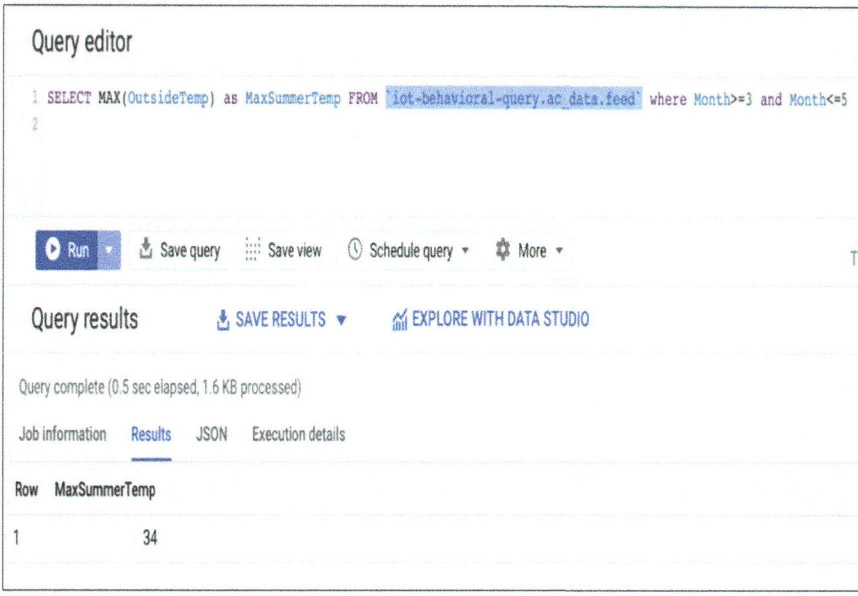

Fig. 6 Historical query to find maximum temperature in the summer season (March–May)

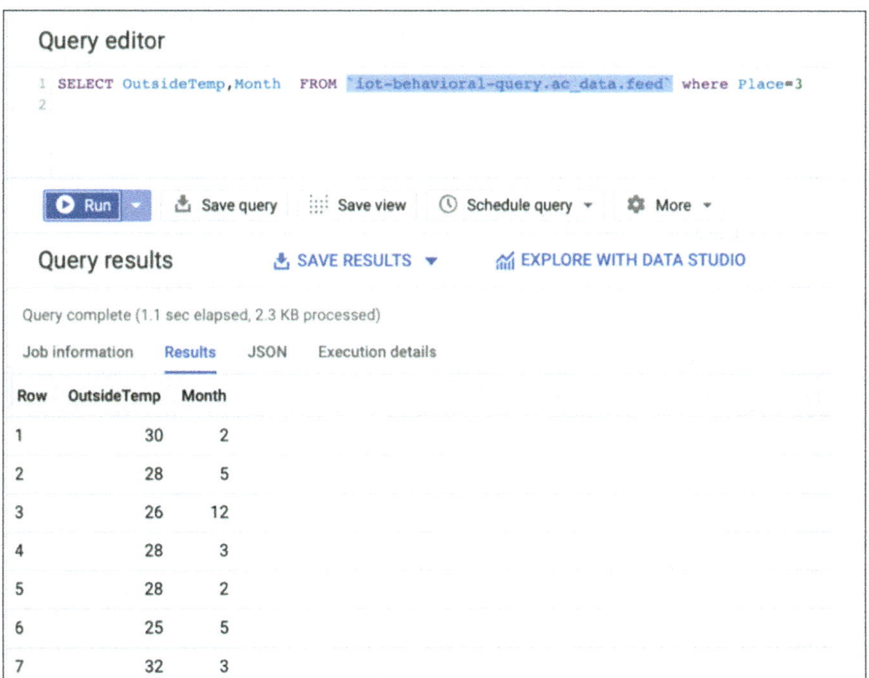

Fig. 7 Historical query to find the outside temperature at a particular place for each month in the past one year

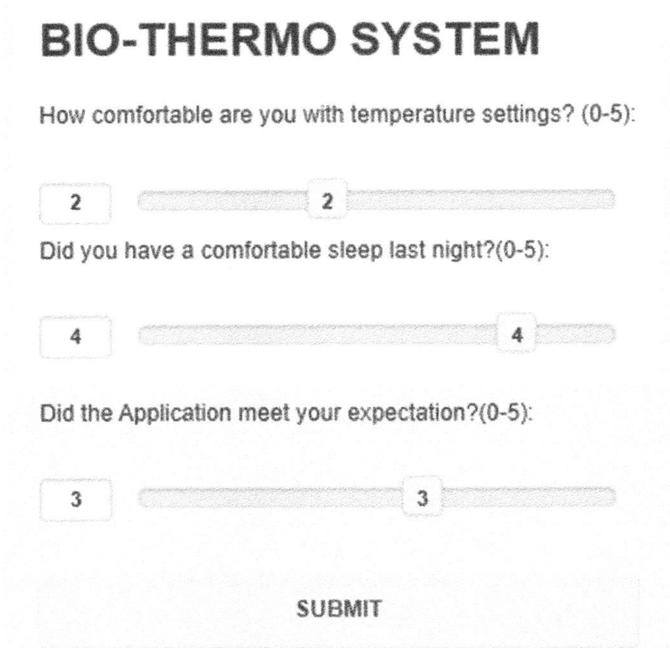

Fig. 8 Case-based application interface for feedback ranking

then the decision must be recalculated. If the score is greater than 10, then the case along with the decision is updated in the case database, and no change is done to the actuated temperature on the AC.

The performance metric of the system is based on the feedback received. A successful actuation would mean that the number of actions re-computed is less and number of retrievals from the knowledge base is also less. Throughout the night, the feedback metric will be used to actuate the AC with respect to the thermoregulation theory in the knowledge base. In a situation where the user wants a particular temperature to be set, a manual override option has been given.

5 Conclusion

The case-based expert system is novel because it helps to store previously witnessed cases, which correspondingly reduces the necessity to recompute actions for the system. Hybridisation of a case-based and expert reasoning system is a promising endeavour, yet to find a strong foothold in smart city applications. The knowledge base is a system of rules which encapsulates the HVAC theory, thermoregulatory theory, and insights from cloud analytics. It results in a customised system for every individual because it takes into consideration body temperature of the person and is able to navigate accordingly when the temperature of the human drops in deep sleep . Behavioural queries and historical queries have been performed in the cloud to gain cognizance about user's habits. The user interface in the application provides a good experience for the user to communicate, query, and give feedback to the system. Overall, the case-based expert thermoregulated AC is a cutting edge technology that understands, relates, and prioritises the thermal comfort of the users.

References

1. J. Zakri, Body Temperature and Sleep. Tuck Sleep (2019). http://www.tuck.com/thermoregulation/
2. H. Kimberly, Thermoregulation. Healthline Media (2017). http://www.healthline.com/health/thermoregulation
3. J.F. Duffy, D.W. Rimmer, C.A. Czeisler, Association of intrinsic circadian period with morningness-eveningness, usual wake time, and circadian phase. Behav. Neurosci. (2001). https://doi.org/10.1037/0735-7044.115.4.895
4. G. Kelly, Body temperature variability (Part 1): a review of the history of body temperature and its variability due to site selection, biological rhythms, fitness, and aging. Altern. Med. Rev. **11**(4), 278–293 (2006)
5. E. Abouzar, K. Iman, V. Marian, G.E.G. Georges, A review on internet of things solutions for intelligent energy control in buildings for smart city applications. Energy Proc. (2017). https://doi.org/10.1016/j.egypro.2017.03.239
6. A.J. Waheb, K.K. Tee, M.R. Roshahliza, N.Z. Siti, S.M. Nurthaqifah, Mohammed B. Zamrizaman, V. Shepelev, S. Alharbi, Design and fabrication of smart home with internet of things enabled automation system. IEEE Access **7**, (2019). https://doi.org/10.1109/ACCESS.2019.2942846
7. G. Selahattin, Z. Nikolay, K.K. Gunes, O. Berna, IoT in action: Design and implementation of a building evacuation service. J. Comput. Netw. Commun. (2017). https://doi.org/10.1155/2017/8595404
8. J. Krishnan, S.K. Vasudevan, An extensive survey on IoT smart gateways, software architecture, in *Related Protocols and Challenges International Conference on Vision Towards Emerging Trends in Communication and Networking (ViTECoN)* (2019). https://doi.org/10.1109/ViTECoN.2019.8899513
9. W. Song et al., An IoT-based smart controlling system of air conditioner for high energy efficiency, in *IEEE International Conference on Internet of Things (iThings) and IEEE Green Computing and Communications (GreenCom) and IEEE Cyber, Physical and Social Computing (CPSCom) and IEEE Smart Data (SmartData)* (2017). https://doi.org/10.1109/iThings-GreenCom-CPSCom-SmartData.2017.72
10. A.M. Ali et al., IoT-based smart air conditioning control for thermal comfort, in *IEEE International Conference on Automatic Control and Intelligent Systems (I2CACIS)* (2019). https://doi.org/10.1109/I2CACIS.2019.8825079
11. L. Tacklim, J. Seonki, K. Dongjun, W.P. Lee, P. Sehyun, Design and implementation of intelligent HVAC system based on IoT and Bigdata platform, in *IEEE International Conference on Consumer Electronics (ICCE)* (2017). https://doi.org/10.1109/ICCE.2017.7889369
12. R. Carli et al., IoT based architecture for model predictive control of HVAC systems in smart buildings. Sens. Open Access J. (2020). https://doi.org/10.3390/s20030781
13. António r, S. Sérgio, D. Helder, P.M. Ferreira, Wireless sensors and IoT platform for intelligent HVAC control. Appl. Sci. (2018). https://doi.org/10.3390/app8030370

14. A. Rajith, S. Sakurai, H. Mine, Real-time optimized HVAC control system on top of an IoT framework, in *Third International Conference on Fog and Mobile Edge Computing (FMEC)* (2018). https://doi.org/10.1109/FMEC.2018.8364062
15. G. Kowshik et al., An inventive and innovative system to detect fall of old aged persons—a novel attempt with IoT, sensors and data analytics to prevent the post fall effects. Int. J. Med. Eng. Inform. (2020). https://doi.org/10.1504/IJMEI.2020.105654
16. F. Roschelle, Identifying varying health states in smart home sensor data: an expert-guided approach. World Multi-Conf. Syst. Cybern. Inform. (2017). https://eecs.wsu.edu/~cook/pubs/sci17.pdf
17. C. Iuliana, O. Geman, An approach of a decision support and home monitoring system for patients with neurological disorders using internet of things concepts. WSEAS Trans. Syst. (2014). https://www.wseas.org/multimedia/journals/systems/2014/g045702-416.pdf
18. M.S. Suh, W.C. Jhee, Y.K. Ko, A. Lee, A case-based expert system approach for quality design. Expert Syst. Appl. (1998)
19. C. Lifan, D. Yu, W. Hongyan, Z. Yi, Z. Jing, Case-based reasoning expert system for concept design of economical car, in *Proceedings of the IEEE International Vehicle Electronics Conference (IVEC99)*
20. K. Joshi, Expert systems and applied artificial intelligence. http://www.umsl.edu/~joshik/msis480/chapt11.htm
21. Q. Xi, Z. Jin, H. Wei, W. Lixin, Y. Yi, W. Jw, Integrated configuration management based on system engineering. EasyChair (2020). https://easychair.org/publications/preprint/hDD7
22. D. Jyoti, B. Udhav, A study of mammogram classification using AdaBoost with decision tree, KNN, SVM and hybrid SVM-KNN as component classifiers. J. Inform. Hiding Multimed. Signal Process. http://bit.kuas.edu.tw/jihmsp/2018/vol9/JIH-MSP-2018-03-004.pdf
23. B. Luca, C. Giovanni, D.D. Palma, M. Gianfranco, M. Antonio, Embedded wireless sensor network system for cultural heritage monitoring, in *Fourth International Conference on Sensor Technologies and Applications* (2010)
24. M. Calum, IoT Device Management: What is it and Why do You Need it? (2019). https://www.iotforall.com/what-is-iot-device-management/

Prediction of Solar Power in an IoT-Enabled Solar System in an Academic Campus of India

Dhruvraj Singh Rawat and Kumar Padmanabh

Abstract Solar power is available in abundance in India. Not only it helps in minimizing carbon emission, but it is also very cost effective. There are multiple arrays of solar panels installed in the academic and the residential buildings of the LNM Institute of Information Technology (LNMIIT) campus Jaipur, Rajasthan. These installations are helping LNMIIT to reduce electricity bills. However, the solar panel is not reliable because LNMIIT does not know how much power these installations would produce at a particular instant of time. Hence, solar power is used in non-trivial applications. Production of power from such a system depends upon the weather condition of the moment. We argue that if the weather can be predicted, then the prediction of solar power is also possible. This paper proposes to make the solar power more reliable by predicting the power output. In this paper, many existing algorithms are customized for prediction and a mechanism is proposed to dynamically select the best performing algorithm. The accuracy of the proposed prediction model has been more than 92%.

Keywords Solar power · Machine learning · Solar power prediction

1 Introduction

Uninterrupted power supply is required for the economic development of any country. Government can build capacity according to the gross domestic product (GDP). A co-author of this paper has earlier proved that since due to specific demography and

This work was done when Kumar Padmanabh was working as Associate Professor in LNMIIT.

D. S. Rawat (✉)
LNM Institute of Information Technology, Jaipur, India
e-mail: dhruvrajsingh.rawat.y15@lnmiit.ac.in

K. Padmanabh
EBTIC, Khalifa University Campus, Abu Dhabi, UAE
e-mail: padmanabh@ieee.org

M. Paprzycki et al. (eds.), *Intelligent Systems, Technologies and Applications*,
Advances in Intelligent Systems and Computing 1353,
https://doi.org/10.1007/978-981-16-0730-1_28

urbanization in India, the peak demand of electricity is exponentially proportional to the GDP [1], whereas in the rest of the world, the peak demand is linearly proportional to the GDP. It has been proven that due to this exponential nature of the peak demand of electricity, the government cannot allocate exponentially increasing fund to build such capacity. Usage of non-renewable source of energy such as solar power is inevitable for uninterrupted power supply. With advancement of technology and government subsidy, solar power is also affordable now.

Solar Power does not contribute to carbon emission or global warming. Paris Agreement-2015 was signed by 195 countries, where it is decided to keep global temperatures below 2 °C above the pre-industrial era. Due to specific demand of electricity in India, While there is an urgent need of solar power in India for the uninterrupted power supply [2]. Hence, solar power could be the major source of clean energy. However, solar power is not uniform throughout the day. It depends upon the weather and other local conditions. Hence, solar power is not reliable enough so that it could be used in critical applications. If the output of an installation is known in advance accurately only then it could be used in all applications.

The power produced by a solar system largely depends upon the variability of solar irradiance and environmental factors [3]. Solar power is a perishable commodity. Reliability of the solar power could be ensured by accurately predicting how much power would be available at a particular moment of time. The objective of this paper is to predict solar power in real-life setup.

This paper is based on the data from an installation of a university campus of LNMIIT in India which has both residential and commercial consumption. Several academic buildings and residential campus houses have rooftop installation of solar panels, as depicted in Fig. 1. Electricity bill of LNMIIT campus has been around 20 million Indian rupees in an year. The management decided to maximize the usage of the solar power to reduce the bill. In the first phase, 15% capacity was built.

It is important to note here that though solar power was produced at 15% of total annual demand of grid power, however there was no corresponding saving in electricity bill. In the daytime when energy consumption was at its peak, when most of academic and administrative work was going on, the management was not confident to use it due to its unpredictable nature. In this scenario, prediction of solar power was scheduled to be used for real-life energy management.

Rest of this paper is organized in following way. In Sect. 2, details about the data and data acquisition would be discussed. Before any prediction algorithm is applied, data need to be cleaned, filtered, and transformed into the required format. This preprocessing of the data is reported in Sect. 3. Outliers degrade the performance of prediction algorithms. Section 4 deals with the outliers, finally result is discussed in Sect. 5.

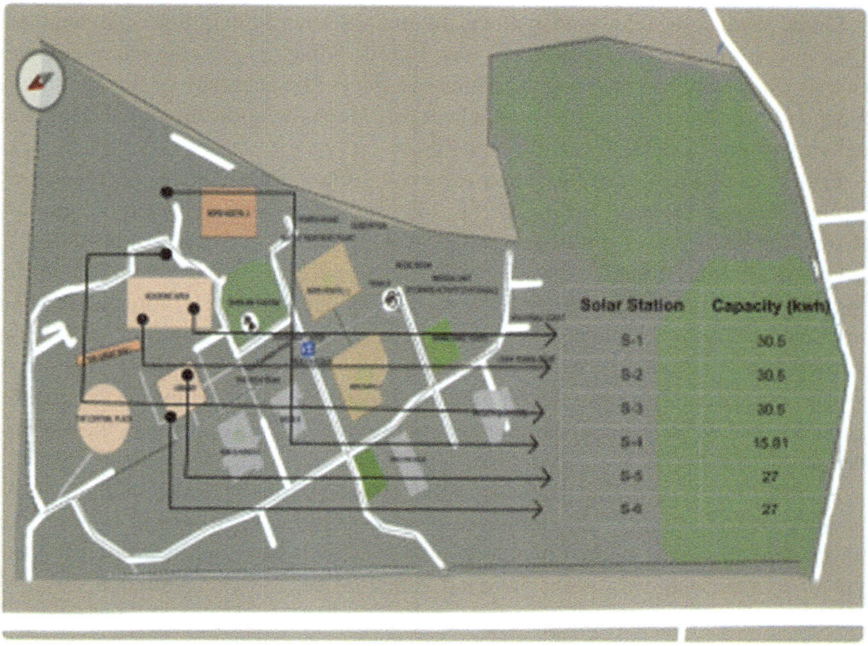

Fig. 1 Location and solar power panels installed at LNMIIT campus

2 Literature Survey

The solar power depends mainly on the projected power on the horizontal surface which is known as irradiance. Since all solar power does not come to the earth, environment above earth surface put an obstruction. Since environmental condition changes dynamically, therefore, solar power also depends upon the instantaneous weather information.

Weather prediction is a very complex system. It requires complex modeling of wind flows in all geography. It requires sophisticated modeling of cloud also. Since it requires very complex and expensive system, hence, there are only about 15 entities of world which do fundamental weather forecasting.

There are broadly two types of models available in the literature: physical models [4, 5] and statistical models [6, 7]. A physical model is basically a mathematical model that describes the initial physical state and dynamic motion of the atmosphere in from of equations. Numerical methods are used to generate solutions to these equations. Numerical weather prediction (NWP) models require high computational power [8]. The statistical models include linear regression models, satellite data-based models, and ANN models using meteorological parameters [9].

Cloud produces shadow and it affects the amount of sunlight falling on solar panel. It varies according to the time of the day. In [10], image processing technique was used to estimate the sunlight in satellite data. Since the accuracy of this technique depends upon the image quality which in turn depends upon the cloud data, this technique is not suitable for commercial use.

Earlier, the co-author of this paper explored different mechanisms of forecasting applicable to predicting load profile [1, 11]. There has been considerable amount of work on solar power prediction using artificial neural network. Authors in [12] proposed an ANN for prediction of daily solar radiation. They obtained correlation coefficient of 0.98 and the mean relative less than 1.5%. The forecast of solar irradiance by using artificial neural network was performed by authors in [13] for performance prediction of a grid-connected PV plants at Trieste, Italy. They also got correlation coefficient in the range 0.98–0.99. It has been demonstrated that the method based on an artificial neural network provides better results than the alternative classical methods.

Authors in [14] predicted irradiance using estimated time of arrival method (ETA) by using neural network. The influence of aerosol in the environment on the solar power prediction was studied by authors in [15]. In all these applications, different type of weather data was used. There individual influence was not studied.

The work mentioned in these literature has sophisticated mathematical and statistical models, however they could not be used in real life because they are not pragmatic. This paper presents mechanism to predict solar power with an objective to integrate in real-life scenario where predicted power would be used to schedule other sources of energy.

3 The Data from Installations of LNMIIT

There are six installations of solar power and they were not physically integrated and fed to the grid. Rather, they were used in isolation. However, the third party who helped the installation managed to get all data on their sensorweb platform [16]. These data from six different arrays of panels were available online. There was no API available there. We have chosen the Web site site scrapping mechanism to harvest all relevant solar data.

There is an "Inspect Element" in the browser which gives access to the source code, the images and CSS that forms its design, the fonts, icons and the JavaScripts. It also gives information on how much time required to launch that page. Inside the "Inspect Element" the option in the network tab "GetAnalysisChart" containing a JSON response is provided. Subsequently, we saved this response as HTTP Archive Format (.HAR) for further analysis. The HAR, is a JSON-formatted archive file format for logging of a web browser's interaction with a site. It provides input having following details:

- How long it takes to fetch DNS information.
- How long each object takes to be requested.
- How long it takes to connect to the server.
- How long it takes to transfer assets from the server to the browser of each object.

The inspection element was used to download data using a scrapping tool. Since the solar power depends upon the weather of that location. Hence, corresponding weather information was also needed to build the model. For weather information we used the developer API's to grab the historical data using a python script for the duration for which we have the solar panel data from "Weather Underground" [17]. In this data was available for every half an hour and hence corresponding weather information was pulled out. Since there was no power in the night hence data from 06:00 PM in the evening to 06:00 AM in the morning was removed.

4 Data Prepossessing and Analysis

There are two aspects of data processing important for this paper—(i) The tool and techniques used to transform data in the format required for building models and (ii) the characteristics of data relevant for the models, especially how these data are dependent on each other.

4.1 Tools Used in Data Processing

We used G-Suite Toolbox-HAR Analyzer for prepossessing of data. Using this tool all the unwanted contents were removed from the HAR file and then saved in JSON format. This was repeated for all required time duration. The saved JSON files are then parsed with python script to transform the data into CSV format. It contains the detailed information of solar power production for different panels. Subsequently, the weather data and the solar data were combined together on the same time-line using a python script.

The harvested data contained the information of 180 d starting from October 01, 2016 till March 30 2017. It consists of daily solar power generation and corresponding weather information. The Data from the solar stations was at the duration of 5 min whereas the data for weather was at the duration of 30 min. This led us creating the two datasets:

- The first dataset contains information at every 30 min, thus neglecting the 4 solar reading (for only maximum correlated solar) in 30 min.
- The second dataset contains data fields at every 5 min but the weather information is available only after 30 min thus there was 10 missing weather fields for solar power generation output per hour (for only maximum correlated solar). Thus, this dataset has around 6 times the first dataset.

To fill the missing value in the second dataset we used the interpolation technique. For our purpose we used the "Akima interpolation" [18] which is a continuously deferential subspline interpolation. This interpolation uses the third degree polynomials. In this interpolation, only data from the next neighbor points is used to determine the coefficients of the interpolation polynomial.

4.2 Correlation Analysis of Different Parameters

As described earlier, there were two sets of data: one from the solar installations and other from the weather station. There are four different elements of weather relevant for us: (i) temperature, (ii) wind speed gives notion of dust accumulation, (iii) visibility gives notion of the intensity of solar light, and (iv) humidity which indicates the rain.

Analysis of correlation is very important here to understand which parameters can help predicting the solar power better or which parameter is irrelevant for the prediction model. Therefore, there are two aspects of correlation of our interest: (i) correlation of solar output produced from different installations to understand whose installation is the best for building models and (ii) correlation of solar output to the weather parameter to select and eliminate particular weather parameters.

We have data from several solar stations. Since they are installed in the same geographical area, the production of respective solar powers is ought to be the same. However, the angle of the orientation of the panel with respect to the sun, the shadows from nearby trees, dust accumulation, etc., might affect the production to the individual panel. Therefore, it was important to understand if the solar power produced from them are correlated, so that prediction model based on the data of one solar panel is applicable to the rest of them (Table 1).

It is to be noted that solar installation for academic block is different and there is no correlation. However, other from hostel, mess, and residential buildings are correlated. Hence, the prediction model based on the data of one can be used for others.

Table 1 Correlations of solar output between different solar stations

Solar	S-1	S-2	S-3	S-4	S-5	S-6
S-1	1	0.91	0.94	0.89	0.93	0.99
S-2	0.91	1	0.97	0.97	0.97	0.92
S-3	0.94	0.97	1	0.95	0.99	0.94
S-4	0.89	0.97	0.95	1	0.95	0.89
S-5	0.93	0.97	0.99	0.95	1	0.94
S-6	0.99	0.92	0.94	0.89	0.94	1

Table 2 Correlation of weather parameters with solar output

Humidity	Temperature	Wind speed	Visibility
−0.41	0.40	0.43	0.24

From above correlation tables, following insight can be concluded.

- By removing the unwanted range of values beyond 06:00 AM to 06:00 PM during which majority of sunlight is not available, we were able to increase the average correlation between solar stations from 0.68 to 0.95. This is a significant improvement and hence also demonstrates the impact of outliers on the data.
- The maximum correlation that was found between any two solar panels was 0.99 that was between S-5 and S-3.
- S-3 (Academic Eco 2) has the maximum correlation to the other solar stations. Its average correlation to the rest of the panels is around 0.97.
- Correlation of solar output also demonstrates in general how well fine-tuning to these solar panel was done during deployment.

From the third point, we can say that the solar-3 (S-3) is the best deployment so the result of this panel can be generalized. Hence, we would be using this particular panel as a candidate for further analysis.

Theoretically, the solar power is proportional to the irradiance. We are claiming that solar power is also proportional to the local weather. To prove it, we need to find the correlation between produced power and the weather.

Table 2 shows the correlation between the weather parameters that are used in model building and the solar output from the panel. From the table, it is quite evident that the wind speed has the maximum correlation of 0.43.

Humidity has the next maximum absolute value of 0.41. A negative sign means their is a inverse relation between the solar output and the humidity. The temperature has correlation of 0.40. The Fig. 2 can be referred for the correlation heat map of solar output, time and various weather parameters.

Solar output with respect to time of the day, visibility, wind speed, and temperature is displayed in the three-dimensional plot presented in Fig. 3. Solar output is directly proportional to the visibility. More is the visibility more would be the temperature. Since visibility is maximum at the noon, surface graph in Fig. 3 reveals that both temperature and solar production are maximum at noon.

Thus, analysis of different correlation indicates that the third installation is the best suitable for building models because its orientation with respect to the sun and the possibility of dust accumulation is most generalized. It is also evident that the humidity and visibility play least role in impacting the production of solar power.

Table 3 Performance of algorithms

Algorithms	R^2 score
Adaboost	0.7438
Bagging	0.9453
Bayesian Ridge	0.3313
Hubber	0.3232
Lasso	0.3331
Linear	0.3331
Random Forest	0.9468
Ridge	0.3331
SVM linear	0.31009
SVM RBF	0.6076

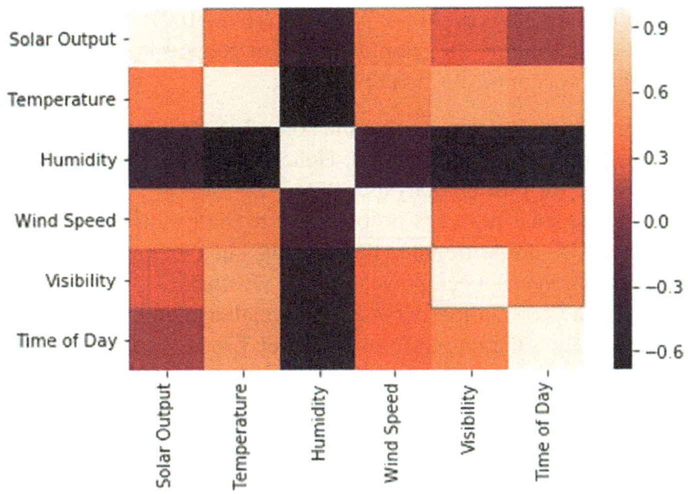

Fig. 2 Correlation heat map of dataset

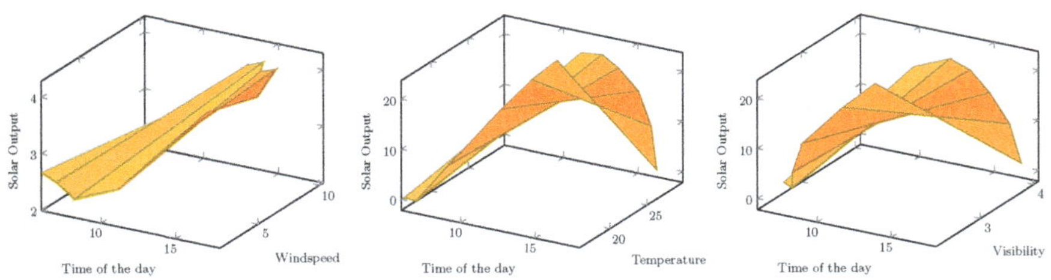

(a) Variation with 'Windspeed'
(b) Variation with 'Temperature'
(c) Variation with Visibility

Fig. 3 Variation of output with respect to wind speed, temperature, and visibility

Table 4 Rank table

Parameter	Rank
Humidity	1
Time	1
Temperature	2
Wind speed	3
Visibility	4

5 The Prediction Model and the Characteristics

In previous section, we have demonstrated through correlation analysis and through different visualization that solar output is not only function of irradiance but also it is also a function of the local weather. We have also demonstrated that solar power is linearly proportional to weather parameter such as wind speed, temperature, visibility, and humidity. In this situation, there are two important aspects of building the models: (i) solar output is linearly proportional to the weather parameters and (ii) there are number of other dependencies. Moreover, we do not know the nature of dependencies of solar power on other parameters hence both multilinear regression and neural network would be investigated to build the model.

5.1 Prediction Using Different Regression Models

Objective of this exercises is to select a particular form of regression model which could give most accurate prediction. Mostly in regression problems R^2 score is widely used as an indicator for quality of prediction. It represents quality of fit of the regression line. Hence, R^2 in this paper score has been used as one of the evaluation metric. Apart from R^2 score, mean absolute error (MAE) can also be used as a metric to estimate how a particular model is performing. Unlike in root mean squared error (RMSE), in which first error is squared and subsequently its square root is considered as error, in MAE mean error of the absolute value is taken into account. Due to square of the error, the RMSE gives more weightage to very large outlier values which is present in very small numbers. Since in our dataset there are variation, MAE would give better result.

It is to be noted that all the results shown here are based on present weather in Rajasthan, a state of India which is considerably far away from the experimentation site. Since there was no weather station available precisely to this location, a more precise weather information would increase the accuracy.

Summary of different regression model and their corresponding R^2 score is given in Table 3.

Table 5 Methodology and comparison of R^2 score

Dataset	Methodology	R^2 score
Normal dataset	67% train, 33% test split	94.3748
Normal dataset	Tenfold cross-validation	90.52
Without wind speed	67% train, 33% test split	92.9
Without wind speed	Tenfold cross-validation	89.17
Without visibility	67% train, 33% test split	93.4586
Without visibility	Tenfold cross-validation	90.24
Without visibility and wind speed	67% train, 33% test split	89.9924
Without visibility and wind speed	Tenfold cross-validation	88.27

As random forest regressor has the best R^2 score using this as the base line we calculated how much an input feature effect the prediction or in other words we ranked the input feature in accordance to their contribution to solar prediction (Table 4).

From the above rank table of different input parameters, it is quite evident that the "Humidity", the "Time of the Day", and the "Temperature" are the most import parameters for predicting the solar output.

5.2 *Evaluation of Random Forest Regressor on Various Dataset*

Since now it is established that random forest regressor gives the best result, it is important to demonstrate the performance of this regressor on different datasets. In order to carryout this exercise, the data was divided in "test data" and "training data" in different samples.

The result of all possible combinations is displayed in Table 5. In all these circumstances, when the model is trained with 67% of data and tested with remaining 33% of test data then it gives the best result. Interestingly visibility does not change the accuracy of prediction. However, windspeed is a big impact on the accuracy. It might be possible that in a geographical area like Rajasthan a desert wind also carries sands with it which is deposited on the solar panel or high wind speed would remove the same from the solar panel.

The same result is presented in Fig. 4 where actual data and predicted data are plotted on the same times scale. The visibility has the same impact here also. Interestingly, when visibility and windspeed both were not considered the drop in R^2 score is not significant. It means just by using temperature and humidity data and time of the day, the solar power can be predicted.

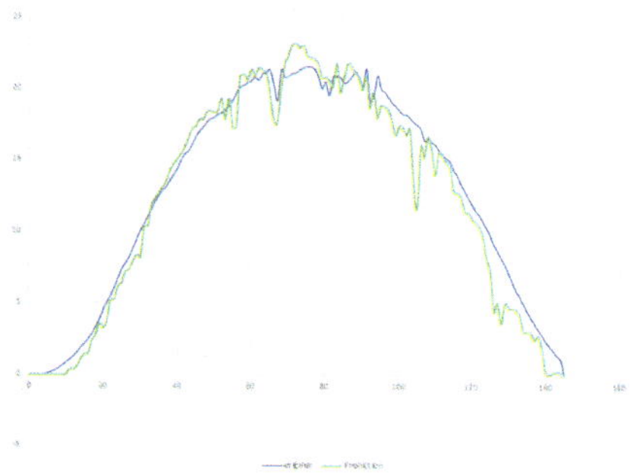

(a) Normal Data Random Forest

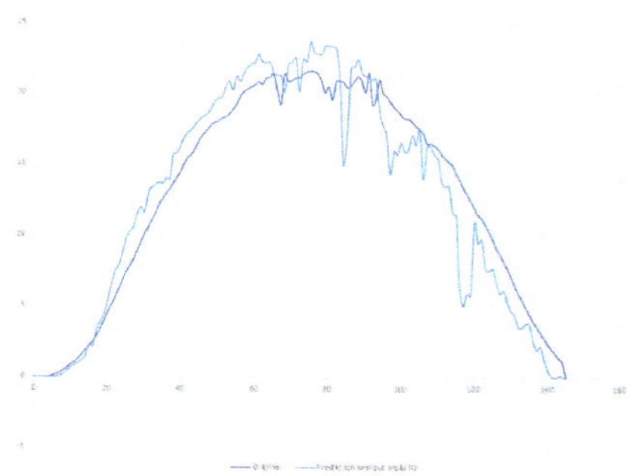

(b) without visibility Random Forest

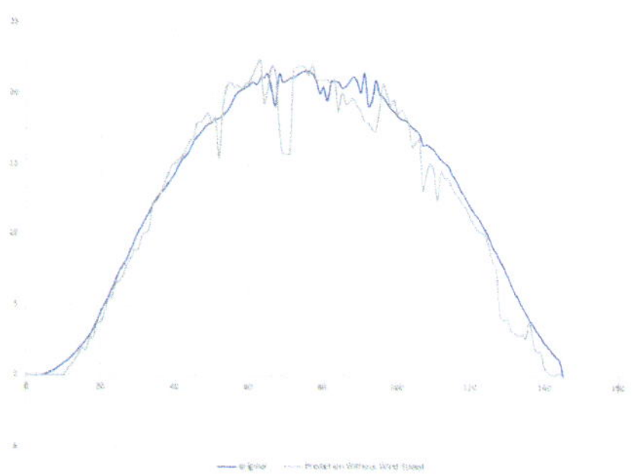

(c) without windspeed Random Forest

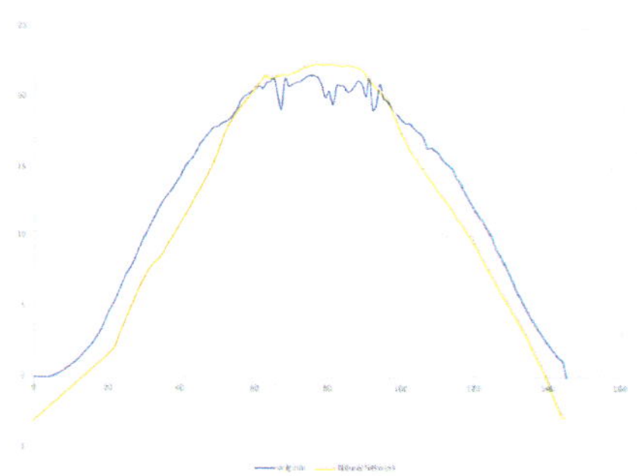

(d) without visibility and windspeed

Fig. 4 Correlation of different solar stations

5.3 Prediction Model Using Neural Network

We also used the feed-forward neural network for predicting the solar output. We have used 10 hidden layers with activation function of rectified linear unit. The learning rate of neural network was 0.01. These parameters are summarized in table below.

Tuning parameter of neural network

S. No.	Argument	Value
1	Hidden layer size	10
2	Activation function	ReLU
3	Solver	Adam optimizer
4	Learning rate	0.01
5	Epos	1000
6	Momentum	10

The result of the neural network-based prediction is displayed in Table 6. Model was trained on various combinations of the weather parameter. Interestingly visibility does not have much impact on the accuracy (R^2 score).

Table 6 Methodology and comparison of R^2 score

Dataset	Methodology	R^2 score
Normal dataset	67% train, 33% test split	87.84
Normal dataset	Tenfold cross-validation	92.02
Normal dataset without wind speed	67% train, 33% test split	86.5181
Normal without windspeed	Tenfold cross-validation	92.67
Normal without visibility	67% train, 33% test split	86.69
Normal without visibility	Tenfold cross-validation	92.30
Normal dataset visibility and windspeed	67% train, 33% test split	85.90
Normal dataset without visibility and wind speed	Tenfold cross-validation	91.63

6 Conclusion

In this research, we tried to develop mechanism to forecast solar power in LNMIIT campus which has multiple solar station. After overcoming with nuances with the quality of data, we develop mechanism to identify and eliminate outliers. We applied various regression mechanisms to forecast solar power and we found that random forest has been giving the best model to predict.

References

1. K. Padmanabh, Load forecasting in India at distribution transformer considering economic dynamics, in *2016 International Conference on,Advances in Computing, Communications and Informatics (ICACCI)* (IEEE, 2016)
2. A.A. Amrutha, P. Balachandra, M. Mathirajan, Model-based approach for planning renewable energy transition in a resource-constrained electricity system—a case study from India. Int. J. Energy Res. **42**(3), 1023–1039 (2018)
3. M. Šúri et al., Potential of solar electricity generation in the European Union member states and candidate countries. Solar Energy **81**(10), 1295–1305 (2007)
4. R. Marquez, C.F.M. Coimbra, Forecasting of global and direct solar irradiance using stochastic learning methods, ground experiments and the NWS database. Solar Energy **85**(5), 746–756 (2011)
5. M. Diagne, M. David, P. Lauret, J. Boland, N. Schmutz, Review of solar irradiance forecasting methods and a proposition for small-scale insular grids. Renew. Sustain. Energy Rev. **27**, 65–76 (2013)
6. F. Wang, Z. Mi, S. Shi, H. Zhao, Short-term solar irradiance forecasting model based on artificial neural network using statistical feature parameters. Energies **5**(5), 1355–1370 (2012)
7. Weather data, https://www.wunderground.com/
8. R.H. Inman, H.T.C. Pedro, C.F.M. Coimbra, Solar forecasting methods for renewable energy integration. Progress Energy Combust. Sci. **39**(6), 535–576 (2013)

9. R. Marquez, C.F.M. Coimbra, Intra-hour DNI forecasting based on cloud tracking image analysis. Solar Energy **91**, 327–336 (2013)
10. C.W. Chow et al., Intra-hour forecasting with a total sky imager at the UC San Diego solar energy testbed. Solar Energy **85**(11), 2881–2893 (2011)
11. M.J. Singh, P. Agarwal, K. Padmanabh, Load forecasting at distribution transformer using IoT based smart meter data from 6000 Irish homes, in *2016 2nd International Conference on Contemporary Computing and Informatics (IC3I)* (IEEE, 2016)
12. W. Gu et al., Modeling, planning and optimal energy management of combined cooling, heating and power microgrid: a review. Int. J. Electric. Power Energy Syst. **54**, 26–37 (2014)
13. Marc Beaudin, Hamidreza Zareipour, Home energy management systems: a review of modelling and complexity. Renew. Sustain. Energy Rev. **45**, 318–335 (2015)
14. R.A. Guarnieri, E.B. Pereira, S.C. Chou, Solar radiation forecast using artificial neural networks in South Brazil, in *Proceedings of the 8th ICSHMO* (2006), pp. 24–28
15. R. Perez et al., Forecasting solar radiation? Preliminary evaluation of an approach based upon the national forecast database. Solar Energy **81**(6), 809–812 (2007)
16. http://www.solarweb.com
17. https://www.wunderground.com/
18. Hiroshi Akima, A new method of interpolation and smooth curve fitting based on local procedures. J. ACM (JACM) **17**(4), 589–602 (1970)

Author Index

M. Paprzycki et al. (eds.), *Intelligent Systems, Technologies and Applications*, Advances in Intelligent Systems and Computing 1353,
https://doi.org/10.1007/978-981-16-0730-1

The manufacturer's authorised representative in the EU is Springer Nature Customer Service Centre GmbH, Europaplatz 3, 69115 Heidelberg, Germany. If you have any concerns regarding our products, please contact ProductSafety@springernature.com

Printed and bound by CPI Group (UK) Ltd, Croydon, CR0 4YY
30/09/2026
02225292-0010